Monitoring Induced Seismicity

Edited by
Cezar I. Trifu

Previously published in *Pure and Applied Geophysics* (PAGEOPH), Volume 167, Nos. 1–2, 2010

Birkhäuser

Editor
Cezar I. Trifu
Engineering Seismology Group Inc.
20 Hyperion Court
Kingston, Ontario K7K 7K2
Canada
trifu@esg.ca

ISBN 978-3-0346-0305-8 e-ISBN 978-3-0346-0306-5
DOI 10.1007/978-3-0346-0306-5

Library of Congress Control Number: 2009939764

Cover illustration: Based on a graphic from the article „Microseismic monitoring of a controlled collapse in solution mining field II at Ocnele Mari, Romania“ by C-I. Trifu and V. Shumila

Printed on acid-free paper

Springer Basel AG is part of Springer Science+Business Media

www.birkhauser.ch

Contents

Pure Appl. Geophys. 167 (2010), 1–3
© 2009 Birkhäuser Verlag, Basel/Switzerland
DOI 10.1007/s00024-009-0009-0

Pure and Applied Geophysics

Introduction

The occurrence of seismicity associated with the active extraction of minerals and hydrocarbons, the decommissioning of mines including solution mining fields, the storage of radioactive waste and hydrocarbons, as well as energy generation from geothermal and water reservoirs has long been amply documented. Stress redistribution as a result of carrying out these activities in the presence of complex regional and local geology cannot be eliminated. Consequently, the effort must be directed to a better understanding of the underlying processes for the management of the hazard and risk associated with these operations while maintaining an increased productivity. The study of induced seismicity has continuously evolved over the past couple of decades, as underlined by both the number and complexity of applied studies required to satisfy the increased economic demands, assure the safety of the workforce and equipment, and protect the environment.

A considerable effort has been put into the development of passive monitoring technology, specialized commercial hardware and software products and services being currently available for a wide range of applications. The recording of substantial high-quality seismic data has stimulated the work on theoretical and practical aspects related to these applications, involving not only seismological knowledge, but also elements of rock mechanics, as well as an understanding of mining, geotechnical, and petroleum engineering. By periodically documenting such studies, the journal of *Pure and Applied Geophysics* has largely contributed to raising the profile of induced seismicity research and its credibility in addressing the challenging targets ahead of it set up by the socio-economic conditions.

The present volume is the third organized on this topic by the editor after *New Trends in Seismological Research: Studies on Seismicity Induced by Petroleum, Mining and Geothermal Activities* (eds. C–I. Trifu and M. Fehler), Tectonophysics *289*, 1998, and *The Mechanism of Induced Seismicity* (ed. C–I. Trifu), Pure Appl. Geophys. *158*, 2001. It contains 12 original studies authored by 30 specialists from Australia, Canada, France, Germany, Israel, Poland, South Africa and the United States. Through the results obtained, these studies recommend the application of passive seismic monitoring for direct, real-time observations, towards a better understanding and control of the phenomena associated with the respective applications. The volume offers reference material on the work status in this domain for both the specialists and graduate students.

Three studies look into applications of cavern monitoring. To validate the use of passive seismics for the monitoring of underground caverns in salt rock, Mercerat *et al.* focus on the analysis of induced seismicity at a French salt mine as recorded by four three- and three one-component geophones in surface boreholes. About 2,000 events occurred during 2 years—some, due to two controlled pressure experiments—exhibit tensile fracture components but no pure shears, the result of progressive cavern damage through dissolution. Trifu and Shumila study a controlled collapse in a salt solution mining field in Romania using a staggered array of 36 one-component geophones in 12 boreholes, which recorded 2,400 events during 9 months. Event mechanism solutions derived by moment tensor inversion exhibit mainly normal and strike-slip shears, and up to 30% explosional failure components, indicative of rock fragmentation and roof caving, with highest deformation rate in areas of detected collapses. To ensure the hazard and risk management associated with post-mining areas, Contrucci *et al.* monitor the effect of stress changes due the presence of extensive underground cavities resulting from coal and iron ore exploitation. They develop quantitative processing of microseismic characteristics based on event locations and seismic energy estimates calibrated by setting small blasts in known locations through field experiments carried out at several instrumented sites.

Somewhat related to the latest study, two analyses are directed to coal mining. Seismicity occurred in the Ruhr area of Germany, associated with longwall mining activity, is investigated by BISCHOFF *et al.* to determine a relationship for peak ground velocity with distance, showing that induced stresses are released rapidly by many small events. Fault plane solutions of events related to a single longwall are consistent with steeply dipping fracture planes parallel to the face and nearly vertical dislocation in the direction of the goaf. FRITSCHEN examines the seismicity induced by coal mining at a mine in Germany and outlines no correlation with extracted coal volume as large events occur only in certain areas, while other areas have no seismicity. Shear and non-shear events occur simultaneously, although at different depths, with shear seismicity located in bands coinciding with the strike of larger regional faults in the area and thus explained as due to the interaction of mining induced and tectonic stresses.

Safety of mining personnel has energized the following three investigations. The study by VALLEJOS and MCKINNON applies the modified Omori's law to seismic aftershock sequences recorded at four Canadian hard rock mines in an effort to develop a real-time re-entry protocol after a major seismic event. They identify a characteristic time that marks the transition from the highest to lowest event rate change, and propose this time as appropriate to re-enter an underground area of the mine affected by a large blast or rockburst. KGARUME *et al.* are also looking into the parameters of large event aftershock sequences in order to establish the conditions for workforce evacuation and re-entry at deep gold mines in South Africa. They found that the aftershocks satisfy the GR scaling with $b \sim 1$, follow the modified Omori law, while the magnitude difference between the main shock and largest aftershock increases for main shocks with $M_L > 3$, and aftershock density with distance suggests dynamic stress triggering. Finally, LUO *et al.* employ 16 three-component geophones in four deep boreholes to examine the mine seismicity at Lihir open pit in Papua New Guinea, located inside an old volcano where outbursts of hot water and steam are a major safety concern for the operations. Of more than 17,000 events recorded in 6 weeks, 75% were interpreted as rock fractures, while 12% exhibited harmonic vibrations due to geothermal activity inside the pit, likely associated with mining.

Two other studies are aimed at techniques employed in seismic tomography and the monitoring of unconsolidated rocks. As such, DEBSKI compares the seismic tomography results obtained through the use of the probabilistic (Monte Carlo), damped least-squares and optimization techniques as applied to a data set recorded at a copper mine in Poland. He concludes that Monte Carlo allows for a relatively easy evaluation of the uncertainties, producing more robust images as compared to the choice of the damping parameters, and is more reliable, particularly under noisy conditions. WUST-BLOCH demonstrates that brittle failure within unconsolidated material can be seismically monitored at close range by analyzing crack development in a sandstone cliff for which event locations are shown to cluster, while event decay over time suggests material undergoes strain change. Polarization of peak waveform amplitude is used to estimate spatial stress distribution, which is consistent with mapped tensile cracks and in accordance with model rock mechanics results.

The final two analyses are dedicated to oil/gas and water reservoir applications. DINSKE *et al.* propose a model for the evaluation of the hydraulic fracture and gas reservoir parameters from microseismic data that combines geometry and diffusion-controlled processes. The models lead to similar geometrical estimates of the spatio-temporal dynamics of the hydraulic fracturing resulting from both gel and water treatments at Cotton Valley, Texas, while linking the difference in the permeability of hydraulic fractures to the volume of proppant. DURA-GOMEZ and TALWANI investigate seismicity induced after the impoundment of Koyna and Warna water reservoirs in India that included 19 events with $M \geq 5.0$ in seven episodes at non-repeating locations. Pore pressure and its daily rate change derived from now available daily lake levels indicate that first episodes are primarily associated with the initial filling of the reservoirs as pore pressure diffused along a saturated, critically stressed network of faults and fractures.

Seismic technology provides a monitoring tool which can be employed effectively in the assessment and mitigation of seismic hazard, similar to its use by civil engineers in establishing design requirements and building codes. The wide range of specific analyses included in the present collection of studies expresses clearly how powerful and resourceful passive seismic monitoring has become to so many applications. Seismic data are routinely evaluated for a variety of development activities specific to each application domain. Engineering practice is continuously improved based on trials and analyses of the ground response and stress levels, confirmed through the monitoring of seismicity. Seismic data are also used for the calibration of numerical models by adjusting model parameters until a good correlation is obtained between observed seismicity and expected fracture modelling results. Furthermore, seismic monitoring allows the identification of major seismically active geological structures for strategic planning. The scientific and engineering communities are better equipped now than ever before to provide the necessary answers for the hazard and risk management of human activities that alter the stress distribution in the upper crust based on a formalized, quantitative and thus objective assessment.

Acknowledgments

This volume was made possible due to numerous authors who are acknowledged for their interest, quality of work, commitment and resourcefulness during the revision process. Moreover, this volume benefited a great deal from the dedicated effort of numerous reviewers whose time, expertise, and objectivity have provided the authors with useful feedback to ensure that they circulate a sound and effective message. I would like to express my most sincere appreciation to Doug Angus, David Beck, Marco Bohnhoff, Johann Braunmiller, Ming Cai, Martin Chapman, Dave Collins, Diane Doser, Cliff Frohlich, Jean-Robert Grasso, Leigh House, Paul Hsieh, Andrew King, Lindsay Linzer, Art McGarr, Steve McKinnon, Wayne Pennington, Scott Phillips, Eliza Richardson, James Rutledge, Robert Shcherbakov, Vladimir Shumila, Jan Sileny, David Simpson, Shahriar Talebi, and Ben Thompson for their insightful comments and suggestions made throughout the review process. Renata Dmowska, the Editor-in-Chief for Topical Issues with Pure and Applied Geophysics is particularly acknowledged for her interest in having this volume organized and assistance throughout.

Cezar I. Trifu
Editor
Engineering Seismology Group Inc.,
1 Hyperion Court, Kingston, ON K7K 7G3, Canada
E-mail: trifu@esg.ca

Pure Appl. Geophys. 167 (2010), 5–25

DOI 10.1007/s00024-009-0008-1

Pure and Applied Geophysics

Induced Seismicity Monitoring of an Underground Salt Cavern Prone to Collapse

E. D. Mercerat,[1] L. Driad-Lebeau,[2,3] and P. Bernard[1]

Abstract—Within the framework of a large research project launched to assess the feasibility of microseismic monitoring of growing underground caverns, this specific work focuses on the analysis of the induced seismicity recorded in a salt mine environment. A local seismic network has been installed over an underground salt cavern located in the Lorraine basin (Northeast of France). The microseismic network includes four 3-components and three single component geophones deployed at depths between 30 and 125 m in cemented boreholes drilled in the vicinity of the study area. The underground cavern under monitoring is located within a salt layer at 180 m depth and it presents a rather irregular shape that can be approximated by a cylindrical volume of 50 m height and 180 m diameter. Presently, the cavern is full of saturated brine inducing a significant pressure on its walls ($\sim$2.0 MPa) to keep the overburden mechanically stable. Nevertheless some small microseismic events were recorded by the network and analyzed (approximately 2,000 events in 2 years of recording). In October 2005 and April 2007, two controlled pressure transient experiments were carried out in the cavern, in order to analyze the mechanical response of the overburden by tracking the induced microseismicity. The recorded events were mainly grouped in clusters of 3–30 s of signal duration with emergent first arrivals and rather low frequency content (between 20 and 120 Hz). Some of these events have been spatially located by travel-time picking close to the actual cavern and its immediate roof. Preliminary spectral analysis of isolated microearthquakes suggests sources with non-negligible tensile components possibly related to fluid-filled cracks. Rock-debris falling into the cavern from delamination of clay marls in the immediate roof is probably another source of seismic excitation. This was later confirmed when the most important seismic swarms occurred at the site during May 2007, accompanied by the detachment of more than 8 $\times$ 10^4 m^3 of marly material on top of the cavern roof. In any case, no clear evidence of classical brittle ruptures in the most competent layers of the overburden has been observed during the analyzed period. Current work is focused on the discrimination of all these possible mechanisms to better understand the damage processes in the cavern overburden and to assess its final collapse hazard.

Key words: Induced seismicity, monitoring, salt cavern, collapse hazard.

1. Introduction

A large research project within the GISOS[1] program has been launched in order to identify and evaluate the potential of geophysical (microseismic, hydroacoustic) and geotechnical (strainmeter, tiltmeter) techniques for monitoring growing underground caverns due to salt dissolution (solution mining). A part of this program focuses on two complementary research areas: (1) the validation of microseismic monitoring techniques in salt mine environments, and (2) the numerical modeling of deformation and failure mechanisms with its associated induced microseismicity. In this study, we focus on the analysis of microseismic activity induced by the presence of an underground cavern in the Lorraine salt basin (Northeast France).

Previous studies on induced seismicity of growing underground caverns confirm the presence of microseismic events associated with the progressive damage of the overburden until its final collapse (see Mendecki, 1997 for a thoughtful review). Continuous monitoring of such microseismicity may provide crucial information for stability analysis of the area surrounding the cavern. The vast majority of such studies focused on underground workings excavated in hard crystalline rocks at important depths (greater than 300 m) and consequently relatively high ambient

[1] Equipe de Sismologie, Institut de Physique du Globe de Paris, 4 Place Jussieu, Tour 24-14, Case 89, 75005 Paris, France. E-mail: mercerat@ipgp.jussieu.fr

[2] INERIS, Ecole des Mines de Nancy, Parc de Saurupt, 54000 Nancy, France.

[3] GEODERIS, 1 rue Claude-Chappe, BP 25198, 57075 Metz, France.

[1] Research Group for the Impact and Safety of Underground Workings involving French institutions INERIS, BRGM, INPL and Ecole des Mines de Paris.

stresses (McGarr *et al.*, 1989; Gibowicz *et al.*, 1991; McGarr, 1992b; Urbancic and Young, 1993; Trifu *et al.*, 1995; Knoll *et al.*, 1996). More specifically, induced microseismic investigations have been carried out in abandoned salt caverns used for hydrocarbon storage to discriminate between regional and mining-induced seismicity (Fortier *et al.*, 2006), as well as to monitor the post-sealing behavior of brine production caverns (Maisons *et al.*, 1997).

There are few published microseismic studies on relatively shallow (less than 200 m) salt mine environments, and therefore interesting challenges are encountered in the present study. Some information is found in Branston (2003), where microgravity and microseismic techniques are analyzed as potential tools to monitor underground salt caverns at comparable depths. Apart from that, we found other sources of useful information for improving our understanding of seismic signatures of rockslide failures in recent microseismic investigations of unstable mountain slopes (Suriñach *et al.*, 2005; Brückl and Mertl, 2006; Spillmann *et al.*, 2007), ground vibrations generated by rocks and debris flows (Huang *et al.*, 2004, 2007), and pre-collapse sinkhole identification in the Dead-Sea area (Wust Bloch and Joswig, 2006).

Throughout the entire Lorraine salt basin, a competent layer known as Beaumont dolomite is found at around 100 m depth. It is composed of a dolomite and an anhydrite layer which present classical elastobrittle behavior with extraordinarily high strength values for typical sedimentary rocks. The presence of this layer favors the creation of large salt caverns in the region. It is believed to be responsible for the whole overburden stability and its fracturing may produce the entire overburden collapse (Buffet, 1998; Nothnagel, 2003). Therefore, it is mandatory to assess damage progression in this particular layer and microearthquakes are expected to be one of the key observations. Hence a local microseismic network was installed over the site test of Cerville-Buissoncourt (Lorraine, France), where a stable brine-filled cavern is found. When the salt exploitation intensifies, the spatial and temporal evolution of microearthquake locations and magnitudes can be recorded and further analyzed to identify, and eventually quantify, the damage progression in the overburden. These experimental results are expected to be correlated with geomechanical modeling to confirm back-analysis results of rock strength parameters (Mercerat *et al.*, 2007; Souley *et al.*, 2008).

In this paper, we present the first results of the microseismic analysis, and discuss some possible causes for the induced seismicity recorded at the test site since January 2005. Although salt dissolution has moved northward from the actual cavern position and the whole overburden is mechanically stable, some microseismic activity has been recorded by the network. It can be associated with localized ruptures within fine anhydritic layers present in the salt deposit and/or block rearrangements caused by material degradation and induced stress-loads. Moreover, in October 2005 and April 2007, two transient pressure experiments were carried out in the brine-filled cavern, in order to check the instrumentation layout and to calibrate other monitoring techniques deployed on the site. The first relevant seismic swarms occurred on the site during May 2007. It was afterwards confirmed by borehole logging that around 8×10^4 m^3 of cavern roof material had fallen in the cavern. Here we suggest that the tremor-type events recorded during the May 2007 swarms are the microseismic signature of these rock-fall avalanches.

After a brief introduction of the test site of Cerville-Buissoncourt in terms of geological and salt exploitation contexts, we present the microseismic monitoring system deployed at the site. Secondly, we discuss the general characteristics of the recorded seismicity during the studied period. Next, we compare hypocenter locations of isolated microearthquakes recorded in 2005 and tremor-type events belonging to May 2007 swarms. Spectral source analysis is then carried out on some chosen microearthquakes with clear P- and S-wave arrivals. We conclude with a discussion concerning the possible origin of the tremor-type events and the correlation with other monitoring techniques, such as brine pressure and strainmeter data, as well as cavern logs carried out to track cavern roof evolution.

2. *Site of Cerville-Buissoncourt*

2.1. *Geological Context*

The salt concession of Cerville-Buissoncourt (SOLVAY S.A.) is located in the Lorraine region

(Northeast France), 20 km east of the city of Nancy. The salt deposit corresponds to the Keuper epoch (Late Triassic period around 200 Ma) associated with the eastern limits of the Paris basin. It extends more than 200 km in the EW direction and 100 km in the NS direction. In the region of Cerville-Buissoncourt, the salt deposit lies between 180 and 220 m depth. This makes it economically feasible for its exploitation by solution mining. The total exploitable thickness varies from 60 to 80 m and the salt layer is intercalated with thin layers of argilites and anhydrites (tens of centimeters to a meter thick). From a tectonic point of view, it is a quite stable region and there is no evidence of the presence of faults or strong discontinuities close to the study area.

2.2. *Brine-filled Cavern and Overburden Characteristics*

The brine-filled cavern under study is actually located within the salt layer at about 180 m depth and presents an irregular shape that, in first approximation, can be considered a cylindrical cavern of 180 m of diameter and 50 m of maximum height. It was formed by intensive salt dissolution during the period 1997–2003, at the southernmost end of two parallel lines of solution mining wells that reach the bottom of the salt deposit. A plan view of the instrumented site, and a schematic geological cross section of the area (perpendicular to the exploitation lines) are shown in Figs. 1 and 2, respectively.

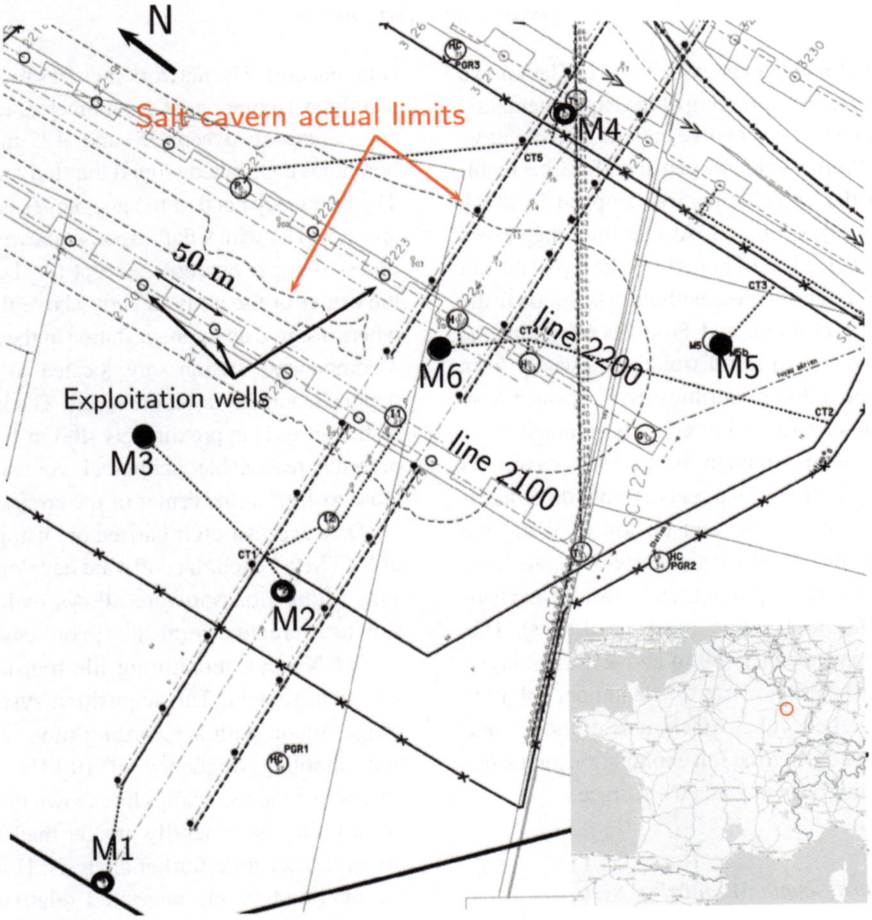

Figure 1
Plan view of the instrumented site of Cerville-Buissoncourt showing microseismic stations M1–M6, the exploitation wells of line 2,100 and line 2,200, and the actual cavern limits (*dashed line*)

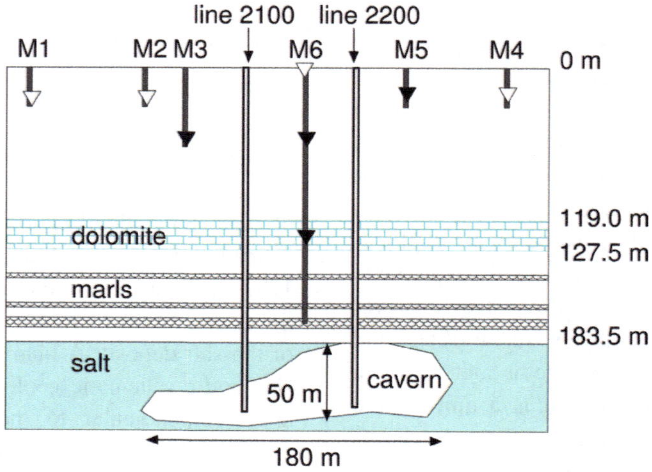

Figure 2
Scheme of the WE vertical section with main lithological units and location of the 3-component (*filled triangles*) and single component (*open triangles*) microseismic stations

Two essential points must be noted: (1) The known cavern geometry at the beginning of the experimentation (fall 2004) will ineluctably evolve until its final collapse, all stages of cavern evolution can therefore be monitored; and (2) the exploitation was stopped in 2004, stopping the cavern evolution. At that time the cavern roof had reached over half of its surface, the top of the salt layer (see Fig. 2). A first phase, without dissolution in the cavern, could then be monitored. From 2006 forward, the operator has undertaken salt dissolution downdip from the cavern at about 200 m northwards; and, since May 2007, significant cavern evolution has been noted.

Typically, the overburden of the salt cavern is characterized by a series of intercalated anhydritic marls lying over the salt deposit, and on top of them, the presence of a highly competent anhydrite-dolomite layer of 8.5-m thickness which presents elasto-brittle mechanical behavior (from laboratory mechanical tests). This layer, located approximately 60 m above the salt layer, corresponds to what is known as the Beaumont dolomite and it represents the level at which most of the seismic damage is expected when the salt exploitation intensifies and the cavern migrates towards the surface.

3. *Microseismic Monitoring System*

A microseismic network was installed in 2004 over the stable underground brine-filled cavern of Cerville-Buissoncourt. The network includes four 3-components and three 1-component seismometers located at different depths (between 35 and 125 m) in cemented boreholes distributed around the study area (see Fig. 1). The frequency band of the geophones is between 28 Hz and 1.5 kHz, with a flat response above 40 Hz. There is one station per borehole, except for the deepest one at the center of the array directly above the cavern (M6), where a single component station at the surface and two 3-components stations are located at 60 and 125 m depth (Beaumont dolomite level). The lateral extension of the array is approximately 400 m by 600 m, and it presents reasonable azimuthal coverage (major gap close to 130° at the center of the array).

Data acquisition is carried out using SYTMISscop of SYTMIS® modular software developed by INERIS. This acquisition software allows real-time transmission to a remote central site (in our case located at the city of Nancy), monitoring file transfers and remote site management. The acquisition system is set up in trigger mode with a recording time window of 0.8 s and sampling frequency of 10 kHz. A preliminary analysis of the recordings has shown that the signal-to-noise ratio was generally greater than 10 allowing, in principle, accurate further analysis. However, most of the qualified events presented relatively low amplitudes ranging between 10^{-7} m/s (ambient noise level) and 10^{-5} m/s. Therefore, a classification of the microseismic events has been undertaken in order to select those for which the signal-to-noise ratio was sufficiently large to ensure the identification of seismic arrivals (P and S waves) and acceptable time picks. This analysis led us to set and adapt the triggering threshold providing an acceptable data quality in terms of signal-to-noise ratio.

4. Recorded Seismicity: General Characteristics and Temporal Evolution

Microseismic activity has been recorded since the installation of the monitoring system in the study area. More than 2,000 triggered event files were recorded between January 2005 and September 2007. After rejection of noise transients, we identified mainly two types of recordings: Isolated microseismic events (less than 10% of the total recorded seismicity), and tremor-type events grouped in clusters of tens of seconds of duration (more than 90%). The frequency content of these relatively long events is between 20 and 120 Hz, and they present an emergent P-wave onset, which is generally immersed in the coda of the previous event, hence it is nearly impossible to pick P waves manually. It must be stressed that the amplitudes are well above the noise level in the study area (less than 10^{-7} m/s). A selection of these two types of events with the corresponding time–frequency analysis is shown in Fig. 3. In addition, small isolated microearthquakes (less than 0.5 s of time duration) were also recorded by the network. For these events, the frequency content is sensibly higher (between 20 and 150 Hz), with P-wave onsets better resolved. This allowed more precise manual picking of first arrivals. Polarization analysis of the waveforms indicates the presence of S waves in some recordings around 40–80 ms after the P-wave arrival. Despite the higher uncertainties, they were also manually picked and utilized to better constrain the spatial location of these events.

In this study it is not possible to quantify the microseismic activity simply by counting the number of triggered event files recorded by the acquisition system, as is classically done in many microseismic monitoring investigations. This is because more than 90% of the recorded activity is composed of swarms of events with time durations between 3 and 30 s (recorded by almost continuous triggering of the acquisition system). Therefore, we easily find triggered event files with more than one individual microearthquake, as well as triggered event files with no identifiable microearthquake. Accordingly, we decided a better way to quantify the microseismic activity was by analyzing the averaged RMS (i.e. sum of squared amplitudes) recorded at three 3-component stations (M3, M5 and M62). These stations are approximately equidistant from the seismogenic zone, i.e. the surroundings of the actual cavern, in order to minimize differential path-effects.

In Fig. 4 we can observe the temporal evolution of the recorded RMS, as well as the cumulative curve, since January 2005 until September 2007. We can identify two notable jumps in the cumulative RMS curve (8 May 2007 and 31 May 2007) and certain small jumps spread throughout the analyzed period.

We will return to some of these events later. It is interesting to note that there is some evolution in the frequency content of the events as a function of time: During 2005 there were hardly any events with spectral dominant frequencies higher than 120 Hz, whereas from 2006 forward, the dominant frequency increases slightly. However, the low frequency limit is stable near 20 Hz. The majority of analyzed events presents dominant frequencies between 40 and 80 Hz.

4.1. Transient Pressure Experiments

Presently, the cavern is full of saturated brine inducing a significant pressure (close to 2.0 MPa) on its walls to keep the overburden mechanically stable. In October 2005 and April 2007, two transient pressure experiments were carried out in the cavern in order to analyze the mechanical response of the overburden and to check over the site instrumentation sensitivity. The transient experiments consisted of a pressure drop of around 0.3 MPa (some 25 m of the brine column height), a stabilization phase, and a final phase at which the pressure was raised to the original 2.0 MPa. The amplitude of the transient was fixed in agreement with SOLVAY S.A. considering safety/industrial reasons. The difference between both experiments was the time duration of each phase. That of October 2005 lasted almost 3 weeks, while the time duration of April 2007 lasted only 4 days. In both cases, an irreversible quasi-static deformation was measured by

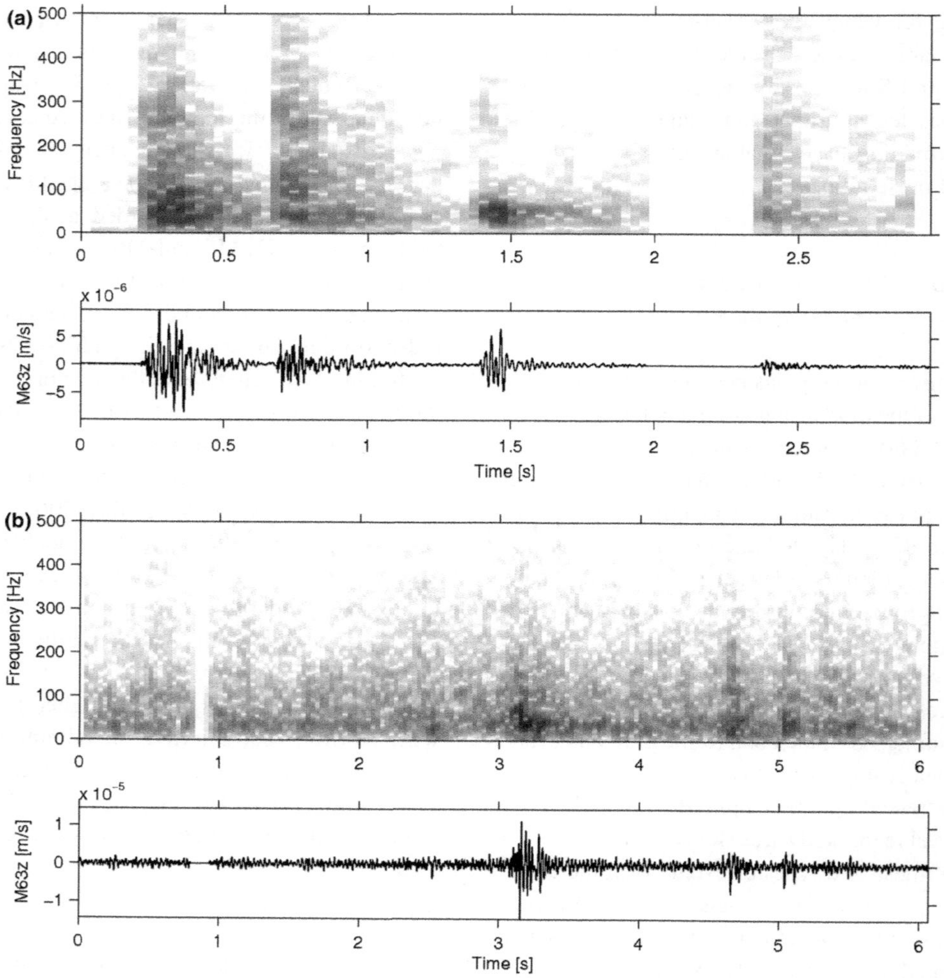

Figure 3
Spectrograms derived from M63 vertical component data using a 0.1 s moving time window: **a** Four isolated events triggered one after the other (2 October 2005), and **b** tremor-type event of more than 6 s duration (22 October 2005). Gaps: No data recorded

the strainmeters placed in the deep borehole M6 at three different depths. The amplitudes were between 1.2 and 1.5 mm at the end of each transient for the deepest strainmeter (129 m depth) just below the dolomite layer. Surprinsingly, the microseismic activity was not significantly increased by the pressure transients with respect to the overall activity, as can be seen in Fig. 5. Nevertheless, we observe almost exclusively tremor-type events in the pressure increase phase compared to the pressure drops, especially for the October 2005 experiment. There may exist a causality between abrupt pressure rises and a triggering of tremor-type events (reactivation of degraded zones in the cavern roof). A more detailed analysis is needed to highlight this assertion. In any case, it was concluded that, at the time of the transient experiments, a pressure drop of 15% ($\sim$0.3 MPa) was not large enough to induce significant ruptures in the overburden (especially in the competent dolomite layer). Experiments with larger pressure drops were not possible for safety reasons (see next section).

4.2. Seismic Swarms of May 2007

It is clear from the cumulative RMS curve of Fig. 4 that the main microseismic activity of the

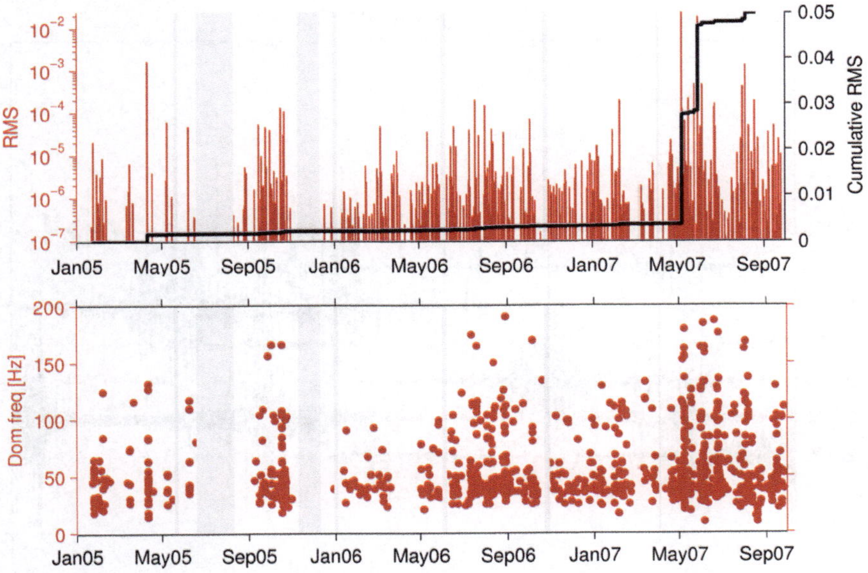

Figure 4
Recorded RMS per day (*fine vertical bars*), cumulative RMS curve (*black line*) and dominant frequency of microseismic events (*dots*) recorded from January 2005 until September 2007. *Vertical gray bars* Acquisition system down

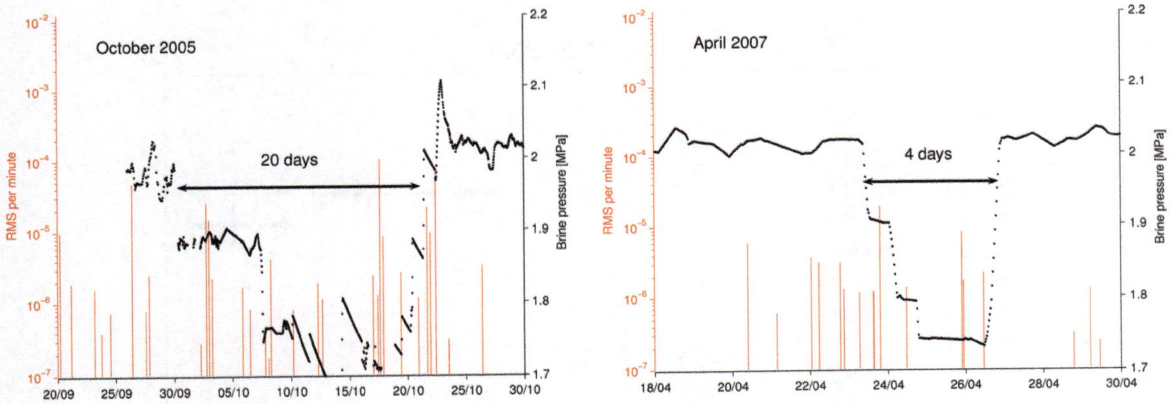

Figure 5
Recorded RMS per minute (*fine vertical bars*) and brine pressure (*black dots*) within the cavern during both transient pressure experiments of October 2005 (*left*) and April 2007 (*right*). There is no clear evidence of seismic activity directly induced by the pressure transients of 0.3 MPa maximum amplitude

analyzed period occurred during May 2007, when more than 1,500 triggered events were recorded by the network during the evenings of 8 May 2007 and 31 May 2007. Each of the crises lasted not more than 10 min, however the energy recorded largely exceeded the cumulative trend by at least one order of magnitude (Fig. 4). The recordings of the vertical component at the deepest station (M63) can be seen in Fig. 6. At that time an abrupt collapse was suspected and consequently any additional pressure experiments in the cavern were canceled for safety reasons. The analysis of these crises presents the same challenges as before: the vast majority of tractable events are immersed in swarms of even longer time durations (on the order of minutes) that are reminiscent of low-frequency tremor-like events

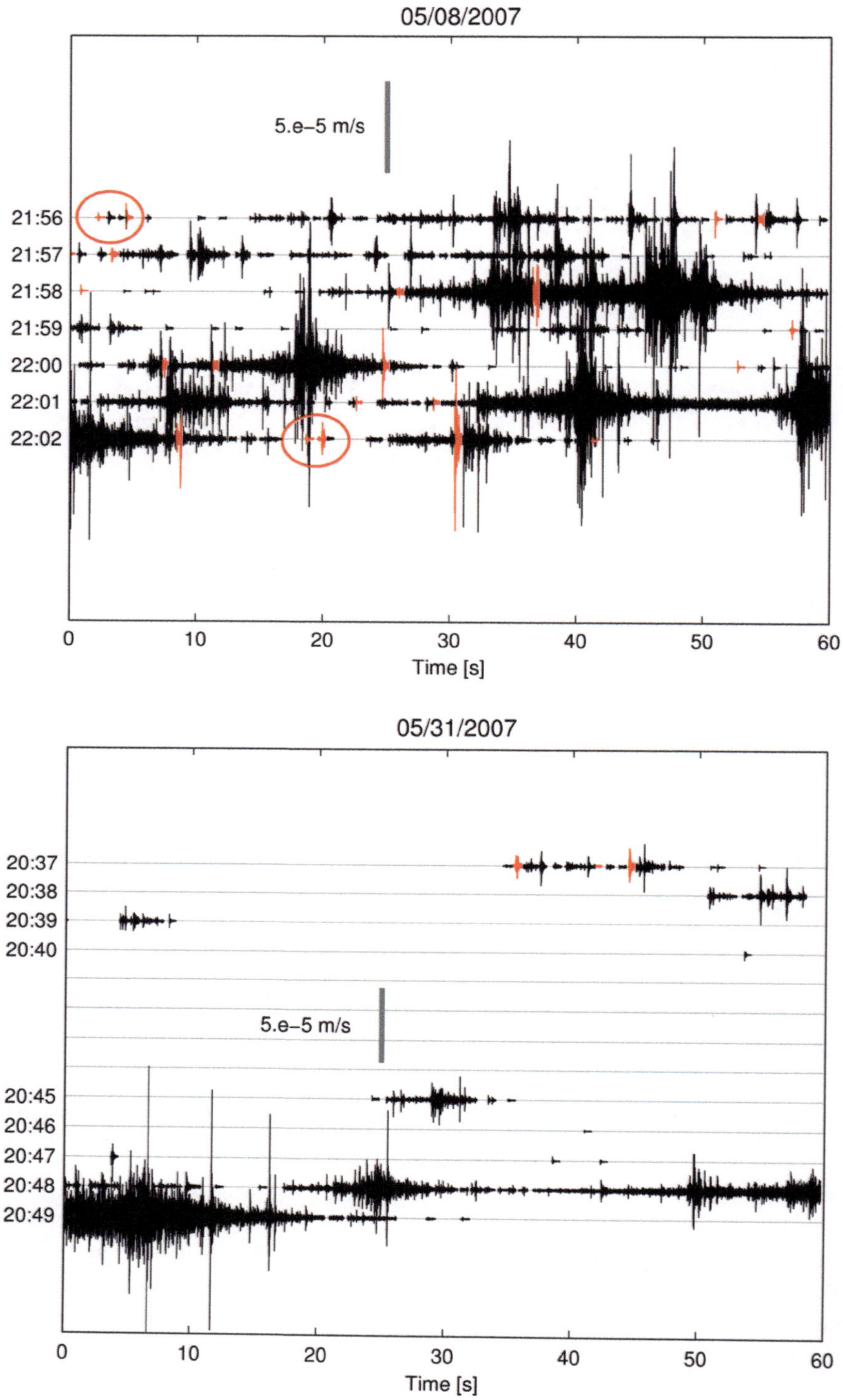

Figure 6
Swarms of 8 May 2007 and 31 May 2007 recorded at M63 (vertical component). In *gray*, some isolated events that could be picked and located. Two ellipses show the individual microearthquakes chosen for spectral source analysis. *Gray horizontal lines* No data recorded

produced by active volcanic systems (WASSERMANN, 2002). Despite these difficulties, some of them could be recognized in several stations and manually picked for further analysis (shown in red in Fig. 6).

5. Hypocenter Determination

5.1. Methodology

Hypocentral determination was possible using the NonLinLoc software (LOMAX, 2006) based on the probabilistic approach of Tarantola and Valette (TARANTOLA and VALETTE, 1982; MOSER *et al.*, 1992), and the Oct-Tree global search approach for hypocentral parameters (LOMAX *et al.*, 2000). This non linear probabilistic location technique has widely been used for regional hypocenter location in highly heterogeneous areas (LOMAX *et al.*, 2001; ZOLLO *et al.*, 2002; HUSEN *et al.*, 2003; PRESTI *et al.*, 2004; LIPPITSCH *et al.*, 2005), and it has been recently incorporated in more local microearthquake studies (SPILLMANN *et al.*, 2007). In contrast with classical linearized algorithms, the complete solution to the inverse problem is a probability density function (PDF) that includes all likely solutions, i.e., hypocentral parameters that are in accordance with both data (picked arrival times), and a priori information, in terms of model (theoretical) and observational errors (TARANTOLA and VALETTE, 1982). Even if these uncertainties are assumed to follow Gaussian distributions in the formulation (TARANTOLA and VALETTE, 1982), the non linear relationship between hypocentral parameters and data, in heterogeneous velocity models, may result in highly irregular and/or multimodal hypocenter PDF's (LOMAX *et al.*, 2000).

For all hypocenter locations done in this study, we define Gaussian uncertainties of 5 ms (standard deviation) for the observational errors at every station, except for the two deepest ones (M62 and M63) where we have used 2 ms (overall better signal to noise ratio). Despite the well-known difficulty in setting realistic model or theoretical travel-time errors (LOMAX *et al.*, 2000; SPILLMANN *et al.*, 2007), we assume that they follow a Gaussian distribution with 2 ms standard deviation for all stations in the network.

5.2. Velocity Model Determination

The velocity model of the monitored area was defined on the basis of vertical seismic profiling data and a geological log of the study area. The first source of information served to constrain the average seismic velocities in each layer, while the second one was used to tightly set the most relevant geological interfaces. Tomographic inversion results from reflection seismic data suggests rather low compressional velocities (less than 2,000 m/s) near the surface (SUFFERT *et al.*, 2006).

In earlier stages of this research, active seismic shots from the tomographic campaign provided the information for network calibration (MERCERAT, 2007). This allowed us to define a 1-D velocity model used in hypocenter locations presented in this study, as well as to characterize anelastic attenuation (constant Q model) for some specific lithologies present at the site (marly materials on top of the dolomite layer). Unfortunately all shots were carried out on the surface lacking penetration to cavern surroundings. We used further data to constrain the velocity model as accurately as possible. Hence, high-resolution seismic sections and borehole tomography results were used to refine the 1-D velocity model provided by the shots points. After careful calibration using a number of well-located shots, the chosen velocity model includes three layers, the first one with a uniform vertical gradient from the surface (V_P = 2,400 m/s) up to 120 m depth (V_P = 3,000 m/s), then the Beaumont dolomite of 10 m thickness (V_P = 5,000 m/s), and finally the marls and salt formations (V_P = 4000 m/s) to the bottom of the model. A constant V_P/V_S ratio of 1.73 is assumed for all layers. We can accept that layer depths are relatively well constrained, however largest uncertainties lie in the absolute values of seismic velocity in each formation. Moreover, possible lateral velocity variations, not taken into account in this model, may occur, especially in the intercalated marls on top of the actual cavern where highly damaged material may be present. In addition, the presence of the brine-filled cavern may mask some seismic phases. All this information should be considered in future investigations to reduce hypocenter location uncertainties for subsequent interpretation.

5.3. Location Results (April–October 2005)

Between April and October 2005, 123 microearthquakes were recorded by the network (isolated or immersed in swarms). Almost 60% of them were hand-picked at more than four stations with a reading accuracy of ~5 ms. For some isolated microearthquakes, after classical polarization analysis, S-wave arrivals were also identified and hand-picked with higher uncertainties (~10 ms) respective to P arrivals.

The hypocenter location results from NonLinLoc can be displayed in many different ways: (1) as a PDF scatter cloud in which densities of dots are proportional to probabilities (impractical when many events are located nearby), (2) as the sum of PDF volumes in color scale proportional to probability, or (3) as the maximum likelihood hypocenter (minimum misfit) with the traditional (Gaussian) estimates of expectation and covariance matrix measures of the PDF. We decided here to use the last option for the location of individual microearthquakes, although to use option (2) while analyzing swarms of events (see next section).

From the hypocentral location results shown in Fig. 7, we estimate an uncertainty between 25 and 40 m in the horizontal hypocentral coordinates, and more than double this for the hypocenter depth, as expected. For the analysis and plotting of hypocenters, following LOMAX *et al.* (2001), we only use the maximum likelihood locations with a maximum semi-axis length of the 68% confidence ellipsoid of less than 60 m, and a maximum station residual of less than 0.05 s. In Fig. 7, all microearthquakes that could be satisfactorily located from the April–October 2005 period (less than 30%) are shown. The vast majority of them have been spatially located close to the actual cavern within the salt layer and in its immediate roof. For this period there is no further evidence of ruptures near, or through, the Beaumont dolomite layer. The spatial correlation with the actual cavern limits in the horizontal plane is quite consistent. On the bottom right of Fig. 7, we observe the time residuals at each station for the located microearthquakes. It is clear that stations corrections (especially at M61 and M63) would improve future hypocentre locations.

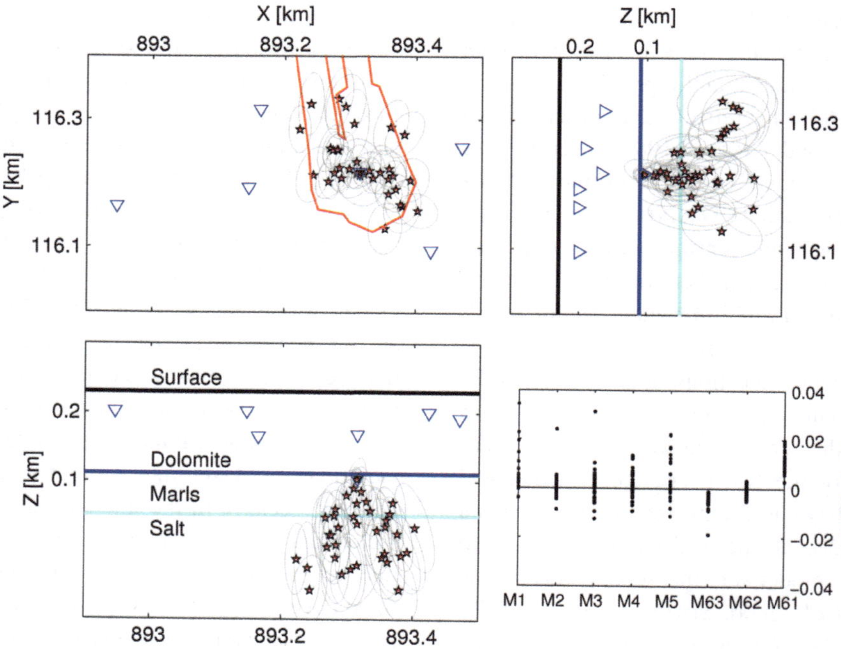

Figure 7
Hypocenter locations April–October 2005 (*stars*) with 68% confidence ellipsoids. Microseismic stations (*triangles*) and cavern outline on the horizontal plane (*solid line*). (*Bottom right*) Time residuals at each microseismic station

5.4. Location Results (May 2007)

From more than 1,000 triggered event files recorded by the network during May 2007, we could identify 134 isolated events recorded at more than four stations. They were hand-picked only for P-wave arrivals, as the S waves were rarely recognizable. Those isolated microearthquakes corresponding to the major seismic swarms of 8 May 2007 and 31 May 2007 are plotted in red in Fig. 6. We used the same observational and theoretical errors as in the 2005 microearthquake hypocenter locations. Consequently, we keep the same criteria for hypocenter location reliability for plotting and analysis. Therefore, only 53 microearthquakes have been satisfactorily located.

The results are shown in two different forms: In Fig. 8 (top) the maximum likelihood locations for the analyzed period are shown, while the sum of hypocenter PDFs computed for the microearthquakes satisfactorily located is shown in Fig. 8 (bottom) in vertical and horizontal slices with color scale proportional to probability. It is argued that the cumulative hypocenter PDFs may separate seismogenic from inactive zones (PRESTI *et al.*, 2004; SPILLMANN *et al.*, 2007). The location results strongly suggest that the activity during May 2007 was concentrated below the southmost end of the exploitation line 2,200. Apart from this, the hypocentral depths are mainly located close to the top of the salt layer, i.e. the base of the intercalated marls. However, depth determination is considerably more inaccurate as can be inferred from the vertical extension of the high probability patches in Fig. 8. Some isolated hypocenters are localized near the dolomite layer. This may indicate the beginning of brittle damage in the competent dolomite layer. Although these microearthquakes present slightly higher frequency content, we cannot discard the hypothesis that this is merely due to their proximity to station M63 (much less anelastic attenuation compared to microearthquakes generated near the cavern roof).

Temporal evolution of the recorded seismicity during May 2007 shows interesting patterns that can be further analyzed. In Fig. 9, we can see all satisfactorily located microearthquakes in a color scale representing their occurrence during May 2007. The majority of them correspond to the first tremor-like swarm of 8 May 2007 (in light blue in Fig. 9). Regarding the locations of microearthquakes preceding or immersed in the other day of high microseismic activity (31 May 2007), they cluster near the southernmost end of line 2,200 (in dark red in Fig. 9). Apart from these, there are some isolated microearthquakes located northwards between both days of high activity. It is clear that this clustering is not fortuitous and a more detailed analysis by waveform similarity (SPOTTISWOODE and MILEV, 1998) and/or multiplet approach (POUPINET *et al.*, 1994; GOT *et al.*, 1994) could provide better insight into the mechanisms of roof detachment.

5.5. Source Analysis

Microearthquake source analysis borrowed traditional techniques from regional seismological studies, of which the assumption of shear (double-couple) mechanisms is well founded. The source spectral models of BRUNE (1970) and MADARIAGA (1976) have been widely used in mining-induced seismicity for estimating source size, stress drop and fault slip (MCGARR *et al.*, 1989; GIBOWICZ *et al.*, 1991; MCGARR, 1992b; URBANCIC and YOUNG, 1993; KNOLL *et al.*, 1996). However, the application of such models in the mining environments must be done with caution, because the presence of mechanisms with explosive or implosive components are not excluded (ŠÍLENÝ, 1989; WONG *et al.*, 1989; GIBOWICZ *et al.*, 1991; MCGARR, 1992a). Furthermore, it was found that source sizes extracted from such models are sometimes too large compared to areas affected by rockbursts in mines (GIBOWICZ *et al.*, 1991; CAI *et al.*, 1998). Further work (MCGARR *et al.*, 1989; TRIFU *et al.*, 1995) suggested that seismic source complexity may well explain the differences between the source size extracted from models and the area actually damaged. An alternative model for extensive microearthquake sources based on energy considerations was developed by CAI *et al.*, (1998) who considered the normal stress acting on the crack surface as the key parameter in controlling the size of seismic ruptures. However, the application of this kind of model requires further parameters than those determined directly from seismograms, such as elastic moduli of the source environment, the crack normal stress and the surface energy of the fracture;

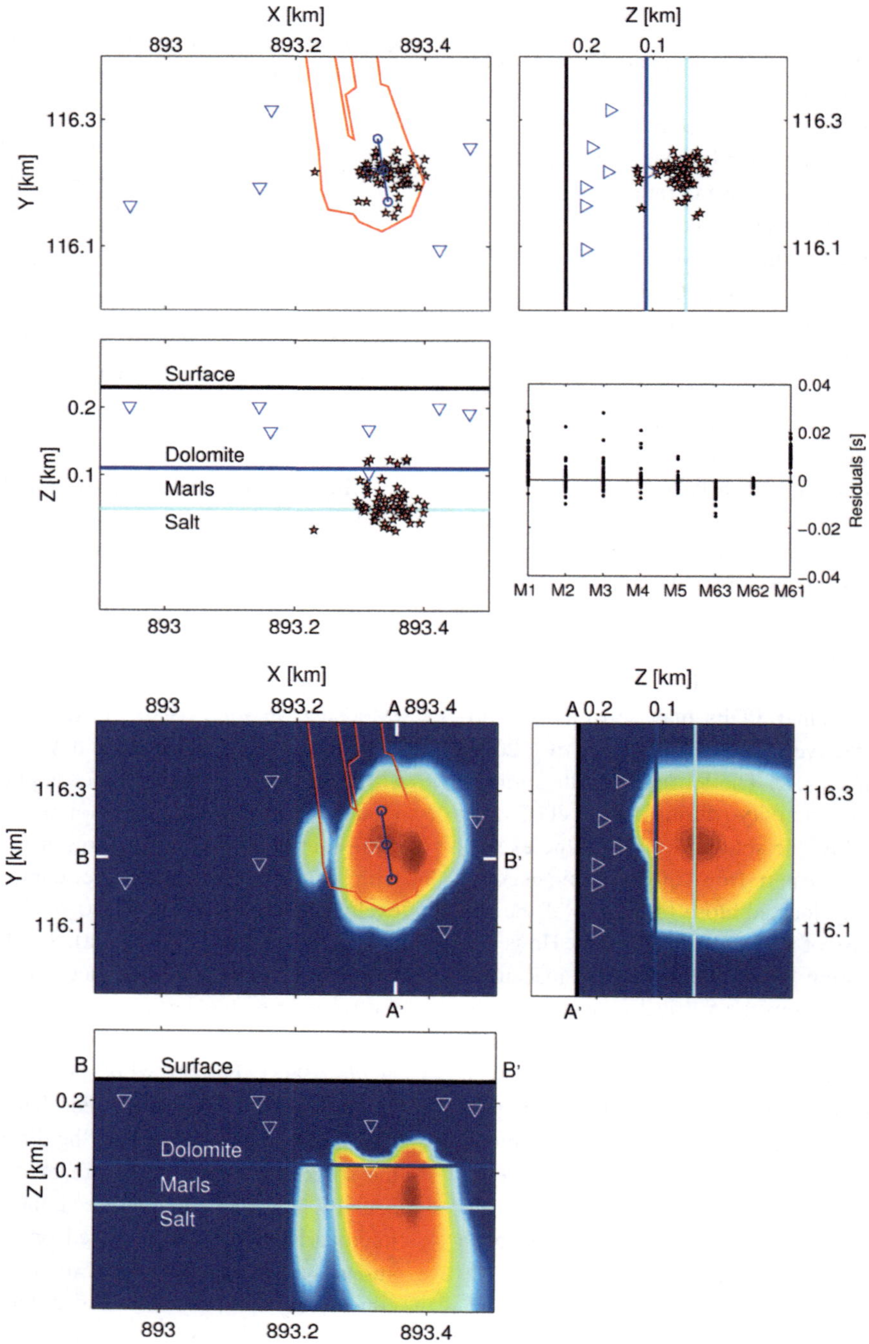

Figure 8

Hypocenter locations of May 2007: (*Top*) maximum likelihood hypocenters (*stars*) and time residuals at each station, (*Bottom*) PDF sums of the previous probabilistic locations. The three exploitation wells of line 2,200 where significant roof evolution was observed are marked on the horizontal plane (*blue circles*)

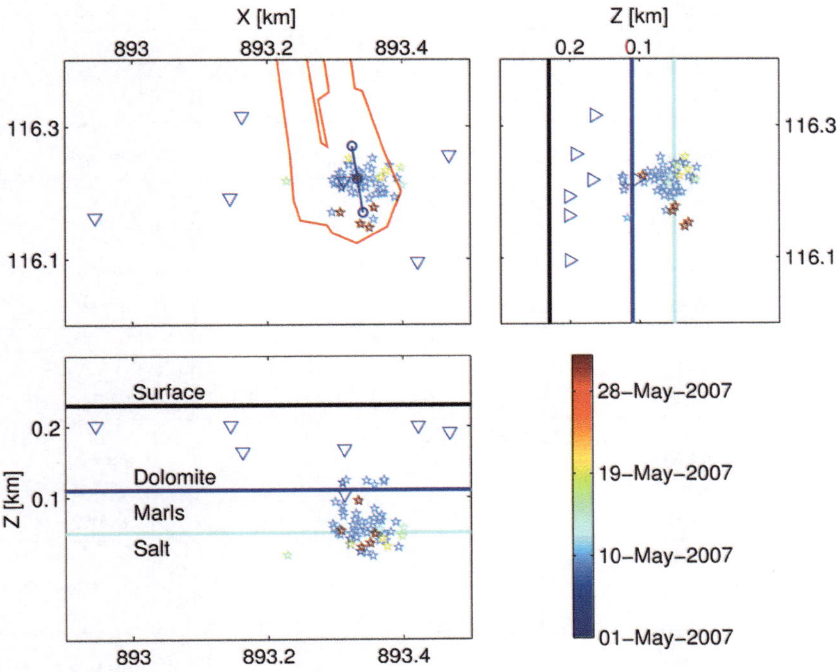

Figure 9
Hypocenter locations of May 2007. *Color scale* represents occurrence times

parameters difficult to obtain in most in situ applications. Walter and Brune (1993) showed that the spectral shape of a pure tensile or a shear crack does not differ from each other, however the low frequency amplitude from the displacement spectra of P wave (Ω_p) and S wave (Ω_s) averaged over the focal sphere differs significantly, mainly due to differences in radiation patterns ($\Omega_s/\Omega_p \sim 2.1$ for pure extensive crack, and $\Omega_s/\Omega_p \sim 7.1$ for pure shear crack).

Therefore, a straighforward manner to identify the presence of tensile components in microearthquake sources is simply based on calculating this ratio. In order to evaluate the magnitude and approximate size[2] of microearthquakes triggered during the analyzed period, we estimate the seismic moment M_0 and the moment magnitude M_w associated from corrected spectra using the formulas introduced by Hanks and Kanamori (1979),

$$M_w = \frac{2}{3}\log_{10}(M_o) - 6, \qquad (1)$$

with

$$M_o = \frac{4\pi\rho V_{p,s}^3 r\Omega_{0(p,s)}}{R_{p,s}} \text{ in [Nm]}, \qquad (2)$$

where ρ is the density, $V_{p,s}$ the wave velocities, r the hypocentral distance (maximum likelihood location), and $R_{p,s}$ the mean radiation patterns (absolute values).

Some isolated microearthquakes with clear S arrivals are selected and rotated to analyze the SH–SV spectra, and to compare them with those of P waves. In Fig. 10, two microearthquakes recorded in the four 3-component stations are presented. After correcting for geophone response in order to recover the low frequency energy below 40 Hz, we converted to acceleration spectra using $A_{0(p,s)} = (2\pi f_c)^2\Omega_{0(p,s)}$, where f_c is the corner frequency. Finally, we corrected for anelastic attenuation with $Q_p = Q_s = 20$ using a simple exponential model independent of frequency, in order to better determine the high-frequency amplitude level of acceleration spectra. The Q_p value was estimated between 10 and 30 from spectral

[2] In contrast to the moment magnitude, the source size and stress drop are strongly model-dependent, we only attempt here to obtain a rough estimation of these quantities (order of magnitude).

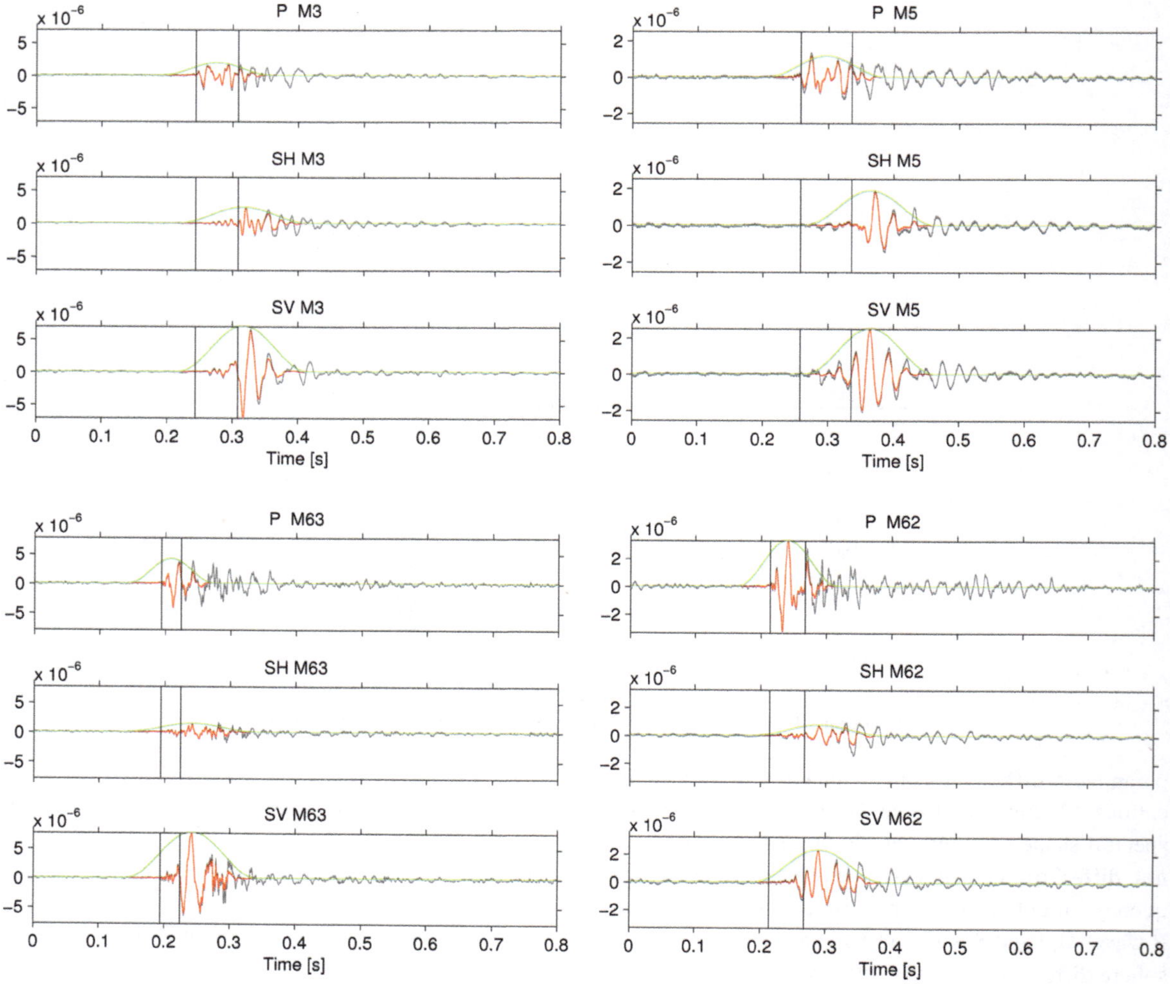

Figure 10
Example of isolated microearthquake recorded on 8 May 2007 at 3-component stations M3, M5, M63 and M62. *Bold lines* denote windowed portions of P and S waves used for spectral analysis. P–S picked arrival times shown as *vertical lines*

analysis of active seismic data recorded at the site in October 2004 (Mercerat, 2007). The assumption $Q_P = Q_s$ does not have a major influence on the $\Omega_{p,s}$ determination. In Fig. 11, we can see two examples of corrected acceleration spectra at the four 3-component stations. Differences in amplitude levels are surely related to differences in radiation patterns. In any case, we estimate an uncertainty of 50 and 20% in the high-frequency amplitude level and the corner frequency determinations, respectively.

From the results shown in Table 1, we notice that: (1) All analyzed microearthquakes have moment magnitudes smaller than -1, (2) all of them present Ω_s/Ω_p lower than 4, suggesting the presence of extensive components in the source mechanisms, and (3) the static stress drops (Brune, 1970) are rather low compared to classical mining-induced seismicity in crystalline rocks (Gibowicz *et al.*, 1991; Feignier and Young, 1992; Urbancic and Young, 1993; Oye *et al.*, 2005), however we are close to the lower limit of stress drops reported in previous studies of microseismicity associated with fluid injection at similar depth and lithologies (Talebi and Boone, 1998; Talebi *et al.*, 1998). It must be stressed that if we used Madariaga's model assuming a rupture propagation velocity of 0.9 Vs, we would obtain

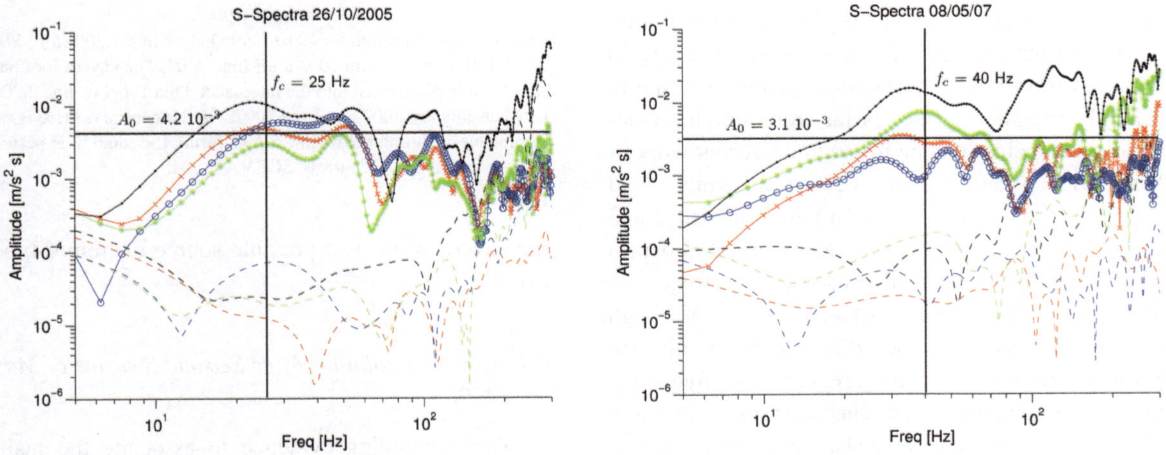

Figure 11
Corrected S-wave spectra (vector sum of SH-SV amplitudes) for microearthquakes 051026_09382470 (*left*), and 070508_21560422 (*right*, traces shown in Fig. 7) recorded at M63 (x-), M62 (o-), M5 (*-), M3 (.-). Noise spectra of 0.1 s before P-wave arrivals (*dashed lines*)

Table 1

Source parameters for selected events from October 2005 and May 2007

Events	A_0 (10^{-3} m/s)	f_c (Hz)	M_o (10^7 Nm)	M_w	r_0 (m)	$\Delta\sigma$ (bar)	Ω_s/Ω_p
051002_18065527	20.0 (1 ± 0.5)	90 (1 ± 0.2)	2.84 ± 1.64	−1.1 ± 0.2	8.5 ± 1.5	0.202 ± 0.095	3.0
051002_18065645	3.0 (1 ± 0.5)	50 (1 ± 0.2)	1.38 ± 0.79	−1.3 ± 0.2	15.3 ± 2.7	0.017 ± 0.008	2.0
051026_09382470	4.2 (1 ± 0.5)	25 (1 ± 0.2)	7.74 ± 4.45	−0.8 ± 0.2	30.6 ± 5.4	0.012 ± 0.006	4.0
051017_15270917[a]	4.9 (1 ± 0.5)	55 (1 ± 0.2)	1.86 ± 1.07	−1.2 ± 0.2	13.9 ± 2.4	0.030 ± 0.014	2.6
070508_21560276[a]	6.0 (1 ± 0.5)	110 (1 ± 0.2)	0.57 ± 0.32	−1.6 ± 0.2	6.9 ± 1.2	0.074 ± 0.035	2.1
070508_21560422[a]	3.1 (1 ± 0.5)	40 (1 ± 0.2)	2.23 ± 1.28	−1.2 ± 0.2	19.1 ± 3.4	0.014 ± 0.007	2.5
070508_22021850[a]	0.9 (1 ± 0.5)	120 (1 ± 0.2)	0.07 ± 0.04	−2.1 ± 0.2	6.4 ± 1.1	0.012 ± 0.006	1.6
070508_22021948[a]	3.0 (1 ± 0.5)	35 (1 ± 0.2)	2.82 ± 1.62	−1.1 ± 0.2	21.8 ± 3.8	0.012 ± 0.006	3.2

Brune (1970) source model: r_0 source radius, $\Delta\sigma$ static stress drop. Physical parameters: $Q_p = Q_s = 20$, $V_p = 3{,}000$ m/s, $V_s = 2000$ m/s, $\rho = 2500$ kg/m^3, $R_p = 0.44$, $R_s = 0.60$ [mean radiation patterns (Boore and Boatwright, 1984)]

[a] Events immersed in swarms

source sizes 56% smaller and stress drops eight times higher than those obtained by Brune's model (Madariaga, 1979). To conclude relative to the applicability of any specific source model to microearthquakes recorded at Cerville-Buissoncourt is beyond the scope of this paper.

6. Discussion

Between January 2005 and May 2007, the brine-filled cavern entered a rather stable period during which the overburden did not deform considerably beyond some millimetric subsidence occurring after both transient pressure experiments. The microseismic activity recorded can be related to some isolated fragile ruptures in the insoluble layers found within the salt deposit, and to delamination of the clayley marls on top of the salt layer with a noticeable extensive component in the source mechanism, followed by block drops and debris flows within the cavern. The fact that the immediate cavern roof is mainly composed of weak marly material in contact with brine may also lead to the possibility of detaching aseismically, and consequently all recorded microseismicity would be caused by rock falls and debris flow within the cavern. In this case, all isolated microearthquake locations should concentrate near the cavern floor.

Note that the current cavern has a rather irregular shape and its base is partially filled by non-soluble rock debris that was found in the salt layer during cavern formation by salt dissolution. The signal expected from falling blocks of degraded material in a highly irregular surface is far from simple, nonetheless some clues can be found in the recent work of Wust Bloch and Joswig, (2006) in which field experiments of falling blocks in both brine-filled and empty superficial salt caverns (less than 30 m depth) were carried out. They have used a time–frequency analysis to discriminate between them. Although their conditions are rather different from ours (in particular cavern depth and pressure), the frequency range and general signal characteristics are quite alike (emergent first arrivals, frequencies below 75 Hz and tens of seconds of signal duration). Very similar seismic activity also has been recorded by local seismic monitoring networks in Presall Brine Field (Blackpool, England) and in Northridge (Cheshire, England) to monitor abandoned underground salt mines (Branston, 2003). The hypothesis concerning the origin of this activity (progressive delamination of clayley marls and block drops) is quite similar to the one described in this work. In addition, Fortier *et al.*, (2006) studied the microseismicity of brine-filled caverns used for hydrocarbon storage in the Geosel-Manosque field (south of France), where they found evidence of small brittle ruptures followed by falling blocks within underground caverns of more than 300 m height that generate clear monochromatic resonant waves. In our case, it must be stressed that the cut-off frequency of the array sensors (40 Hz), tuned to register microseismic activity generated once the solution mining intensifies and the cavern migrates towards the surface, is just too high to neatly record resonant frequencies corresponding to cavitations.

On the other hand, recent microseismic investigations of unstable mountain slopes (Suriñach *et al.*, 2005; Brückl and Mertl, 2006; Spillmann *et al.*, 2007), as well as experimental studies on ground vibrations caused by various rocks and debris flows in channel beds (Huang *et al.*, 2004, 2007), show signal characteristics comparable to the ones recorded at Cerville-Buissoncourt in terms of frequency content (<50 Hz), time duration (between 5 and 20 s) and moment magnitude ($-2 < M_w < 0$). These results reinforce the idea of debris flow along the cavern walls as a possible source of tremor-type events.

Figure 12 ▶
Cavern shape evolution (*colored lines*) below line 2,200 (*top*) and line 2,100 (*bottom*) from 2002 until June 2007. The cavern roof has significantly progressed into the intercalated marls below line 2,200 between January 2007 and June 2007. On the contrary, there is no significant evolution below line 2,100 during the same time period (Courtesy SOLVAY S.A.)

6.1. *Cavern Evolution After Seismic Swarms of May 2007*

The concluding evidence to associate the main seismic activity recorded in the site with roof detachment followed by rock debris falling in the brine-filled cavern, arrived after the correlation of direct field observations with hypocenter location results for microearthquakes from the seismic swarms of May 2007 (Fig. 8). Subsequent to borehole control measurements of the cavern top level carried out in June 2007, SOLVAY S.A. confirmed up to 25 m of cavern roof migration towards the surface below the southernmost three wells of exploitation line 2,200, just on top of the actual cavern position (Fig. 12). It can therefore be conjectured that the isolated, relatively high-frequency, events correspond to small ruptures in the cavern roof (progressive overburden damage) and in the competent anhydritic levels within the salt layer, while the low frequency tremor-type events correspond to block drops and debris flows within the brine-filled cavern. This idea is sketched in Fig. 13.

This evolution is related to the superposition of the effect of partial salt dissolution conducted about 200 m from the instrumented cavern, and on the other side, the natural roof evolution by marl degradation in contact with the saturated brine. These two effects cannot be discriminated as long as the cavern remains under the indirect influence of the running dissolution. These observations were confirmed by regular well-log measurements: the cavern roof continued to evolve in accordance with recorded microseismic activity and the latest logs (new raise of 6 m of the roof in January 2008). In all probability, this stage of cavern evolution may be comparable with the initial phase of the collapse process which will be monitored

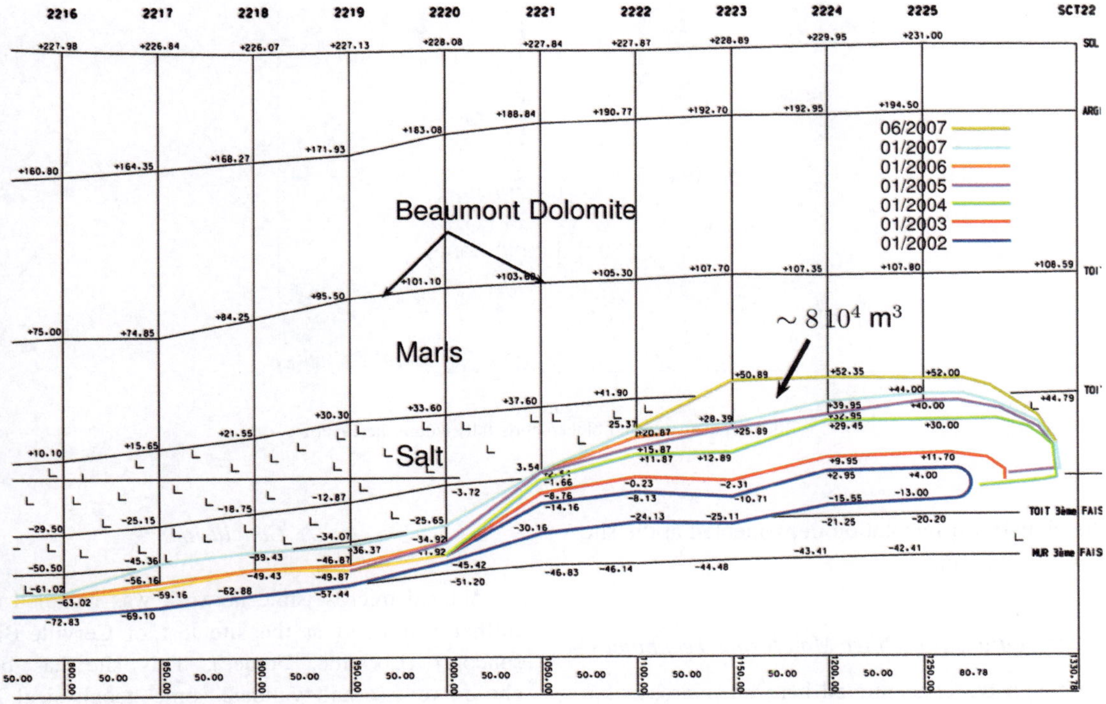
2216
2217
2218
2219
2220
2221
2222
2223
2224
2225
SCT22
06/2007
01/2007
01/2006
01/2005
01/2004
01/2003
01/2002
Beaumont Dolomite
Marls
Salt
~ 8 10^4 m^3
TOIT 3ème FAIS
MUR 3ème FAIS

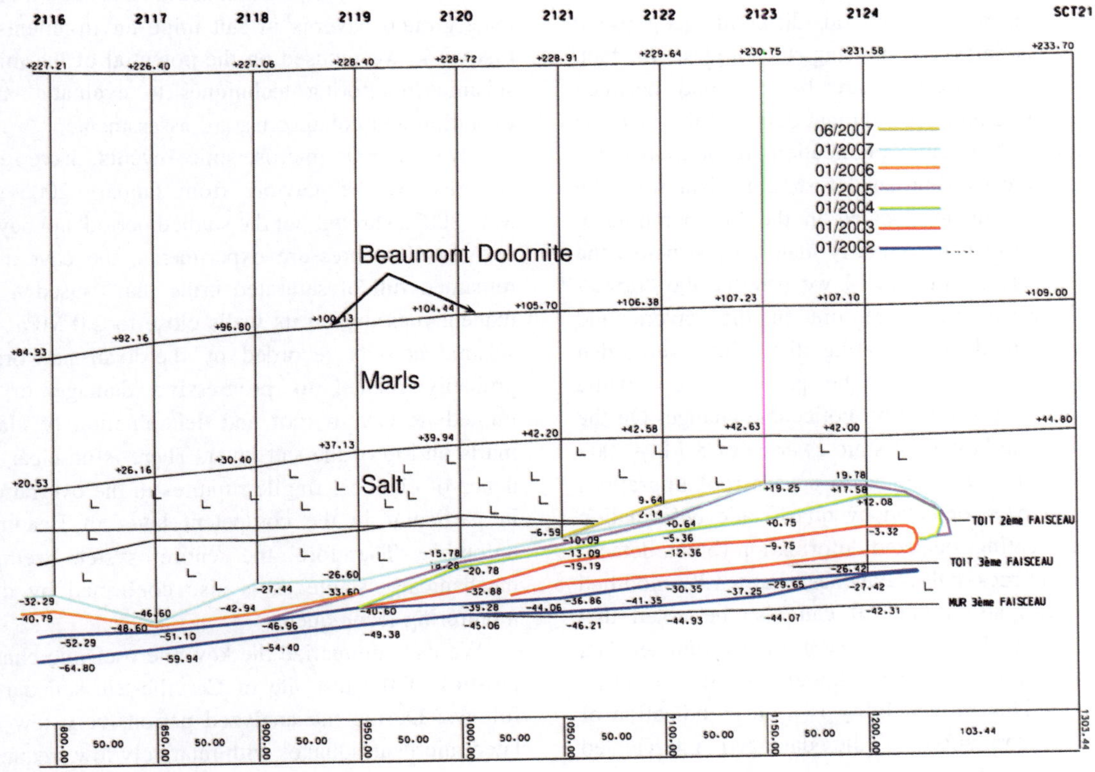
2116
2117
2118
2119
2120
2121
2122
2123
2124
SCT21
06/2007
01/2007
01/2006
01/2005
01/2004
01/2003
01/2002
Beaumont Dolomite
Marls
Salt
TOIT 2ème FAISCEAU
TOIT 3ème FAISCEAU
MUR 3ème FAISCEAU

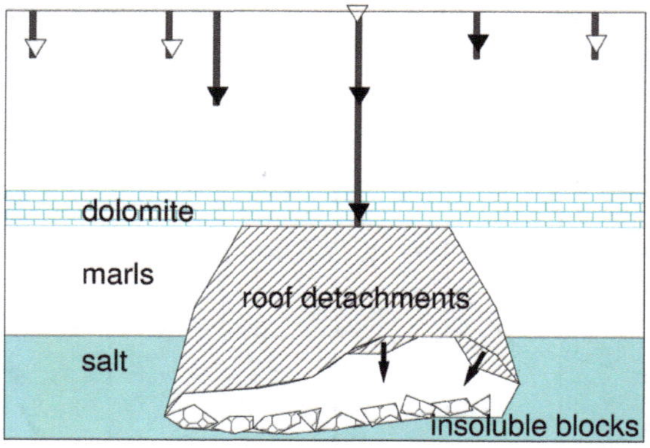

Figure 13
Roof detachments and block/debris falls within the cavern

by the global instrumentation implemented at the site of Cerville-Buissoncourt.

6.2. *Correlation with Other Monitoring Techniques*

Although we concentrated here on microseismic data analysis, the site of Cerville-Buissoncourt is being used to test several different geophysical and geotechnical monitoring techniques. In fact, interesting correlations are being found between microseismic and geotechnical data and they will be the object of a future publication. In particular, we present here overburden deformation data from the deepest strainmeter located in the M6 borehole at 129 m depth (in weak marly material just below the dolomite layer). In Fig. 14 we observe the correlation between brine pressure in the cavern and deformation data indicating that the overburden immediately reacted to the pressure drop, while the seismicity showed no noticeable change. On the contrary, the largest seismic swarms of 8 May 2007 and 31 May 2007 related to cavern roof migrations have not been detected by quasi-static deformation data, indicating localized deformation (block detachments and rock-fall avalanches) that hardly modified the overall stress field. It can also be noted that there was negligible recovery in the strainmeter data after restoration of the original 2.0 MPa of brine pressure. This irreversibility is a clear indication of plastic deformation in the damaged intercalated marls just on top of the cavern roof.

7. *Conclusion*

A local microseismic network was installed and further calibrated at the site test of Cerville-Buissoncourt (Lorraine, France). This site has been chosen to test and validate some geophysical and geotechnical techniques that could be used to monitor underground caverns in salt mine environments. In this work, we focused on the potential of the microseismic monitoring techniques to evaluate cavern evolution and collapse hazard assessment.

Over 2,000 microseismic events have been recorded by the network from January 2005 until May 2007. Throughout the studied period and beyond two transient pressure experiments, the cavern has remained full of saturated brine that caused a permanent pressure on its walls close to 2.0 MPa. The seismic activity recorded by the local network is probably related to progressive damage of the immediate cavern roof and delamination of clayey marls on top of the salt layer. There is no clear evidence of classical fragile ruptures in the overburden, in particular in the competent layer of Beaumont dolomite. Therefore the entire system remains mechanically stable as is also confirmed by other monitoring techniques.

We can summarize the key microseismic characteristics of the test site of Cerville-Buissoncourt as follows: During the analyzed period, very few isolated microearthquakes with relatively low frequency content (20–120 Hz) and rather emergent first

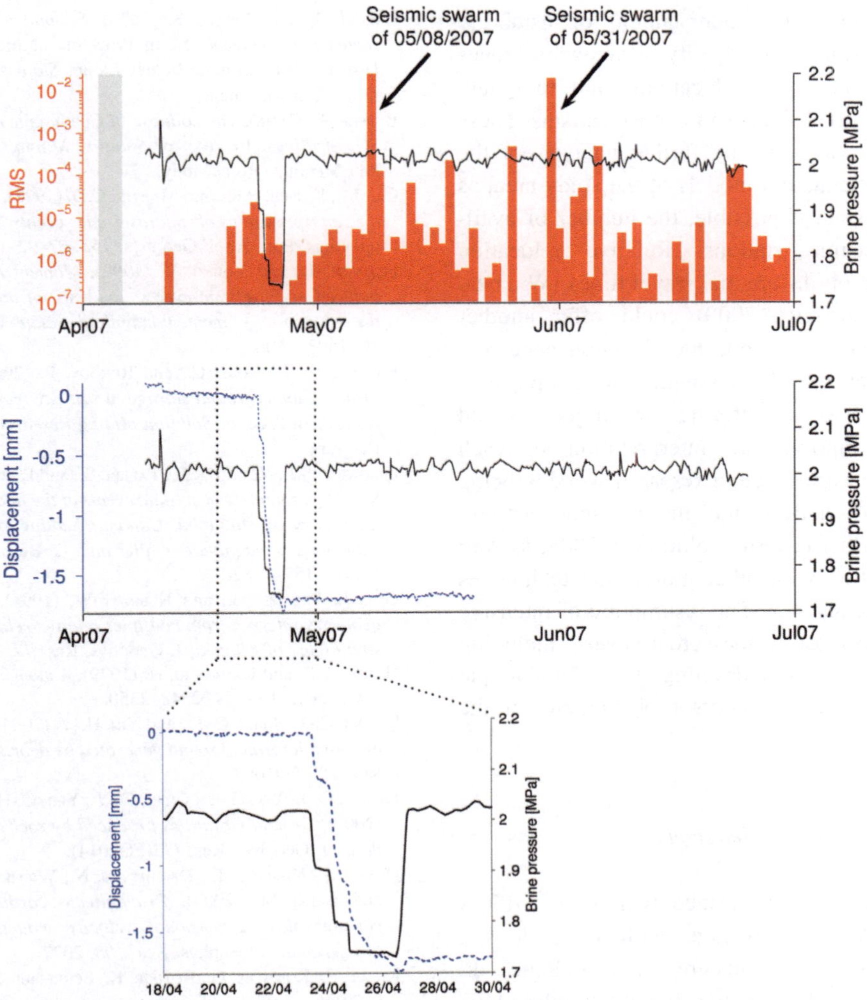

Figure 14
Recorded RMS per day (*vertical bars*), brine pressure (*solid line*) and vertical displacement from strainmeter (*dashed line*) just below the dolomite layer from April 2007 until July 2007. Main seismic swarms of May 2007 are marked by *arrows*

arrivals were recorded. All hypocenter locations concentrate in the actual cavern surroundings within the salt layer and the intercalated marls on top of it. From spectral source analysis, typical moment magnitudes M_w are lower than -1 and the presence of extensive components in the source mechanisms is suggested. In contrast, we found a large majority of tremor-type events with tens of seconds of duration, presenting amplitudes one order of magnitude greater than isolated microearthquakes. We hypothetize that individual microearthquakes may reflect small ruptures of fragile layers within the salt layer and on the actual cavern roof, while tremor-type events are the consequence of large block falls and debris flows within the brine-filled cavern. The largest seismic swarms occurred at the site in May 2007. Despite larger uncertainties, we could pick and locate some isolated microearthquakes immersed in long tremor-type events. The hypocenter location results correlate fairly well with the information supplied by SOLVAY S.A. of more than 25 m of cavern roof migration towards the surface below the southernmost end of line 2,200.

Throughout this work, we highlighted the potential of microseismic monitoring as a useful tool to assess overburden damage and salt cavern evolution.

In order to improve the understanding of instability phenomena, we need to identify seismogenic zones by more accurate spatial locations, and to obtain better insight into physical source mechanisms. These goals could be achieved by further analysis of the velocity and attenuation models of the study area, as well as increasing, if possible, the number of available microseismic stations. Relative relocation methods based on differential travel times (WALDHAUSER and ELLSWORTH, 2000) could offer another approach to increase hypocenter location accuracy. Apart from that, correlations with other geophysical and geotechnical monitoring techniques could significantly improve the interpretation of each microseismic swarm. Actual research work is being carried out to analyze the microseismic activity induced by further cavern evolution in 2008, as well as the correlation with other monitoring techniques deployed at the test site. The resumption of intensive salt dissolution close to the actual cavern, including depressurization by brine draining, will offer a unique opportunity to record precursor phenomena of the final overburden collapse.

Acknowledgments

This work has been performed within the GISOS framework. We acknowledge with gratitude the financial support of Ministère de l'Écologie, du Développement et de l'Aménagement Durables (MEDAD). The authors also thank SOLVAY S.A. for the experimental site, the technical support and the data for site characterization. Comments and remarks from two anonymous reviewers and the guest editor Cezar Trifu helped us to improve the final manuscript.

REFERENCES

BOORE, D.M. and BOATWRIGHT, J. (1984), *Average body wave radiation coefficients*, Bull. Seismol. Soc. Am. *74*(5), 1615–1621.

BRANSTON, M. (2003), *The application of time-lapse microgravity to investigate and monitor subsidence related to salt dissolution*, Ph.D. Thesis, Keele University, England.

BRUNE, J.N. (1970), *Tectonic stress and the spectra of seismic shear waves from earthquakes*, J. Geophys. Res. *75*(26), 4997–5009.

BRÜCKL, E. and MERTL, S. (2006), *Seismic monitoring of deep-seated mass movements*. In Proc. of International Symposium *Disaster Mitigation of Debris Flows, Slope Failures and Landslides*, Tokyo, Japan.

BUFFET, A. (1998), *The collapse of Compagnie des Salins SG4 and SG5 drillings*. In Proc. of *Solution Mining Research Institute*, Fall Meeting, Rome, Italy.

CAI, M., KAISER, P.K., and MARTIN, C.D. (1998), *A tensile model for the interpretation of microseismic events near underground openings*, Pure Appl. Geophys. *153*, 67–92.

FEIGNIER, B. and YOUNG, R. (1992), *Moment tensor inversion of induced microseismic events: Evidence of non-shear failures in the $-4 < m < -2$ moment magnitude range*, Geophys. Res. Lett. *19*, 1503–1506.

FORTIER, E., MAISONS, C., and RENOUX, P. (2006), *Seismic monitoring of underground storage in salt cavity in a seismo-tectonic context*. In Proc. of *Solution Mining Research Institute*, Brussel, Belgium.

GIBOWICZ, S.J., YOUNG, R.P., TALEBI, S., and LAWRENCE, D.J. (1991), *Source parameters of seismic events at the underground research laboratory in Manitoba, Canada: Scaling relations for events with moment magnitude smaller than -2*, Bull. Seismol. Soc. Am. *81*(4), 1157–1182.

GOT, J., FRÉCHET, J., and KLEIN, F.W. (1994), *Deep fault plane geometry inferred from multiplet relative relocation beneath the south flank of Kilauea*, J. Geophys. Res. *99*, 15375–15386.

HANKS, T.C. and KANAMORI, H. (1979), *A moment magnitude scale*, J. Geophys. Res. *84*, 2348–2350.

HUANG, C.-J., SHIEH, C.-L., and YIN, H.-Y. (2004), *Laboratory study of the underground sound generated by debris flows*, J. Geophys. Res. *109*(FO1008).

HUANG, C.-J., YIN, H.-Y., CHEN, C.-Y., YEH, C.-H., and WANG, C.-L. (2007), *Ground vibrations produced by rock motions and debris flows*, J. Geophys. Res. *112*(FO2014).

HUSEN, S., KISSLING, E., DEICHMANN, N., WIEMER, S., GIARDINI, D., and BAER, M. (2003), *Probabilistic earthquake location in complex three-dimensional velocity models: Application to Switzerland*, J. Geophys. Res. *108*, 2077.

KNOLL, P., KOWALLE, G., ROTHER, K., SCHREIBER, B., and PASKALEVA, I. (1996), *Analysis of microtremors within the Provadia region near a salt leaching mine*, Pure Appl. Geophys. *147*(2), 389–407.

LIPPITSCH, R., WHITE, R., and SOOSALU, H. (2005), *Precise hypocentre relocation of microearthquakes in a high-temperature geothermal field: The Torfajokull central volcano, Iceland*, Geophys. J. Int. *160*, 370–387.

LOMAX, A. (2006), *An on-line guide to software for Probabilistic, Non-Linear, Global-Search Earthquake Location in 3D Media*, Anthony Lomax Scientific Software, Mouans-Sartoux, France.

LOMAX, A., VIRIEUX, J., VOLANT, P., and BERGE, C., *Probabilistic earthquake location in 3D and layered models: Introduction of a metropolis-gibbs method and comparison with linear locations*. In (C.H. Thurber and N. Rabinowitz, Eds.), *Advances in Seismic Event Location*, pp. 101–134 (Amsterdam, Kluwer 2000).

LOMAX, A., ZOLLO, A., CAPUANO, P., and VIRIEUX, J. (2001), *Precise, absolute earthquake location under Somma-Vesuvius volcano using a new 3D velocity model*, Geophys. J. Int. *146*, 313–331.

MADARIAGA, R. (1976), *Dynamics of an expanding circular fault*, Bull. Seismol. Soc. Am. *66*(3), 639–666.

MADARIAGA, R. (1979), *On the relation between seismic moment and stress drop in the presence of stress and strength heterogeneity*, J. Geophys. Res. *84*(B5), 2243–2250.

MAISONS, C., FORTIER, E., and VALETTE, M. (1997), *Induced microseismicity and procedure for closure of brine production caverns*, Pure Appl. Geophys. *150*, 585–603.

MCGARR, A. (1992a), *An implosive component in the seismic moment tensor of a mining-induced tremor*, Geophys. Res. Lett. *19*(15), 1579–1582.

MCGARR, A. (1992b), *Moment tensors of ten Witwatersrand mine tremors*, Pure Appl. Geophys. *139*, 781–800.

MCGARR, A., BICKNELL, J., SEMBERA, E., and GREEN, R. (1989), *Analysis of exceptionally large tremors in two gold mining districts of South Africa*, Pure Appl. Geophys. *129*(3-4), 295–307.

MENDECKI, A.J., *Seismic Monitoring in Mines* (Chapman and Hall, London, UK 1997).

MERCERAT, E.D. (2007), *Sismicité induite et modélisation numérique de l'endommagement dans un contexte salin*, Thèse de doctorat (Ph.D. Thesis), LAEGO, Institut National Polytechnique de Lorraine, France.

MERCERAT, E.D., SOULEY, M., DRIAD-LEBEAU, L., and BERNARD, P. (2007), *Evaluation of overburden damage in a salt mine environment by a coupled continuum-discrete numerical modelling approach*. In Proc. of *Solution Mining Research Institute*, Basel, Switzerland.

MOSER, T., VAN ECK, T., and NOLET, G. (1992), *Hypocenter determination in strongly heterogeneous earth models using the shortest path method*, J. Geophys. Res. *97*, 6563–6572.

NOTHNAGEL, R. (2003), *Modélisation des instabilités en Mécanique des Roches: Application à l'exploitation de la concession de Drouville*, Thèse de Doctorat (Ph.D. Thesis), École des Mines de Paris, France.

OYE, V., BUNGUM, H., and ROTH, M. (2005), *Source parameters and scaling relations for mining-induced seismicity within the Pyhäsalmi Ore Mine, Finland*, Bull. Seismol. Soc. Am. *95*(3), 1011–1026.

POUPINET, G., ELLSWORTH, W., and FRÉCHET, J. (1994), *Monitoring velocity variations in the crust using earthquake doublets: an application to the Calaveras fault, California*, J. Geophys. Res. *89*, 5719–5731.

PRESTI, D., TROISE, C., and NATALE, G.D. (2004), *Probabilistic location of seismic sequences in heterogeneous media*, Bull. Seismol. Soc. Am. *94*, 2239–2253.

SOULEY, M., MERCERAT, E.D., DRIAD-LEBEAU, L., and BERNARD, P. (2008), *A large scale continuum-discrete numerical modelling: Application to overburden damage of a salt cavern*. In *Post-Mining 2008*, Nancy, France.

SPILLMANN, T., MAURER, H., GREEN, A.G., HEINCKE, B., WILLENBERG, H., and HUSEN, S. (2007), *Microseismic investigation of an unstable mountain slope in the Swiss Alps*, J. Geophys. Res. *112*(B07301).

SPOTTISWOODE, S. and MILEV, A. (1998), *The use of waveform similarity to define planes of mininginduced seismic events*, Tectonophysics *289*, 51–60.

SUFFERT, J., GRANDJEAN, G., and BITRI, A. (2006), *Etude d'une cavité saline par méthodes sismiques*, Rapport de stage BRGM-EOST. RP-54531-FR.

SURIÑACH, E., VILAJOSANA, I., KHAZARADZE, G., BIESCAS, B., FURDADA, G., and VILAPLANA, J.M. (2005), *Seismic detection and characterization of landslides and other mass movements*, Natural Hazards and Earth System Science *5*(6), 791–798.

TALEBI, S. and BOONE, T.J. (1998), *Source parameters of injection-induced microseismicity*, Pure Appl. Geophys. *153*(1), 113–130.

TALEBI, S., BOONE, T.J., and EASTWOOD, J.E. (1998), *Injection-induced microseismicity in Colorado shales*, Pure Appl. Geophys. *153*(1), 95–111.

TARANTOLA, A. and VALETTE, B. (1982), *Inverse problems = quest for information*, J. Geophys. *50*, 159–170.

TRIFU, C.-I., URBANCIC, T.I., and YOUNG, R. (1995), *Source parameters of mining-induced seismic events: An evaluation of homogeneous and inhomogeneous faulting models for assessing damage potential*, Pure Appl. Geophys. *145*, 461–470.

URBANCIC, T.I. and YOUNG, R.P. (1993), *Space-time variations in source parameters of mining-induced seismic events with* $M < 0$, Bull. Seismol. Soc. Am. *83*, 378–397.

ŠÍLENÝ, J. (1989), *The mechanism of small mining tremors form amplitude inversion*, Pure Appl. Geophys. *129*(3-4), 309–324.

WALDHAUSER, F. and ELLSWORTH, W. (2000), *A double-difference earthquake location algorithm: Method and application to the northern Hayward Fault, California*, Bull. Seismol. Soc. Am. *90*(6), 1353–1368.

WALTER, W.R. and BRUNE, J.N. (1993), *Spectra of seismic radiation from a tensile crack*, J. Geophys. Res. *98*(B3), 4449–4459.

WASSERMANN, J., *IASPEI New manual of seismological observatory practice*, Chapter 13: *Volcano Seismology*, pp. 42. (P. Bormann ed.,) (GeoForschungsZentrum Potsdam 2002).

WONG, I., HUMPHREY, J., ADAMS, J., and SILVA, W.J. (1989), *Observations of mine seismicity in the Eastern Wasatch Plateau, Utah : A possible case of implosional failure*, Pure Appl. Geophys. *129*(3-4), 369–405.

WUST BLOCH, G.H. and JOSWIG, M. (2006), *Pre-collapse identification of sinkholes in unconsolidated media at Dead Sea area by 'nanoseismic monitoring' (graphical jackknife location of weak sources by few, low-SNR records)*, Geophys. J. Int. *167*(3), 1220–1232.

ZOLLO, A., MARZOCCHI, W., CAPUANO, P., LOMAX, A., and IANNACCONE, G. (2002), *Space and time behavior of seismic activity at Mt. Vesuvius volcano Southern Italy*, Bull. Seismol. Soc. Am. *92*, 625–640.

(Received September 2, 2008, revised March 5, 2009, accepted April 29, 2009, Published online November 25, 2009)

Pure Appl. Geophys. 167 (2010), 27–42

DOI 10.1007/s00024-009-0013-4

Pure and Applied Geophysics

Microseismic Monitoring of a Controlled Collapse in Field II at Ocnele Mari, Romania

C-I. Trifu[1] and V. Shumila[1]

Abstract—Several decades of faulty exploitation of salt through solution mining led to the creation of an underground cavern containing several million cubic meters of brine. To eliminate the huge hazard near a densely inhabited area, a technical solution was implemented to resolve this instability concern through the controlled collapse of the roof while pumping the brine out and filling the cavern with sterile. To supervise this, an area of over 1 km^2 was monitored with a staggered array of 36 one-component, 15 Hz geophones installed in 12 boreholes about 160–360 m deep. A total of 2,392 seismic events with M_w −2.6 to 0.2 occurred from July 2005 to March 2006, located within an average accuracy of 18 m. The *b*-value of the frequency-magnitude distribution exhibited a time variation from 0.5 to 1 and from there to 1.5, suggesting that the collapse initiated as a linear fracture pattern, followed by shear planar fragmentations and finally a 3-D failure process. The brunching ratio of seismicity is indicative of a super-critical process, except for a short period in mid-February when temporary stability existed. Event relocation through the use of a collapsing technique outlines that major clusters of seismicity were associated with the main cavern collapse, whereas smaller clusters were generated by the fracturing of smaller size nearby caverns. It is shown that one-component recordings allow for stable and reliable point source event mechanism solutions through automatic moment tensor inversion using time domain estimates of low frequency amplitudes with first polarities attached. Detailed analysis of failure mechanism components uses 912 solutions with conditional number CN < 100 and a correlation coefficient $r^2 > 0.5$. The largest pure shear (DC) components characterize the events surrounding the cavern ceiling, which exhibit normal and strike-slip failures. The majority of mechanism solutions include up to 30% explosional failure components, which correspond to roof caving under gravitational collapsing. The largest vertical deformation rate relates closely to the cavern roof and floor, as well as the rest of the salt formation, whereas the horizontal deformation rate is most prominent in areas of detected collapses.

[1] Engineering Seismology Group Inc., Kingston, ON K7K 7G3, Canada. E-mail: trifu@esg.ca

1. Introduction

The salt deposit of Ocnele Mari is located in the sub-Carpathian hills of southern Romania. About 8 km long, it runs from east to west, with a width of 3.5 km and a thickness of up to 400 m, dipping about 20° to the north (Fig. 1). Salt has been exploited through dissolution in four fields which entered in production sequentially, beginning with 1954. Operations in Field II were initiated in 1969 and led to the extraction of 13.5 million tones of salt until a major collapse occurred in March 1991. This event underscored the risks associated with the presence of large dissolution chambers close to a densely populated area, prompting the decision to shut-down the field (Zamfirescu *et al.* 2007a).

Cavernometry measurements carried out in 1993 outlined the presence of a massive cavern formed by the complete dissolution of the inter-chamber pillars of 6 of the 15 wells, containing 5.5 million m^3 of compressed brine. Another major collapse took place towards the northern edge of the cavern in September 2001, when a part of the roof caved in, forming a quasi-circular crater of almost 200 m in diameter and leading to the spill of 1.7 million m^3 of brine. The ground material plunged from the surface, however, acted like a cork, stopping a larger outflow of brine. A second major event occurred in July 2004, during which the original sinkhole was enlarged, accelerating thus the demand for a solution to this mechanical instability problem.

A description detailing how the controlled collapse of the cavern roof was designed and carried out is presented by Zamfirescu *et al.* (2007b). In short, a fragmentation process was sought by cutting open a part of the cavern roof, pumping the brine out and

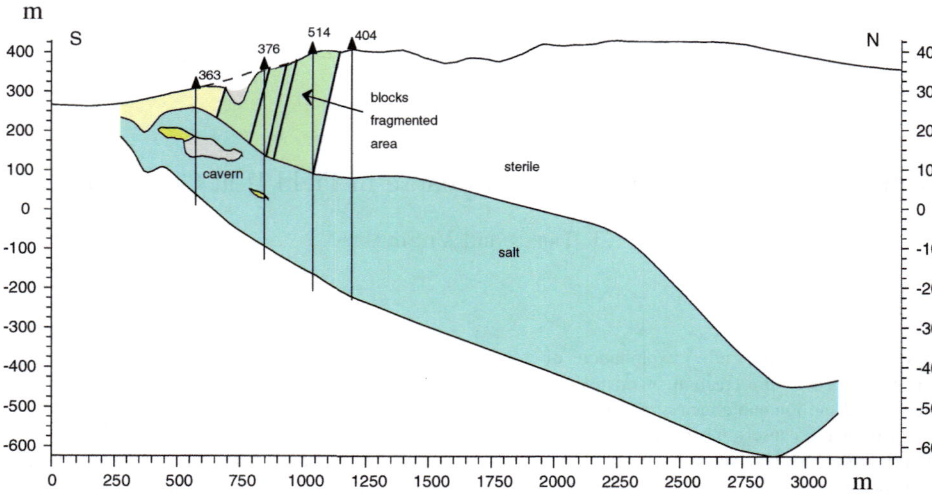

Figure 1
Cross-section of the Ocnele Mari salt formation

introducing sterile in the cavern all simultaneously to keep overall hydraulic pressure stable. A subsurface seismic array consisting of one-component geophones was installed around the cavern to continuously monitor an area of about 1 km^2 to ensure the safety of the workforce and equipment employed, as well as to offer quantitative information about the stress redistribution, indicative as to how the collapse proceeded.

It is for the first time that passive microseismic technology is used for the monitoring of a controlled roof collapse in a solution mine field. This study documents the seismic array design and undertakes an analysis of the seismic data recorded between July 2005 and March 2006. The mechanisms of the seismic events induced during the controlled collapse are subsequently investigated using the seismic moment tensor inversion from one-component recordings. The reliability of such inversions is analyzed for the sensor array employed, after which the spatial distribution of different failure components is examined. Based on the derived mechanism solutions, an estimate of the seismic deformation field is finally obtained.

In addition to the microseismic array, topographic marks were installed to provide settlement data. Various other geotechnical measurements were regularly carried out during the controlled collapse, such as surface crack orientation and size, rate of water inflow, brine pressure, etc. The study of the relationship between changes in these parameters and microseismicity will be the subject of a future analysis.

2. *Microseismic Array*

A monitoring system was specifically designed to identify, locate and report in real-time on the occurrence of microseismic activity as a result of stress redistribution during various stages of the controlled collapse. The system included a staggered array of 36 sensors installed in 12 boreholes, three sensors per hole (Fig. 2). Sensors are one-component, omnidirectional 15 Hz geophones with a sensitivity of 43.3 V/m/s, in stainless steel cases with polyurethane jacket cables to withstand the very corrosive medium. The boreholes were drilled vertically and were cased to avoid accidental closure. Each three-unit geophone string was installed attached to a support pipe and grouted in place by pouring the cement through this pipe directly to the bottom of the hole to ensure a tight, airless grouting.

The array geometry was designed to provide an average event location accuracy of 20–30 m throughout the entire monitoring volume, while being capable of identifying events with magnitudes as low as -1.5. As such, the boreholes were rather evenly

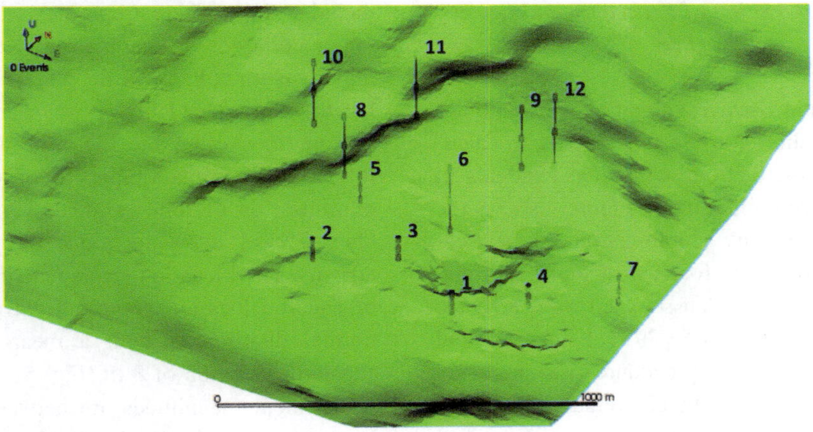

Figure 2
Array geometry; numbers correspond to boreholes and scale bar is 1,000 m

spaced at approximately 200 m, avoiding though any placement too close to the sinkhole. The depth of each hole was designed to ensure that the bottom sensor is located within the salt layer, thus reaching at times close to 400 m below surface.

A 24-bit Paladin seismic recorder manufactured by Engineering Seismology Group was placed on top of each hole, along with external amplifiers, lightning protection, power supply and 2.4 GHz radio Ethernet communication. Paladin is a web-enabled device capable of operating simultaneously in a continuous and triggered mode at up to 20 kHz sampling frequency with an effective dynamic range of 120 dB. For this application, data sampling was carried out at 2 kHz. GPS units employed for time synchronization offered a better than 1 μs accuracy. Radio signals were collected at a central control station either directly or through up to two repeaters, depending on the local topography. Example waveforms are shown in Fig. 3.

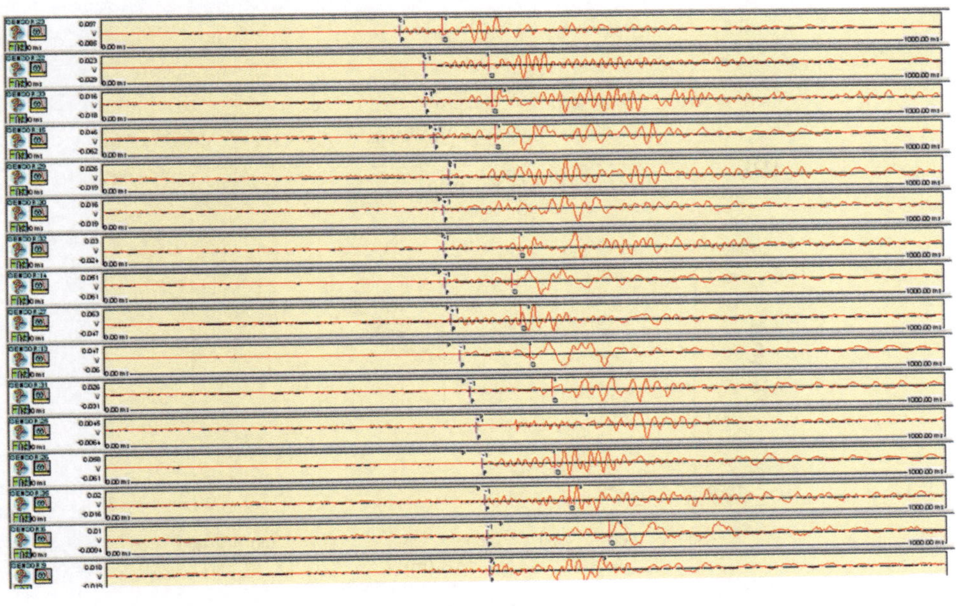

Figure 3
Example waveforms recorded for a seismic event which occurred at Ocnele Mari

3. *Seismic Data, Location of Seismicity and Seismic Regime*

Constant P- and S-wave velocities were initially estimated from calibration blasts and used to calculate event locations. Stations with time residuals in excess of 20 ms were filtered out. Figure 4 shows the difference between any combination of source-receiver straight raypaths for each actual event and the time difference between associated wave arrivals. The results indicate $V_P = 2933 \pm 11$ m/s and $V_S = 1{,}700 \pm 8$ m/s. This procedure was preferred to the direct absolute distance versus time plot, because it is independent from the origin times and increases the statistics.

A total of 2,392 events were recorded between July 7, 2005 and March 11, 2006. Seismic activity included three outburst episodes associated with two minor collapses in October and November, and the major cavern collapse at the end of December. Event locations employed first P-wave arrivals at 5 to 31 sensors (most often 8 to 13) and S-wave arrivals at ¼ to ½ that many in a Simplex algorithm with L1 norm minimization of time residuals (Press *et al.* 1989). Only 7% of event locations did not use S-wave arrivals. Total location errors (square root of the sum of squared location coordinate errors) fit closely a normal distribution with a mean of 18 m and a standard deviation of 8 m (Fig. 5).

Moment magnitudes are between −2.6 and 0.2. Frequency-magnitude distribution exhibits a liner trend between M_w −1.5 and −0.3 with a *b*-value of 1.5 (Fig. 6). The distribution of event locations is presented in Fig. 7a along with local structures. The top

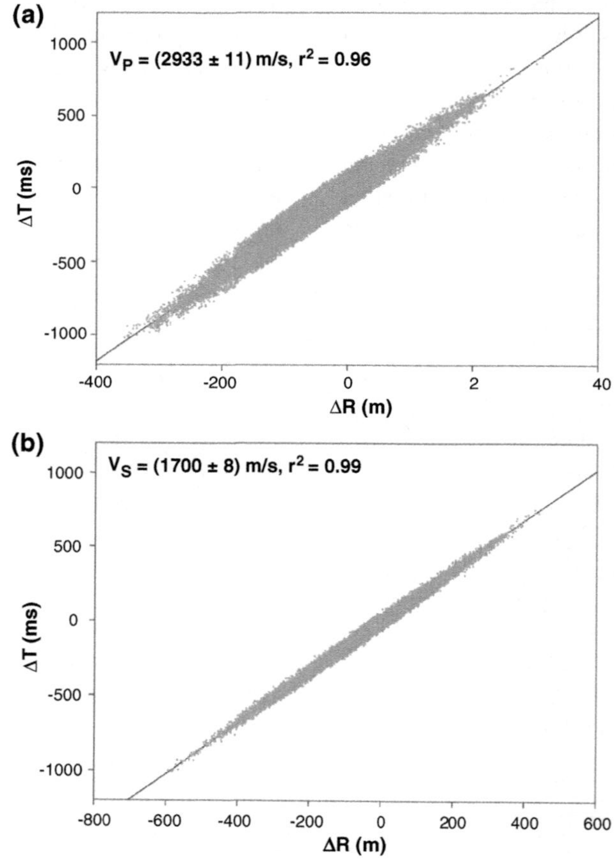

Figure 4
P- (**a**) and S-wave (**b**) velocity estimates

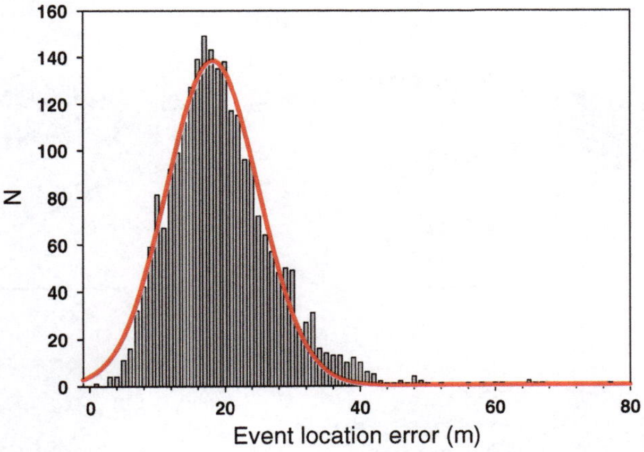

Figure 5
Distribution of event location errors; the curve corresponds to a normal distribution with a mean of 18 m and a standard deviation of 8 m

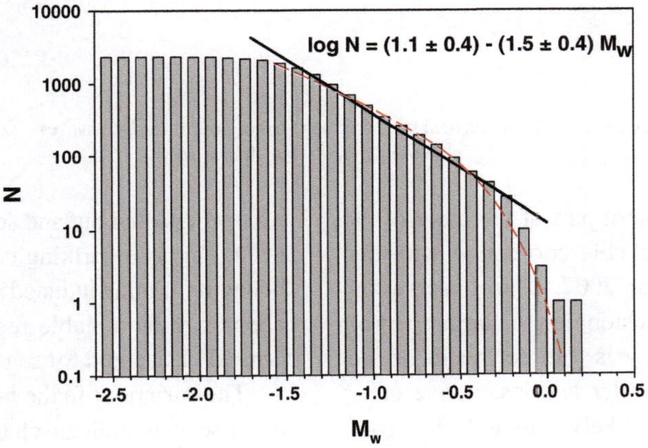

Figure 6
Cumulative frequency-magnitude distribution

surface corresponds to the local topography, whereas the lower surfaces correlate to the roof and the floor of the original large underground salt cavern. Most of the seismicity induced by the controlled collapse appears to be indeed related to the fragmentation and falling of the roof. Deeper seismicity located in the cavern was likely caused by subsequent nearby falling debris during strata readjustment to a new equilibrium.

Uncertainties in wave velocity and misidentification of first wave arrivals due to low signal-to-noise ratios contributed to increased location errors, thus leading to a blurry picture. Assuming that event locations should be better related to local structures and active processes take place in the field, an attempt was made to improve the relative event location accuracy by employing a collapsing technique (JONES and STEWART 1997). This technique considers that event location errors are normally distributed and the event position within the corresponding ellipsoid of uncertainty is random. As such, each event location could be moved within its ellipsoid of uncertainty towards a location that is consistent with locations of other events that fall inside the same ellipsoid. The algorithm optimizes the overall consistency of relocated events with the original event location uncertainties.

The results are presented in Fig. 7b–d and several general characteristics are instantly apparent. The distribution of seismicity is composed of several spatially tight clusters. Most active clusters are

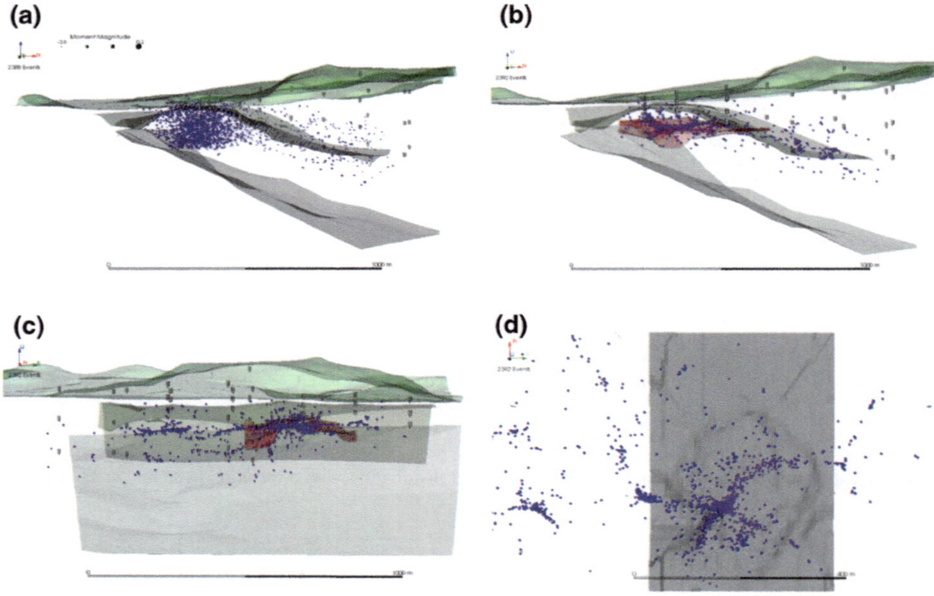

Figure 7
Absolute (**a**) and relative event locations (**b–d**) in vertical cross-section striking north (**a–b**), east (**c**) *–scale bar* 1,000 m, and in north-east plan view *–scale bar* 400 m (**d**)

located in the central-southern part of the monitoring region and exhibit a remarkable correlation with the cavern ceiling surveyed in 2002. This observation strongly supports the conclusion that the larger part of the induced microseismicity is a direct result of the major cavern collapse. Smaller and less active clusters of seismicity are likely related to roof fragmentations of smaller, nearby caverns. These results offer increased confidence in the fact that seismic data contain relevant information for the understanding of the geotechnical processes that took place during the controlled collapse.

In the framework of the Epidemic Type Earthquake Sequence, the average number of off-springs generated by each event is defined as the branching ratio (Helmstetter and Sornette 2002). Figure 8 shows the time variation of the *b*-value and branching ratio using a 2-week moving window with a step of 24 h. The *b*-value is close to 0.5 from July to September 2005, increases and varies around 1 until January 2006, after which it drops slightly in February and then increases to approximately 1.5 in March 2006. The branching ratio consists of values that define a super-critical seismic regime (>1) over the entire period of time in this analysis, except for the mid-February 2006, when the off-springs died out and seismic regime became stable (<1). This is in striking contrast for example with the behavior of mine induced seismicity, which commonly appears to be a stable regime and only rarely super-critical (Trifu and Shumila 2005).

The variations in the *b*-value from 0.5 to 1 and then to 1.5 seem to indicate changes in the general pattern of the fracture process within the monitoring volume. It suggests that the fracture process initiated as a linear fracture pattern ($D = 2b = 1$). Three months later roof fragmentation was well developed and shear fractures dominated ($D = 2$), while towards the end of the reported period fractures occurred within the entire volume ($D = 3$). The branching ratios indicate that once the roof fragmentation process began, the fracturing continued super-critically. Worth noting, it persisted for about another month after the large roof collapse at the end of December 2005.

4. *Stability and Reliability of Seismic Moment Tensor Inversions*

The use of one-component recordings maintains the linear dependence of the low frequency far-field

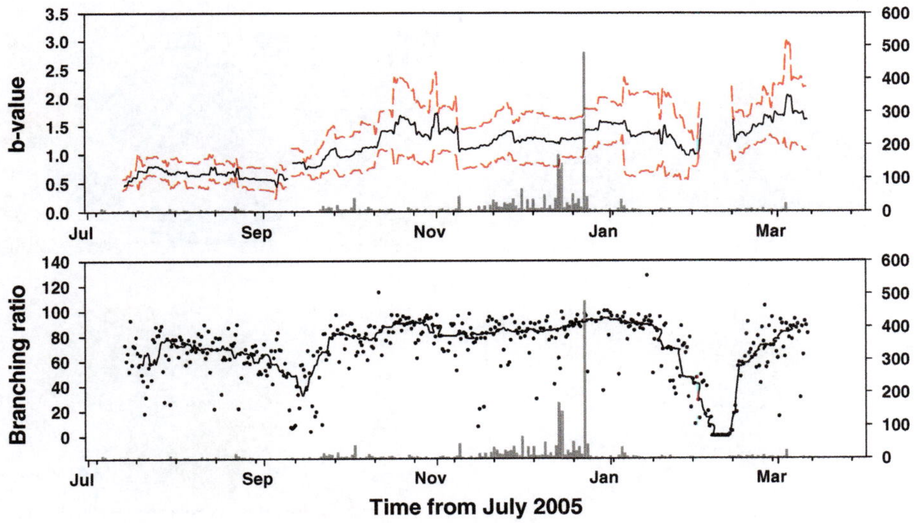

Figure 8
Time variation of the *b*-value (**a**) and branching ratio (**b**). *Dashed lines* in (**a**) denote the standard deviation and *continuous line* in (**b**) represents median estimates

displacements, with first polarities included, on the six unknown components of the symmetrical seismic moment tensor which describes a point seismic source (TRIFU and SHUMILA 2002). As such, evaluation of the event mechanism consists in the determination of the seismic moment tensor through the linear inversion of the above displacements. Worth noting, low frequency displacements can be calculated in the time domain, which allows for an automated estimate of source mechanisms (TRIFU *et al.* 2000). Before discussing the inversion results obtained for the recorded seismicity, the stability and reliability of the seismic moment tensor inversions are analyzed for the array geometry and sensor characteristics employed at Ocnele Mari (Fig. 2).

A measure of the stability of seismic moment tensor inversion is given by the Conditional Number (CN) (DUFUMIER and RIVERA 1997). This is defined as the ratio between the largest and smallest eigen values of the data matrix to be inverted (TRIFU *et al.* 2000). It is equal to one for a perfectly invertible matrix and approaches infinity for a very ill-conditioned matrix. CN can be interpreted as a data noise amplifier that defines the upper bound of the model parameter errors. Since it depends only on the source location and array geometry, expected CNs are calculated and displayed in Fig. 9a, b. Most stable inversions (CN < 10) are expected for the sources located below 200 m elevation. In the close proximity to the major cavern ceiling the anticipated inversion CN is around 15 and increases towards the surface. In the immediate vicinity of most sensor wells CN reaches the highest values and these zones should be considered as 'blind spots' for inversions. Overall, the CN results suggest that reasonably stable moment tensor inversions are expected for the installed sensor array. Average CN estimates for actual inversions are calculated over the same grid and shown in Fig. 9c, d. In general, real CN values are in agreement with expected ones. Slightly higher levels of real estimates in some areas are easily explained by the extrapolation of observed CNs to no-data grid points.

To further analyze the stability and reliability of moment tensor inversions, data were randomly perturbed using a Monte–Carlo approach to simulate increased noise levels. Additionally, tests have been carried out to evaluate the effect of using three-component recordings, as compared with actual one-component recordings, from a potential array of sensors installed in identical locations. Figure 10 summarizes the results obtained for three arbitrarily chosen seismic events located in the vicinity of the major cavern ceiling. Based on actual event

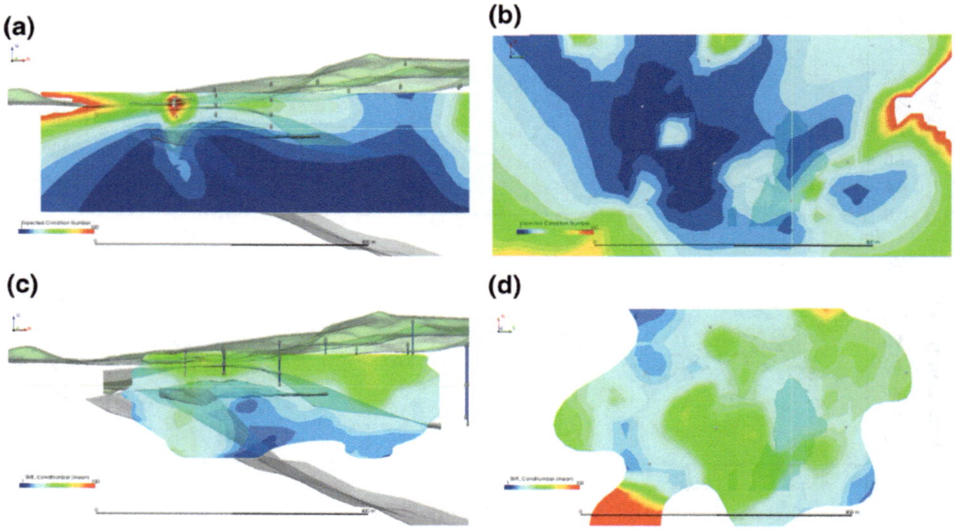

Figure 9
Expected (**a**, **b**) and average (**c**, **d**) condition number (CN, color scale from 1 to 100) in south-north vertical cross-section looking east and north-east plane at 200 m elevation looking down, respectively. *Horizontal* and *vertical thin light lines* show the positions of respective projections. Semi-transparent surface crossed by the *vertical line* in (**b**, **d**) is part of the major cavern that lies above 200 m elevation. *Scale bar* is 800 m

mechanism solutions, theoretical low frequency amplitude levels with first motion polarities were calculated for the P-, SV-, and SH-waves. These data were then perturbed by 10% Gaussian noise in the amplitude domain and random altering of 10% of first polarities.

For each particular moment tensor solution shown in column 1, 500 major double-couple (DC) solutions (JOST and HERRMANN 1989) obtained through random perturbations are displayed superimposed in columns 2 and 3 using a lower hemisphere equal angle projection (AKI and RICHARDS 1980). These results indicate that the use of one-component recordings provided by the geophone array installed at Ocnele Mari offers both stable and reliable seismic moment inversions. Interestingly, the use of three-component recordings would only lead to slightly more stable solutions. Automatic seismic moment tensor solutions have been obtained for 1,518 events. The distributions of Condition Numbers (CN) and multiple correlation coefficients (r^2) for this data set are shown in Fig. 11.

5. *Analysis of Event Mechanisms*

Due to previous massive collapses which occurred in 1991, 2001 and 2004, the salt deposit is expected to contain highly fractured volumes with a heterogeneous local stress distribution. The collapsing of a fractured rockmass is compatible to the presence of both shear and non-shear fracture components. Consequently, it is anticipated that the seismicity generated by the controlled roof collapse will exhibit a large variety of seismic source mechanisms. After filtering out the events with CN > 100 and $r^2 < 0.5$, a data set of 912 solutions is available for further analysis.

Traditional 2D mapping of event mechanisms involves equal area or equal angle projections on the upper or lower hemisphere. This type of representation is adequate for tectonic seismicity with a relatively limited number of solutions and predominantly DC type mechanisms. In case of induced seismicity, however, source distribution is essentially 3D. To represent as much information as possible, each moment tensor solution will be shown as a 3D sphere with nodal planes defined by the major DC component.

A red–green–blue (RGB) 3D color space offers a convenient equivalent for the coloring of the tensional quadrants based on the fact that the absolute values of the decomposition coefficients into isotropic, compensated linear vector dipole (CLVD), and

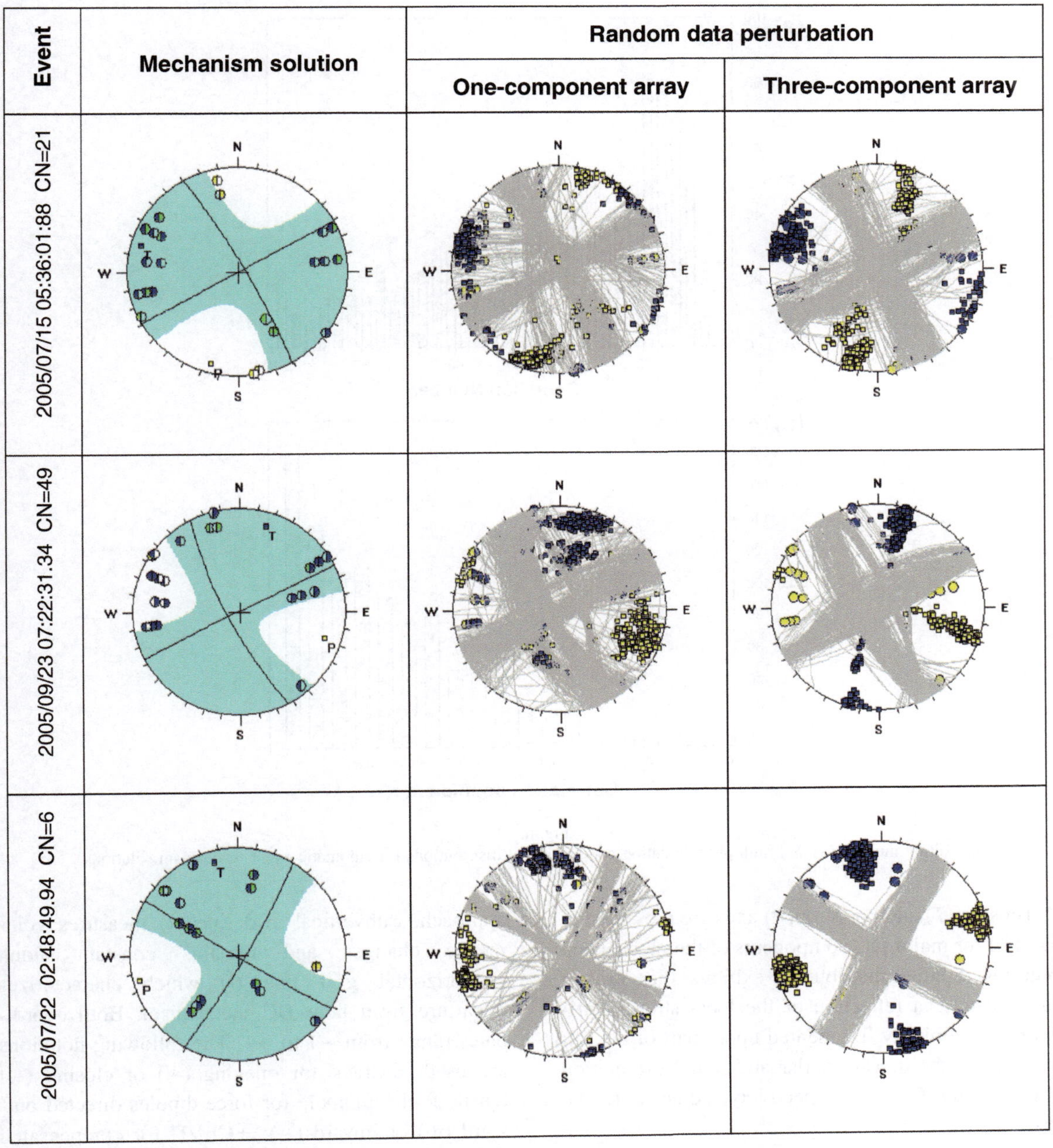

Figure 10
Stability of example moment tensor inversions (column 1) under 10% random perturbations for the one-component seismic array (column 2) and a hypothetical three-component array (column 3) at similar sensor locations

double-couple (DC) failure components vary between 0 and 1 and their sum is equal to one (VAVRYCUK 2001). Since present seismic monitoring arrays can provide online mechanism solutions for thousands of events, the visualization of this amount of information is not trivial. Figure 12 shows the spatial distribution of the derived mechanism solutions. Its relationship with the underlying geological structures is apparent, particularly with the sinkholes caused by previous collapses.

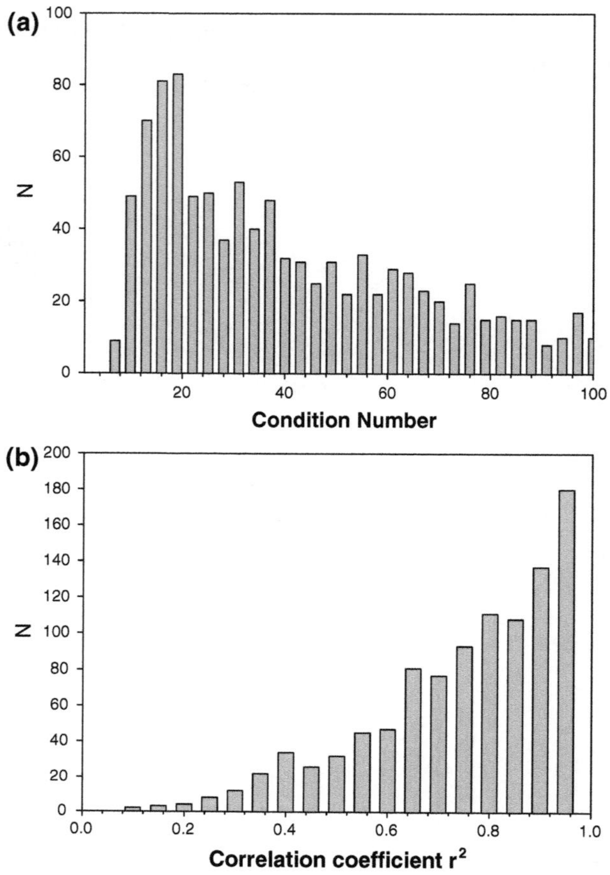

Figure 11
Condition number (**a**) and multiple correlation coefficient (**b**) distributions for automatic event mechanism solutions

Tension (T) and pressure (P) axes are presented in Fig. 12d for major DC components of the mechanism solutions. Spatial distribution exhibits the 'broken glass' pattern, a reflection of the local stress distribution in the highly fragmented upper part of the salt dome. Note the quasi-circular and radial orientation of the T and P axes, respectively, adjacent to the sinkhole.

A summary of the geometry of the shear faulting components and of mechanism distributions is presented in Fig. 13 on the faulting (FROHLICH 2001) and source type (HUDSON *et al.* 1989) diagrams. The faulting type diagram (Fig. 13a) shows that the data set predominantly contains normal and strike-slip events. The mechanism type diagram (Fig. 13b) also includes the location error ellipsoid in t–k coordinates, calculated using a linear error propagation approach. Subvertical grid line (k) measures volumetric changes and remains constant along subhorizontal grid line (t) which characterizes departure from pure DC mechanism. Both coordinates range from -1 to $+1$. The following notations are used: ±Crack for opening (+) or closing (−) tensile fault; ±Dipole for force dipoles directed outward (+) or inward (−); ±CLVD for compensated linear vector dipole with the sign determined by the direction of dominated dipole forces. As expected, the data-set reveals a variety of source mechanism types. Most events lay in the region containing the nodal planes of the P-wave radiation. This is the latitudinal strip between ±Dipole marks. Also, note that majority solutions are in the area where tensile components of failure (slip vector out of faulting plane) are allowed.

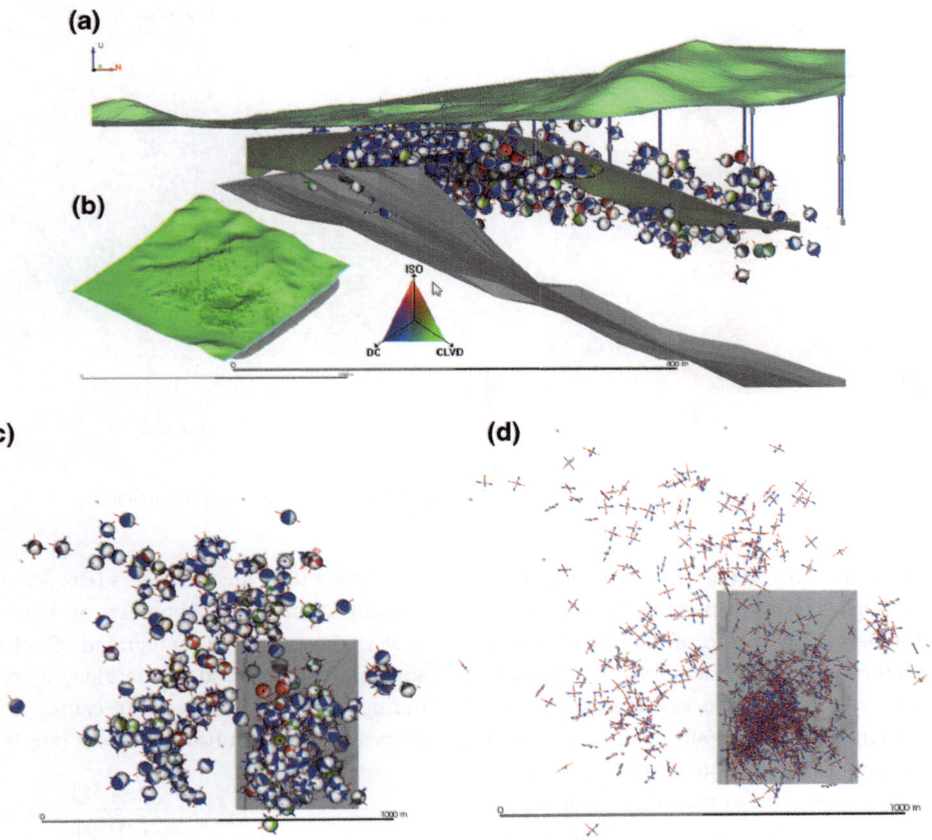

Figure 12
Event mechanism solutions in a cross-section looking east (**a**), with an aerial view of the monitoring area looking north-west (**b**), and in the northeast plan view looking down (**c**) with the dark rectangle showing the location of the cavern which collapsed in 2001. Distribution of T (*blue*) and P (*red*) axes (**d**). Scale bars are 800, 2,000, 1,000 and 1,000 m, respectively

Evaluating the failure components is important for the understanding of the stress re-distribution and geomechanical processes taking place on site. Although Fig. 13 provides an insight into seismicity generation in the area of the controlled collapse, it does not contain information on the spatial distribution of the failure components. To obtain this, the average parameter estimate is calculated over the closest events at each grid point by employing a nearest neighborhood approach with restriction on the search radius.

Spatial distribution of the isotropic failure component evaluated through the decomposition of the moment tensor solution is presented in Fig. 14a, b. Scale is from −10 to 100%. Negative and positive values indicate implosional (contraction) and explosional (expansion) volumetric changes, respectively. As expected for the geomechanical processes which took place in the area under investigation, only a small part of the study volume was subjected to seismic events with implosional components of failure. It can be speculated that such events occurred either on the floor of the cavern under gravitational forces, or in places where the inter-chamber pillars were destroyed by dissolution processes. The results indicate that the vast majority of the mechanism solutions include up to 30% explosional failure components, which correspond to roof caving under gravitational collapsing.

DC components indicating pure-shear failure are shown in Fig. 14c, d. The largest such components correspond to events surrounding the cavern ceiling. Above and below this area the percentage of the DC component decreases. Note that low values that occur towards the map boundary are caused by very sparse sampling.

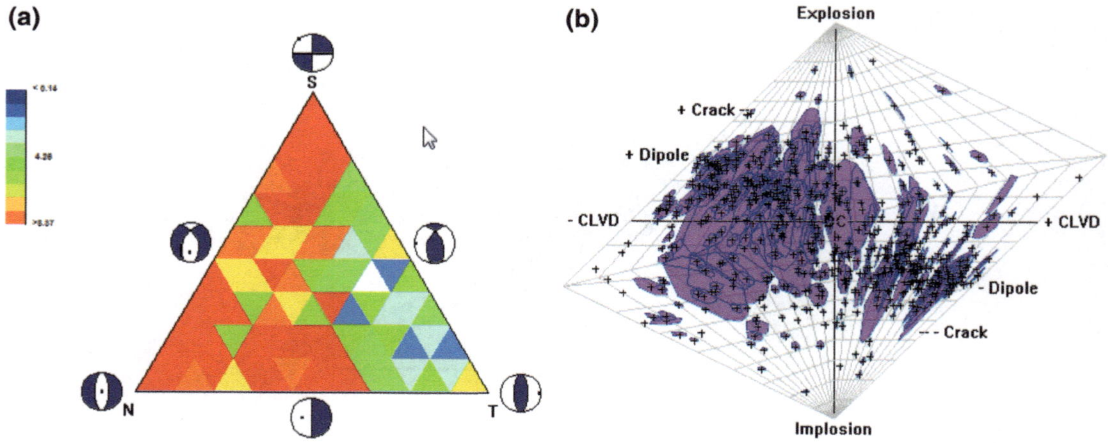

Figure 13
Faulting (**a**) and mechanism (**b**) type diagrams of the derived mechanism solutions

Distribution of the CLVD components is displayed in Fig. 14e, f. Such a failure component could correspond to a rotation of the plane of slip, or off-plane shear. Low percentage CLVD components dominate the mechanism solutions for the vast majority of the monitoring volume. The pronounced zones with a high percentage of CLVD are noted at the periphery of this volume and correspond to poor data sampling.

A generalization of the pure-shear point source model was proposed by VAVRYCUK (2001) in which the slip vector could be out of the faulting plane by an angle α called tensile angle. A positive angle indicates a tensional seismic source, whereas a negative value denotes a compressive source. Pure shear sources will have $\alpha = 0°$, while 90° and $-90°$ correspond to pure tensile and pure compressive sources, respectively. Tensile angle is estimated from the coefficients of the moment tensor decomposition and its spatial distribution is presented in Fig. 14g, h. Discarding extreme values at the boundary of the study volume, it is apparent that regions characterized by tensile/compressional values correlate well with pre-existing geological structures.

6. *Seismic Deformation Field*

When source locations are distributed over an entire volume rather than over a limited number of failure planes, the integrated seismicity can be treated as a flow within the rock mass (RIZNICHENKO 1965). A formalization of this approach by KOSTROV and DAS (1988) shows that the integrated effect of N seismic sources that occurred in the elementary volume ΔV during the time interval Δt causes a deformation described by the following strain rate tensor

$$\dot{\varepsilon}_{ik} = \frac{\sum_{n=1}^{N} M_{ik}^{n}}{2\mu \Delta V \Delta t} \tag{1}$$

where M_{ik}^{n} are the components of the second rank symmetric seismic moment tensor for the n-th event in ΔV, μ is the shear modulus, and i, k are indexes denoting the axes of the coordinate system. Standard approach for strain rate mapping was based on sparse regional seismicity, in which case it was necessary to account for missing small earthquakes. This was done by separating moment tensor geometry (failure components) and size (scalar seismic moment M_o). M_o was then integrated using a recurrence law over the entire expected energy range in the volume of interest (KOSTROV and DAS 1988).

In the current study statistics are very rich. As such, there is no need to use a recurrence law for seismicity extrapolation. Instead, Eq. 1 can be applied directly. Evaluation of the strain rate tensor components is performed over a rectangular 3D grid with 20 m steps in both northing and easting coordinates, and 10 m step in elevation. At each grid point the volume comprising a predefined number of events is used as elementary volume ΔV. Time

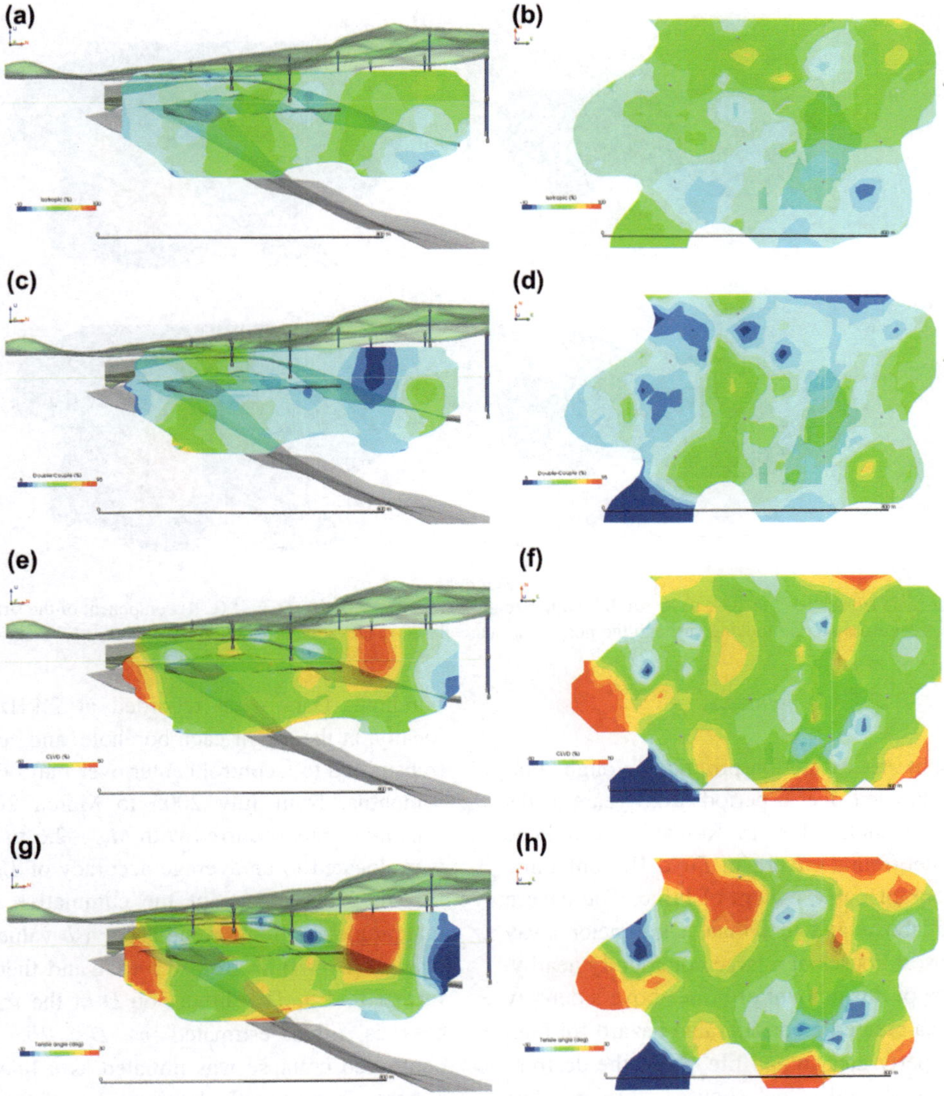

Figure 14
Isotropic—color scale from −10 to 100% (**a**, **b**), pureshear—from 5 to 95% (**c**, **d**), CLVD—from −50 to 50% (**e**, **f**) failure components and tensile angle—from −30 to 30% (**g**, **h**) in the south-north vertical cross-section looking east and the northeast plane at 200 m elevation looking down, respectively. The *horizontal* and *vertical thin light lines* show the positions of the respective projections. *Scale bar* is 800 m

interval between the first and last event occurrence in ΔV is considered as elementary time interval Δt in Eq. 1. Figures 15a, b display the vertical component of the strain rate tensor $(\dot{\varepsilon}_{zz})$.

The highest vertical deformation rate tends to be related to pre-existing geological structures, such as the cavern roof and floor, as well as the rest of the salt formation. The horizontal component of the strain rate tensor is defined as $\bar{u} = \{\dot{\varepsilon}_{xx}, \dot{\varepsilon}_{yy}\}$ and shown in Fig. 15c, d. The vector magnitude corresponds to the color code, while its direction is represented by the unit arrow. Note that the vertical cross-section is looking at the projection of unit direction. As expected, the horizontal deformation rate is most prominent in the areas of detected collapses.

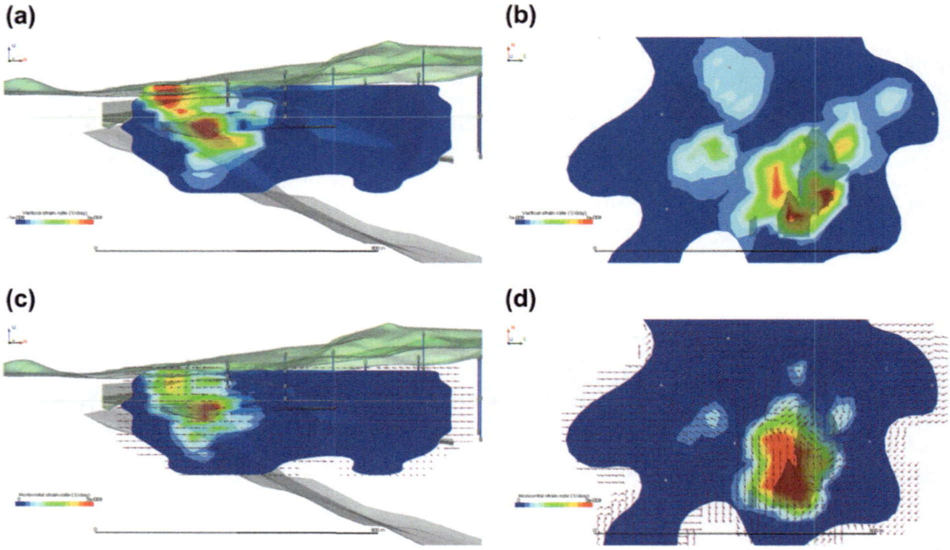

Figure 15
Vertical—color scale from -10^{-9} to 10^{-8} day^{-1} (**a**, **b**) and horizontal—from 0 to 5×10^{-8} day^{-1} (**c**, **d**) component of the strain rate in the south-north vertical cross-section looking east and the north-east plane at 200 m elevation looking down, respectively. *Scale bar* is 800 m

7. *Conclusions*

A deficient salt extraction program through solution mining has led over a period of 25 years to the formation at Ocnele Mari in Romania of a huge artificial underground cavern in Field II, containing many millions of cubic meters of brine. The cavern was located above and less than a kilometer away from the closest houses of a large community nearby. The presence of this amount of brine above a densely inhabited area created a considerable hazard for both personal property and human life. After the destruction of several individual houses through fast developing sinkholes, a major accident occurred in September 2001, followed by a second one in July 2004, accelerating the demand for a solution to this irreversible instability problem. A controlled collapse was proposed to cut open a 'window' in the roof of the underground cavern, pumping simultaneously the brine out and filling the cavern with sterile.

To evaluate the progress of the controlled collapse, microseismic monitoring was implemented within an area of over 1 km^2. Twelve vertical boreholes were drilled from the surface to approximately 160–360 m deep. Three one-component 15 Hz omnidirectional geophones were installed in each borehole. Data were recorded at 2 kHz sampling locally, at the top of each borehole, and continuously transmitted to a control center over radio Ethernet. In 9 months, from July 2005 to March 2006, 2,392 seismic events occurred with M_w −2.6 to 0.2, which were located to an average accuracy of 18 m.

The linear decay of the cumulative frequency-magnitude distribution suggests a *b*-value of 1.5. It changes over time from 0.5 to 1 and then increases further to 1.5. The dimension *D* of the fracture process is often estimated as $D = 2b$. Thus, the controlled collapse was initiated as a linear fracture pattern. Three months later, once roof fragmentation was well developed, shear fractures dominated. Towards the end of the period reported in this study, fractures expanded throughout the entire monitored volume. The analysis of the 'off-spring' seismicity outlines a super-critical process. The relocation of seismicity through a collapsing technique allows the identification of various tight clusters. The most active clusters are closely associated with the main cavern collapse, whereas the less active clusters correspond to the roof fracturing of smaller size, nearby caverns.

Synthetic data with 10% random noise indicate that one-component geophone recordings allow for stable and reliable event mechanism solutions through seismic moment tensor inversion of low frequency displacements calculated in the time domain, with first polarities attached. The use of three-component recordings at similar sensor locations would only marginally contribute to an increase in the stability and reliability of these solutions. A total of 1,518 event mechanisms have been derived for the seismicity which occurred from July and December 2005. Eliminating those characterized by a conditional number CN > 100 and a multiple correlation coefficient $r^2 < 0.5$, a data set of 912 solutions was retained for detailed analysis.

The analysis of spatial distribution of various failure components of the moment tensor solutions reveals interesting correlations with pre-existing geological and man-made structures on site. The tension (T) and pressure (P) axes for the major DC components exhibit a quasi-circular and radial orientation adjacent to the major sinkhole, respectively. The largest pure shear (DC) components characterize the events surrounding the cavern ceiling, which appear to exhibit normal and strike-slip failures.

The analysis of the isotropic failure component outlines that only a small part of the study volume was subjected to seismic events with implosional sources, associated with roof fragments striking the cavern floor under gravitational forces, or the dissolution of inter-chamber pillars. Worth noting, the majority of the mechanism solutions include up to 30% explosional failure components, which correspond to roof caving under gravitational collapsing. Low percentage CLVD failure components, perhaps indicative of shear occurring off the plane of slip, dominate the mechanism solutions within the monitoring volume. The presence of the CLVD and tensile faulting need further investigation. Meanwhile, the presence of high fluid pressure on site satisfies the main physical condition for permissible tensile faulting. Also, salt dissolution processes can produce failure components departing from pure shear.

Moment tensor solutions are further employed to evaluate the strain rate tensor components over a rectangular 3D grid with 20 m steps in both northing and easting coordinates, and a 10-m step in elevation. The highest vertical deformation rate relates closely to the cavern roof and floor, as well as the rest of the salt formation, whereas the horizontal deformation rate is most prominent in areas of detected collapses.

The present study proves that automatic event mechanism evaluation reflects expected trends for controlled salt mine collapsing. Further studies should analyze the evolution of the mechanism solutions over time, since this has the potential to offer indicative elements for detecting the initiation of the above processes and thus contribute to the assessment of the progress and control of this engineering activity. Such an analysis will obviously benefit from information pertaining to the geotechnical measurements carried out on site during the controlled collapse, such as surface crack orientation and size, rate of water inflow, brine pressure, etc.

Acknowledgments

We acknowledge Ervin Medves (ERM SRL) and Florian Zamfirescu (Universitatea Bucuresti) for their efforts towards the implementation of the passive microseismic monitoring technology at Ocnele Mari. Alexandru Danchiv, Monica Andrei, Marius Mocuta (Universitatea Bucuresti), Valentin Filip, Marius Mateica (S.C. FlowTex Technology SA), Ian Leslie and Mike Smith (Engineering Seismology Group Inc.) are acknowledged for their sustained contribution to the maintenance of the microseismic system. Additionally, we would like to thank Florian Zamfirescu and Alexandru Danchiv for various discussions on technical aspects related to this project.

References

AKI, K. and RICHARDS, P.G., *Quantitative Seismology: Theory and Methods* (W.H. Freeman, San Francisco, 1980).

DUFUMIER, H. and RIVERA, L. (1997), *On resolution of the isotropic component in moment tensor*, Geophys. J. Int. *131*, 595–606.

FROHLICH, C. (2001), *Display and quantitative assessment of distributions of earthquake focal mechanisms*, Geophys. J. Int. *144*, 300–308.

HELMSTETTER, A. and SORNETTE, D. (2002). *Sub-critical and super-critical regimes in epidemic models of earthquake aftershocks*, J. Geophys. Res. *107*, 2237–2257.

HUDSON, J.A., PEARCE, R.G., and ROGERS, R.M. (1989), *Source type plot for inversion of the moment tensor*, J. Geophys. Res. *94*, 765–774.

JONES, R.H. and STEWART, R.C. (1997). *A method for determining significant structures in a cloud of earthquakes*, J. Geophys. Res. *102*, 8245–8254.

JOST, M.L., and HERRMANN, R.B. (1989), *A student's guide to and review of moment tensors*, Seism. Res. Lett. *60*, 37–57.

KOSTROV, B.V. and DAS, S., *Principles of Earthquake Source Mechanics* (Cambridge University Press, Cambridge, UK, 1988).

PRESS, W.H., FLANNERY, B.P., TEUKOLSKY, S.A., and VETTERLING, W.T., *Numerical Recipes: The Art of Scientific Computing (Fortran Version)* (Cambridge University Press, Cambridge UK, 1989).

RIZNICHENKO, Y.V., *Seismic rock flow*, In *Dynamics of the Earth Crust*, Nauka, Moskow (in Russian), 1965.

TRIFU, C-I., ANGUS, D., and SHUMILA, V. (2000). *A fast evaluation of the seismic moment tensor for induced seismicity*, Bull. Seismol. Soc. Am. *90*, 1521–1527.

TRIFU, C-I. and SHUMILA, V. (2002), *The use of uniaxial recordings in moment tensor inversions for induced seismic sources*, Tectonophysics *356*, 171–180.

TRIFU, C-I. and SHUMILA, V. (2005). *Seismic hazard assessment in mines using a marked spatio-temporal point process model*, In *Controlling Seismic Risk* (eds. POTVIN, Y., and HUDYMA, M.), Australian Centre for Geomechanics, Perth, Australia, pp. 461–467.

VAVRYCUK, V. (2001), *Inversion for parameters of tensile earthquake*, J. Geophys. Res. *106*, 16339–16355.

ZAMFIRESCU, F., MOCUTA, M., CONSTANTINESCU, T., NITA, C., and DANCHIV, A., *The main causes and processes of instability evolution at Field II of Ocnele Mari – Romania*, In *Proc. Solution Mining Research Institute Spring Meeting*, 28 April – 2 May, 2007a, Basel, Switzerland, 6 pp.

ZAMFIRESCU, F., MOCUTA, M., DIMA, R., CONSTANTINESCU, T., and NITA, C., *A technical solution for the collapse fragmentation of the Field II cavern – Ocnele Mari, Romania*, In *Proc. Solution Mining Research Institute Spring Meeting*, 28 April – 2 May, 2007b, Basel, Switzerland, 8 pp.

(Received October 21, 2008, revised May 20, 2009, accepted May 22, 2009, Published online November 21, 2009)

Pure Appl. Geophys. 167 (2010), 43–62
© 2009 Birkhäuser Verlag, Basel/Switzerland
DOI 10.1007/s00024-009-0005-4

Pure and Applied Geophysics

Management of Post-mining Large-scale Ground Failures: Blast Swarms Field Experiment for Calibration of Permanent Microseismic Early-warning Systems

I. Contrucci,[1] E. Klein,[1] P. Bigarré,[1] A. Lizeur,[1] A. Lomax,[2] and M. Bennani[1]

Abstract—In France, decades of coal and iron-ore mining have left extensive underground cavities beneath or in the vicinity of urban areas. This poses an environmental challenge for society. To ensure post-mining risk management and public safety, wherever remediation is not possible, numerous real-time microseismic monitoring systems are being installed. The objective is to detect remote rock mass fracturing processes, precursory events and acceleration phases for appropriate and timely action. Although no consistent collapse has occurred in any of the monitored areas yet, single 3-D probes record many microseismic events of very low amplitude which create difficulties in the quantitative data analysis. The development of specific quantitative processing has therefore become a major issue in our research work. For that purpose, a field experiment was carried out on six of the instrumented sites. It consisted of sequences of small blasts in mine pillars which were accurately controlled in terms of the location, orientation and energy of the explosive source. The data analysis was used to calibrate parameters (velocity model, 3-D sensor orientation, etc.) for reliable 3-D localization and to develop an empirical law to estimate the source energy from the sensor energy. This work now enables us to analyze real microseismic events with a considerably better level of accuracy and to obtain enough information and confidence to discuss these data in terms of site stability.

Key words: Mine collapse, risk management, microseismicity, early warning system, 3-D location, velocity model.

1. Introduction

In the Lorraine area of eastern France, decades of iron-ore mining from 1850 to 1997 have left vast underground cavities beneath or in the vicinity of urban areas. At present, these residual voids are estimated at 500 million m^3 and represent 40,000 km of underground galleries. This now poses a societal and environmental challenge: Major collapses took place in the 1990s in the southern part of the Lorraine iron-ore basin in the cities of Auboué, Moutiers and Roncourt. These events occurred a few months after the mine closure and the progressive rise of the water level in the underground working caused by the halt of the de-watering system (Didier, 2008).

These large-scale ground failure events prompted a request from both the government and local authorities for a management strategy to be set up to prevent and control post-mining risks. A methodology was thus established in order to assess hazard zones and rank them according to their vulnerability due to human surface infrastructure and activity. The high risk zones are secured either by reducing the hazard or by using *in situ* monitoring. As it is rarely possible to deal with the hazard itself, public safety is often ensured through real-time microseismic monitoring systems that are installed in the zones in question.

For that purpose, INERIS deployed an innovative microseismic monitoring platform that was tested and validated during the Terres Rouges experiment in 1997 (Senfaute *et al.*, 2000; Couffin *et al.*, 2003; Bennani *et al.*, 2004). This platform is designed to detect rock-mass fracturing that first affects the old mine workings before reaching the overburden and finally the surface. More than thirty real-time microseismic monitoring networks have been installed since 1998 in the Lorraine iron-ore basin by the

[1] Institut National de l'Environnement Industriel et des Risques (INERIS), Ecole des Mines de Nancy, Parc de Saurupt, CS 14324, 54042 Nancy Cedex, France.
E-mail: Isabelle.Contrucci@ineris.fr; Emmanuelle.Klein@ineris.fr; Pascal.Bigarre@ineris.fr
[2] ALomax Scientific, 161 Allée du Micocoulier, 06370 Mouans-Sartoux, France. E-mail: anthony@alomax.net

 Reprinted from the journal

CENARIS[1] at INERIS. These networks have been designed following a methodology based on the geological context, the dimensions of the zone being monitored, the hazards involved and the technical installation and maintenance constraints due to urbanization. The most vulnerable buildings are instrumented with geotechnical sensors linked to the microseismic system to ensure a smart triggering scheme (KLEIN *et al.*, 2008).

Although no major collapse has occurred in any of these areas yet, small microseismic events are often recorded. These events, with low amplitudes ($\sim 10^{-6}$ mm/s) and negative magnitudes raise real quantitative analysis problems, especially in terms of localisation and spatial–temporal breakdown which are essential in understanding instability mechanisms (DRIAD *et al.*, 2005). Although publications exist on the microseismic monitoring of working mines (e.g., SENFAUTE *et al.*, 1997; DRIAD *et al.*, 2005; LI *et al.*, 2007) or on the long-term analysis of microseismicity in abandoned mines (MILLER *et al.*, 1989; OGASAWARA *et al.*, 2002), none of them describe the applied methods, tools or data management strategy.

This lack of methods, combined with the flooding of the Nord iron-ore basin scheduled to start at the end of 2005, motivated a large-scale field experiment to facilitate calibration of data processing parameters and to estimate microseismic source parameters. This experiment was performed in six instrumented zones where mine workings were still accessible. It consisted of numerous blast sequences to ensure the high sensitivity of the monitoring devices and to calibrate some of the fundamental numeric data processing procedures using accurately controlled input data.

This article describes the sequence of the 2005 calibration experiment and the methods adopted to confirm the high sensitivity of sensors based on blast data. Details will then be given on the calibration procedure for the 3-D localisation tool and the determination of its main input parameters, especially the velocity model. An empirical law to calculate the source energy from the hypocentral distance, in the geological context of the Lorraine iron-ore basin, is also estimated. Finally, the transfer of these results to the operational point of view will be discussed.

2. *Geological Context of the Lorraine Iron-ore Basin*

The Lorraine iron-ore basin, located on the eastern boundary of the Paris basin, extends for approximately one hundred kilometres from North to South and 20 to 30 km from East to West. The Pont-à-Mousson anticline splits the region into two distinctive zones: the Briey–Longwy–Thionville basin to the North and the Nancy basin to the South. The Lorraine iron-ore basin formations are marine in origin and were subject to very little deformation after their deposit. The basin is nevertheless cut by major faults of several kilometres long oriented NE–SW. The iron-ore series and its cover show a dip of several degrees towards the South-West, except near the major fractures where the local dip can be as much as 10°, as for example at the Audun-le-Tiche fault.

The iron-ore series belongs to the Toarcian and Aalenian stages. The marly carbonated cover belongs to the Bajocian. Overall, the various geological formations found in the basin are as follows (BENNANI and HOMAND, 2004): (1) the Doncourt oolitic Limestones; (2) the Jaumont oolitic Limestones; (3) the upper and lower Polypiers Limestones; (4) the Haut-Pont Limestones; (5) the Ottange Limestones; (6) the Charennes Marls; (7) the iron-ore formation. The geometry, thickness and extent of these series can vary laterally rapidly. Not all these formations are found in all the experimental sites. For example, Doncourt Oolitic Limestones are only found in the west of the iron-ore basin.

The iron-ore formation can be considered as alternating between marls and ores. It is between 0 and 65 m thick, 40 m on average (MONTAGNE *et al.*, 1992). Nine ore-bearing layers have been counted in the entire basin (MAUBEUGE, 1955), named using colours. Usually two to three layers were exploited with the rooms and pillars operating method below urban areas, with pillar extraction outside urban areas. A total of 3.1 billion ton of iron ore were extracted, i.e., about 1.2 billion m^3 over a surface area of 1,700 km^2.

[1] French National Monitoring Centre for Ground and Underground Risks created by INERIS.

3. *Field Experiment Design*

3.1. Experimental Sites

The field experiment took place between October 2005 and May 2006 in the cities of Audun-le-Tiche, Fontoy, Tressange, Nondkeil, Ottange and Moutiers (Fig. 1). These municipalities were already equipped with permanent microseismic monitoring networks; the underground workings were accessible during the experiments on all sites except Moutiers. Each monitoring network includes one or more microseismic stations. Each station consists of three probes equipped with miniature broadband geophones and cemented into boreholes as follows: one 1-D probe on the surface, one 1-D probe about 15 m below the surface and one 3-D probe about 50 m below the surface. The orientation of the 3-D probe was measured at the installation into the borehole. The local coordinate system used is the Lambert 1. The stations are connected to an acquisition unit, which automatically detects, records and transfers data to the monitoring central site of INERIS at Nancy, in quasi real-time via, a secure high bandwidth link.

3.2. Experimental Procedure

Apart from the Moutiers site, where the blasts were made from a vertical borehole, the experiments involved small dynamite blasts in mine pillars at depths of between 50 and 250 m depending on the site. TITADYNE AG 30 dynamite was the explosive used, packed in a cartridge 50 mm in diameter and placed in 76 mm-diameter boreholes 5 m deep. The explosive charges were chosen to produce signals

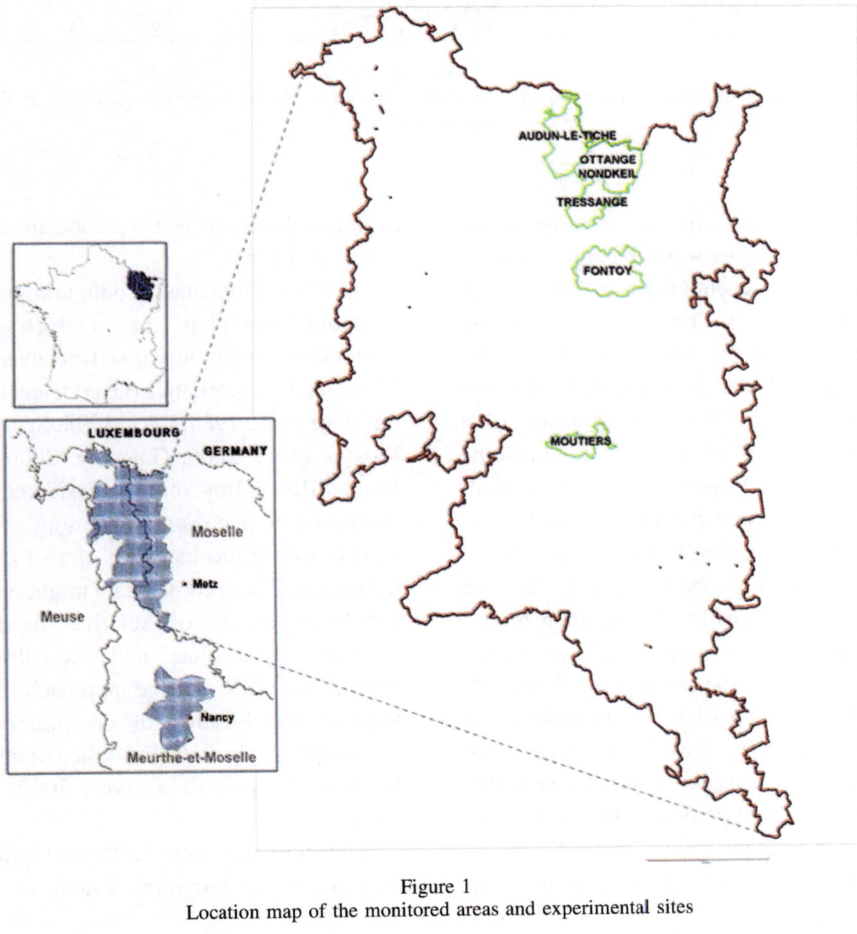

Figure 1
Location map of the monitored areas and experimental sites

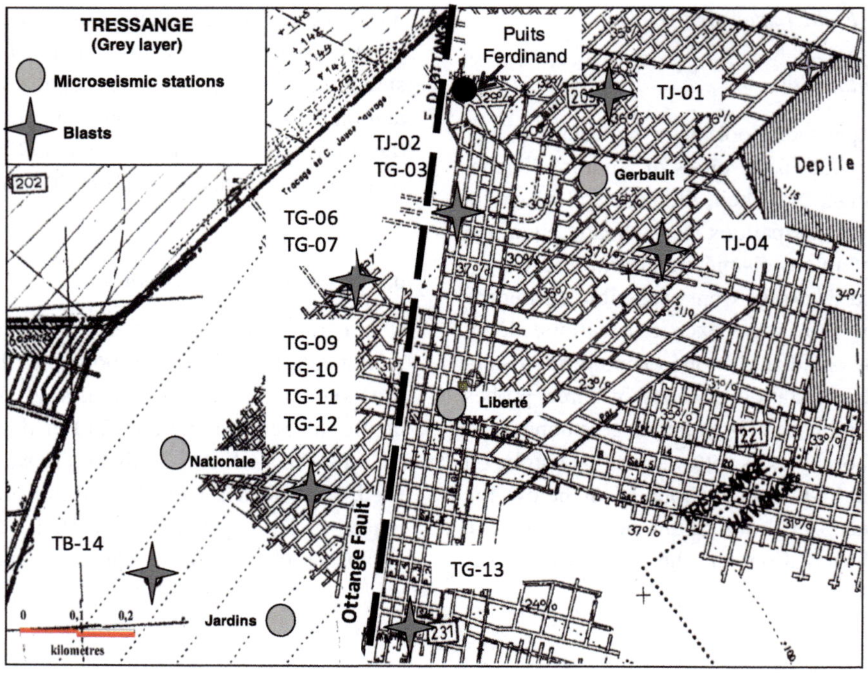

Figure 2
Mining map of the Tressange site. *Grey spots* represent the microseismic stations; *dark crosses* represent the blasts; the Ottange Fault is in a *thick dotted line*

with usable signal-to-noise ratio yet avoiding sensor saturation. Thus, the sources were dynamite charges between 0.5 and 12 kg, depending on the experimental site and its configuration. The blast positions were chosen considering both the best "blast-to-station" distance coverage and the best angular coverage when mine accesses so permitted. Various blast configurations were also used on each site to test the resolution of the numerical processing tools, including 3-D location and source parameter calculations. The first combination, a so-called "single blast", involved one single blast in a borehole (Fig. 3a). The second combination, a so-called "orientation blast," comprised two blasts of equivalent charge in the same pillar placed in perpendicular boreholes (Fig. 3b). Lastly, a combination of orientation and single blasts—"multiplet blast"—involved four blasts using different charges in the same pillar, including one in a borehole perpendicular to the other three (Fig. 3c). When possible, fixed *X* and *Y* coordinate blasts were also reproduced in other exploited levels to test the robustness of the localisation at depth.

The "multiplet blast" configuration was performed to record "multiplets" a term which refers, in classic seismology, to a group of seismic events showing the following characteristics (Geller and Mueller, 1980; Poupinet et al., 1984; Lees, 1998; Slunga *et al.*, 1995; Moriya *et al.*, 2006; Gibowicz, 2006): similar wave form; different time origin; considered as the result of the relaxation of constraint of a single fracture or fault, similar source mechanisms; close location; different magnitude. The recorded data might indeed be used to test the capabilities of "relative" location algorithms compared to "absolute" and classical seismic location algorithms. The relative approach is supposed to improve the location of one order of magnitude compared to absolute location (Rubin *et al.*, 1999; Waldhauser and Ellsworth, 2000; Schaff *et al.*, 2002).

Seventy blasts were performed in total, producing over 1,270 seismograms (Table 1). The recorded

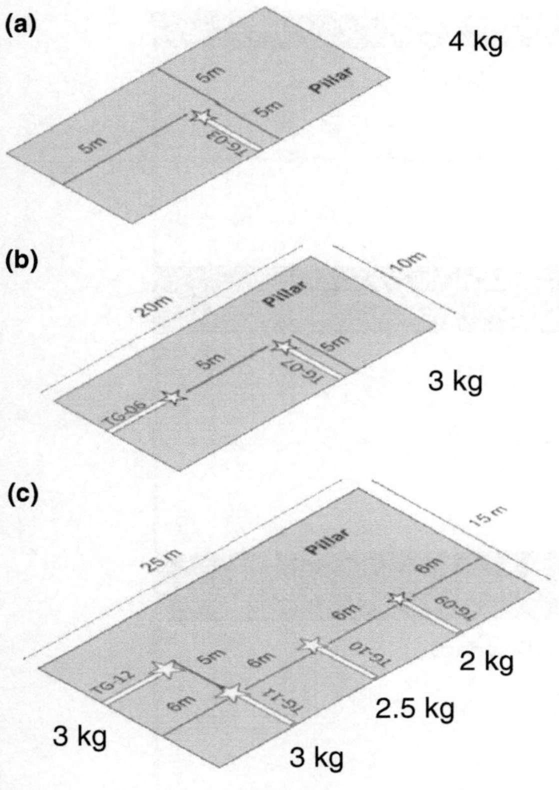

Figure 3
Configuration of **a** a source blast, **b** an orientation blast, **c** a multiplet blast

seismograms show a frequency spectrum of between 30 and 500 Hz for hypocentral distances of less than 300 m (Fig. 4). This frequency range was found to "fit" correctly the frequency range observed from a previous experiment during which caving induced events were recorded (SENFAUTE *et al.*, 2000).

4. Data Processing: Example of the Tressange Site

Microseismic data processing started with a systematic analysis of all the seismograms in terms of amplitude, signal-to-noise ratio and sensor energy to assess data quality. All the recorded 3-D seismograms were then analysed for polarisation to check consistency between measured incident angles and expected values at the 3-D probes, thereby ensuring the correct azimuthal orientation of these probes (Sect. 4.2). The geological formation velocities for the overburden were then estimated from the inversion of both P-wave arrival times and measured polarisation angles, taking into account the multilayered geology of the site. The blasts were relocalised based on the velocity model thus calculated (Sect. 4.3.3). Lastly, an empirical law for calculating source energy was determined based on the sensor distance and energy (Sect. 4.4). This approach is illustrated below with the Tressange check blast data (Fig. 2), except for the source energy calculation in which all the sites are considered.

4.1. Data Quality

A systematic analysis of all seismograms per site and microseismic station in terms of amplitude and energy was undertaken in order to assess data quality. This analysis demonstrated the excellent sensitivity of the monitoring systems, able to detect 1 kg blasts at a distance reaching 300 m. For analysis, only signals with signal-to-noise-ratio higher than ten were considered. Note that in such a "near-field" experiment, where sources generate mainly compressional stress waves, only P waves are observed; the S-wave

Table 1

Table summarising the number of blasts, blast to station distance, charges and signals recorded per site

Site	Number of microseimic stations	Number of blasts	"Blast-to-station" distance		Mass of explosives (kg)	Number of seismograms
			Min (m)	Max (m)		
Fontoy	2	13	100	900	3.0–12.0	130
Tressange	4	13	170	1,000	2.0–5.0	260
Audun-le-Tiche	3	11	80	1,050	2.0–9.0	165
Nondkeil	4	10	110	370	0.5–2.0	140
Ottange	5	18	90	1,300	2.0–10.0	468
Moutiers	4	5	80	480	1.0–5.0	115
TOTAL	22	70	80	1,300	0.5–12.0	1,278

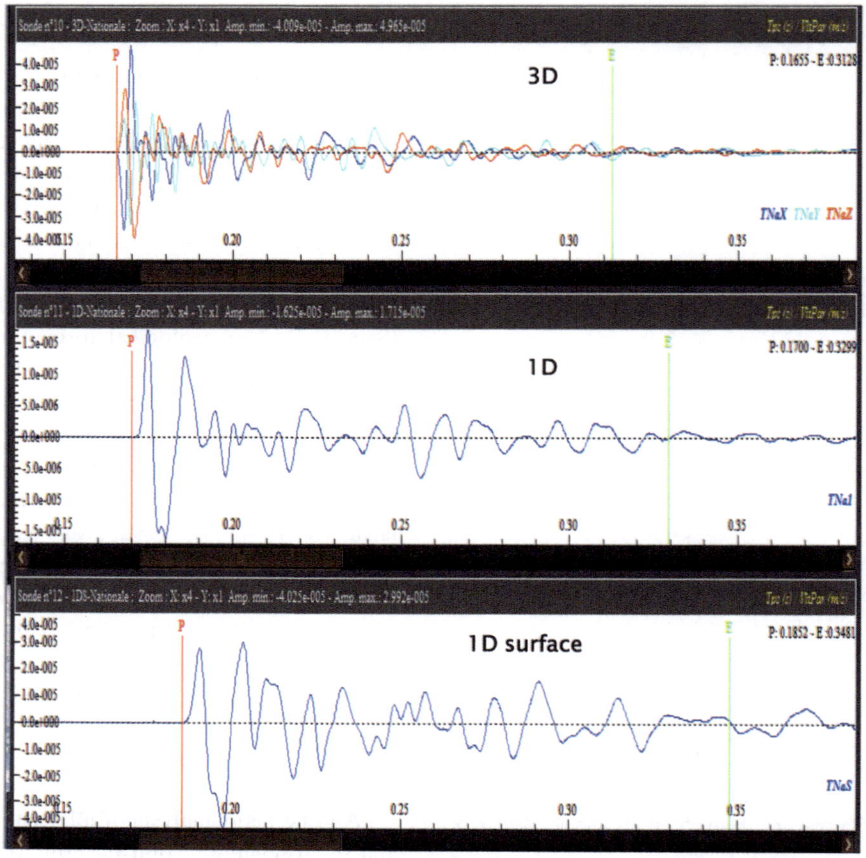

Figure 4
Seismograms of blast number 9 recorded by the microseismic station "Nationale" of the Tressange network (amplitude in mm/s vs. time in s). The 3-D probe is the deepest one, the 1-D surface probe is the shallowest one

energy is not observed on the recorded data. This limitation will be discussed here below.

4.2. Polarisation Analysis

Analysing polarisation by wave rotation is used to calculate incident angles (azimuth and dip) of the ray path at the three-component probe from a 3-D seismogram. The estimated direction of the incident ray path provides important information that can be used along with the measured P- and S- arrival times to efficiently constrain the solution for source location. This approach, which assumes prior knowledge of the orientation of the 3-D probes, is of most importance when localisation relies on only a few microseismic stations (MAGOTRA *et al.*, 1987; ABDUL-WAHED *et al.*, 2001; VOLKER and ROTH, 2003). In the current study, the azimuths of the 3-D probes were measured with a releasable compass during grouting into the vertical boreholes. As the blast positions are known, the orientation measurements can be checked easily by examining the expected and measured directions by a polarisation analysis of each 3-D seismogram. The expected direction is calculated by assuming an isotropic homogeneous medium between the blast and the 3-D probe, i.e., assuming a straight ray path between these two points. This hypothesis is reasonable regarding the short "blast-to-probe" distance and the relative homogeneity of the encountered geology. Note that in such sub-horizontal media, the calculated azimuth does not depend on the velocity model. On the contrary, the measured dip of the ray path may depend significantly on the gradient velocity of the stratified overburden when contrasting velocities between layers are considered.

On the Tressange site, the expected azimuths at the 3-D probes at the Nationale and Liberté stations match the measured azimuths; the mean error, assessed using quality factors such as rectilinearity and planarity factors (SAMSON, 1983), is ~5°. However, for the Gerbault and Jardins probes, a systematic mean misfit of 50° and −200°, respectively, is noted between the expected and calculated azimuths (Fig. 5). These systematic misfits are caused by uncontrolled rotation of 3-D probes during grout injection while the orientation device is released and pulled out. These misfits have been corrected at each station, thus providing true probe orientation. Note that blast 08, located underneath the Liberté station, has a major azimuthal misfit (Fig. 5) because of the inconsistent solution for ray path with a dip close to 90°. Note also that a bias between expected and measured dip values is observed for all stations: the bias is less than 5° for the Liberté, Jardins and Nationale stations (Fig. 6), it is ~9° for the Gerbault station (Fig. 6). It is most likely due to straight ray paths assumption.

4.3. 3-D Localisation of the Blasts

This stage consists in performing blast relocalisation to test the performance of the localisation algorithm thanks to the known blast positions (Sect. 4.3.4). Firstly, it is necessary to build up a velocity model based on the site's geological structure (Sect. 4.3.1). P-wave velocities for each layer are calculated allowing for all available input data, i.e., positions of the blasts, direct P-wave arrival times and dips measured at the 3-D probes (see Sect. 4.2). Note that S-wave arrival times were not available and measured azimuths are of no interest in the velocity calculation. For this purpose, a velocity model optimization program has been developed (Sect. 4.3.3).

4.3.1 Geology of the Tressange site

The Tressange site is located on the eastern part of a fault system extending from Mont-Bonvillers to Ottange, oriented SW–NE to SSW–NNE. Note the presence of the Ottange fault, orientated SSW–NNE, at Tressange (Fig. 2). The fault throw is several metres and the East block is the collapsed block with layers dipping regionally in the order of 3% westwards or WSW (LOPES, 2002). As shown in

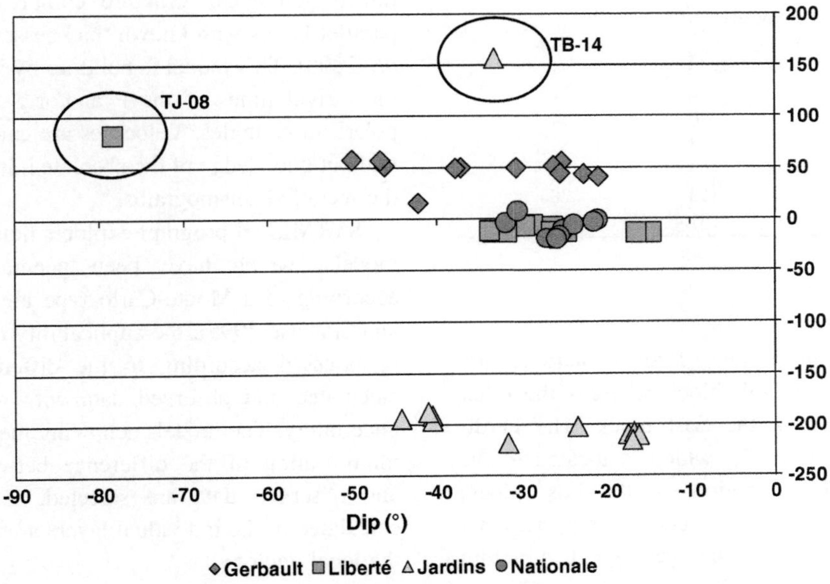

Figure 5
Azimuth error versus dip for the 3-D microseismic probes of Tressange network

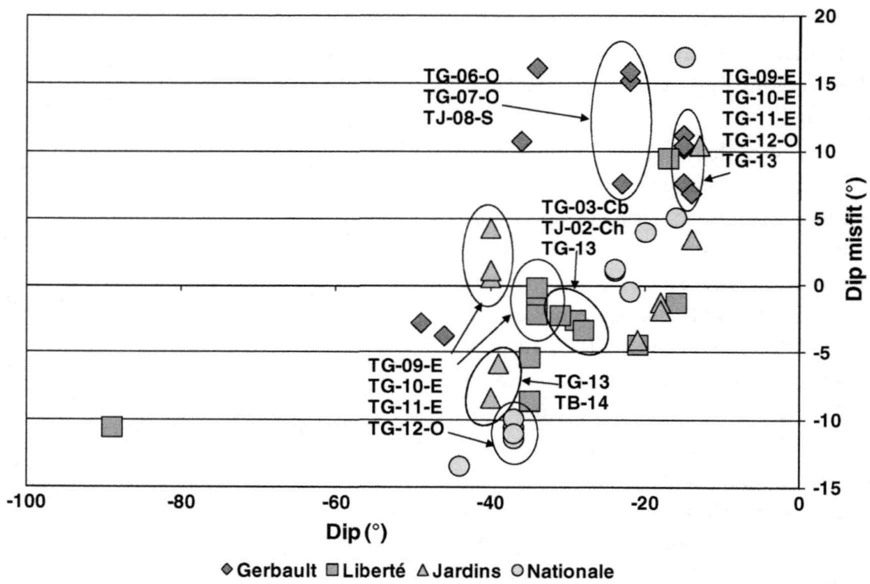

Figure 6
Dip misfit versus measured dip for the 3-D microseismic sensors of Tressange network

Table 2

Geological facies found at the Ferdinand shaft on the Tressange site

Geological facies:	Depth of the layer roof (m)	Layer thickness (m)
Coordinates (Lambert 2 extended) X = 864,456.9 Y = 2,495, 763.8 and Z = 347.51 m		
Overburden	–	–
Jaumont Limestones	0	14
Polypiers Limestones	14	85
Haut-Pont/Ottange Limestones	99	54
Charennes Marls	153	28.4
Iron-ore formation	181.4	–

Fig. 2, the Nationale and Jardins stations are situated in the West fault block whereas the other stations are located in the East block. The Ferdinand mine shaft (Table 2), which is located to the west of the Gerbault station, in the East block, shows that the geological structure comprises six main facies (Table 2). The geological structure adopted for Tressange is based on these different characteristics.

4.3.2 Construction of an Inversion Type Velocity Model

SYTMISvel software was especially developed for automatically adjusting a velocity model to a determined geological structure comprised of inclined parallel layers with known thicknesses (Contrucci *et al.*, 2008). This model is build-up by inversion, using the arrival times of the P and/or S waves and the polarisation angles. Velocities are calculated with or without knowledge of the absolute initial time (T0) of the recorded seismograms.

SYTMISvel program explores numerous velocity models, which have been generated randomly, according to a Monte-Carlo type algorithm (Lomax and Snieder, 1995): the applicability of these models is assessed according to the differences between calculated and observed data with respect to their uncertainty. The models achieving the most effective minimisation of the difference between calculated and observed data are selected, if the velocities calculated in the individual layers are consistent with the local geology.

The differences between calculated and observed data are minimised by using either the least-square misfit function (L2-norm) or the equal differential time misfit function (EDT-norm; PINSKY *et al.*, 2008; LOMAX, 2005).

For the L2-norm, the function to be minimised is expressed as:

$$\mathrm{pdf}_{\mathrm{LSL2}} = k\exp\left\{-\sum_{\mathrm{obs}_i}\frac{[T_{\mathrm{obs}_i}(x)-T_{\mathrm{calc}_i}(x)]^2}{\sigma_{i^2}} - \sum_{\mathrm{obs}_j}\frac{[\theta_{\mathrm{obs}_j}(x)-\theta_{\mathrm{calc}_j}(x)]^2}{\sigma_{j^2}}\right\}$$

with x, the position of the source; T_{obs_i} and T_{calc_i}, the arrival times observed and calculated from observation; θ_{obs_i} and θ_{calc_i} are the polarisation angles observed and calculated; σ_i and σ_j are the uncertainties; k a normalisation constant.

For the EDT-norm, the function to be minimised is expressed as:

$$\mathrm{pdf}_{\mathrm{EDT}} = \frac{1}{K}\left(\sum_{\mathrm{obs}_a,\mathrm{obs}_b}\exp\left\{-\frac{\{[T_{\mathrm{obs}_a}(x)-T_{\mathrm{obs}_b}(x)]-[T_{\mathrm{calc}_a}(x)-T_{\mathrm{calc}_b}(x)]\}^2}{\sigma_{a^2}+\sigma_{b^2}}\right\} + \sum_{\mathrm{obs}_a,\mathrm{obs}_b}\exp\left\{\frac{\{[\theta_{\mathrm{obs}_a}(x)-\theta_{\mathrm{obs}_b}(x)]-[\theta_{\mathrm{calc}_a}(x)-\theta_{\mathrm{calc}_b}(x)]\}^2}{\sigma_{a^2}+\sigma_{b^2}}\right\}\right)$$

with T_{obs_a} and T_{calc_b}, the arrival times for observations a and b; T_{calc_a} and T_{calc_b}, the calculated travelling times; θ_{obs_a} and θ_{obs_b} are the polarisation angles observed for observations a and b; θ_{calc_a} and θ_{calc_d} are the polarisation angles calculated; σ_a and σ_b are the respective uncertainties; K, a normalisation constant.

This minimization process uses a global approach since the travel times are nonlinearly related to the velocities in the individual layers. The program input comprises mainly the following information: (1) The method used to randomly generate sets of velocity models is either the traditional Monte-Carlo method which defines randomly the entire velocity model or the "Metropolis" Monte-Carlo method in which each model tested in a Metropolis loop is close to the last model selected (a random walk similar to the simulated annealing method); (2) the geological structure and the observed data, i.e., the number of geological layers and the velocity range for P and S waves; the thicknesses of the layers and their possible dips; the positions of the microseismic stations and blast points; the arrival times of the P and/or S waves for each blast and each of the 1-D and 3-D probes, as well as associated picking errors and the incidence angles observed for each blast at each 3-D probe and associated errors. The program output principally comprises the difference between calculated and observed arrival times and incidence angles for each blast and each of the considered probes and the probability of occurrence of each velocity model.

4.3.3 Velocity Model for the Tressange Site

The thicknesses of the model's layers are based on the geological structure observed on the log of the Ferdinand Shaft (Table 2). Several tests were carried out in order to determine the best velocity model with the layer thicknesses indicated in Table 2. The minimisation between calculated and observed values was achieved by using the L2-norm. Regardless of the configuration of the data used for the inversion, significant misfits were indeed obtained when the EDT-norm was employed. This norm does not seem to be adapted to the geometry of the problem and the dimensions of the geological structure studied: the double difference of the arrival times and angles affects the information contained in our data, and thus generates aberrant velocity values.

The best velocity model is determined in two main steps. The first step consists in inverting all of the data, i.e., all of P-wave arrival times and incidence angles of all blasts. It allows identification of the data with a large misfit between calculated and observed values

(Fig. 7a, b, c). Figure 7a shows that time misfits higher than 0.04 s are observed for blast 1 on the Gerbault station, and blasts 1, 4 and 14 on Liberté station. The high misfit observed for blast 1 on the Gerbault station is probably related to a local heterogeneity in the vicinity of the blast point. This hypothesis seems reasonable since the hypocentral distance is, in that case, short. On the contrary, the high time misfits observed for blasts 1, 4 and 14 on Liberté station are probably due to the large hypocentral distances. These high distances can induce significant signal attenuation, leading to high picking errors. These misfits and errors can also be emphasized by anisotropic wave propagation. Figure 7b shows three significant azimuth misfits. For blast 8, at Liberté station, the misfit is due to the fact that the blast is located directly below the Liberté station. Indeed, there exists an infinity of azimuth solutions for a dip close to 90°. For blasts 4 and 14, the azimuth misfits observed respectively at Gerbault and Nationale stations, are difficult to explain since the hypocentral distances are short in both cases. These misfits are however lower than 10%. They can be neglected since they do not depend on the velocity model. Regarding the dip (Fig. 7c), the greatest misfits are observed for large hypocentral distances, relatively large error in dip for near-horizontal rays in the assumed, constant velocity layered model. As a general rule, the highest misfits, in terms of both polarisation angles and P-arrival times, are observed for blasts performed on the network border. This is probably due to signal attenuation with distance. Also, the boundary ray paths cross geological sequences that are not intersected by other ray paths: The information collected from the boundary ray paths is not crosschecked, thus creating high misfits. In other words, this first stage demonstrates that the velocity model has to be calculated with high signal-to-noise input data. Thus, the second step in the velocity model calculation was carried out by eliminating the "aberrant" errors or high misfits (Fig. 8) associated with blasts 1, 4 and 14.

The step 2 model significantly reduces misfits which exist between calculated and measured values for P-wave arrival times as well as for the polarisation angles of the incident ray (Fig. 8). Thus, time misfits vary in a maximum range of ±0.02 s instead of 0.07 s after the first step inversion.

The velocities obtained by inversion, for each layer, are indicated in Table 3. The results show that removing aberrant input data considerably reduces the standard deviations, except for Jaumont Limestones which layer is clearly not constrained enough by the input data. Figure 9, which presents the distribution of velocities for each layer, gives additional details on these results. The velocity peak of the Polypiers Limestones layer shows that this layer is very well constrained. For the other layers, the peaks are less well-defined (Fig. 9), and thus associated errors are greater (Table 3).

These calculated velocities can also be compared with velocity measurements made in the laboratory by Homand and Dagallier (2004). These measurements provide an accurate idea of expected velocities in the formations in question and the contrast between layers. They show that P-wave velocities vary between 2,710 and 4,215 m/s for the Polypiers Limestones, between 3,900 and 5,375 m/s for the Ottange/Haut-Pont Limestones, between 1,965 and 2,680 m/s for the Charennes Marls and between 3,270 and 3,450 m/s for the iron-ore formation. These values are consistent with those obtained in this study. Note that the step 2 velocity model shall be selected for relocating the blasts.

4.3.4 3-D Localisation Algorithm

The localisation module implemented in SYTMIS-auto software used in this study, is based on the combination of the microseismic wave arrival times as well as the polarisation angles, in order to determine the hypocenter with the maximum likelihood. Indeed, the integration of polarisation angles enables location of an event with few probes, i.e. one 3-D probe and one 1-D probe if only P waves are detected or a single 3-D probe if P and S waves are recorded (Magotra *et al.*, 1987; Abdul-Wahed *et al.*, 2001; Volker and Roth, 2003). The implemented localisation algorithm is based on a probabilistic approach to solve the inverse problem (Tarantola and Valette, 1982). It consists in maximising the probability density function (pdf) of the hypocenter at a given point using EDT or L2 norm. This is done by minimising the misfit between measured and calculated values, i.e., between observed and calculated

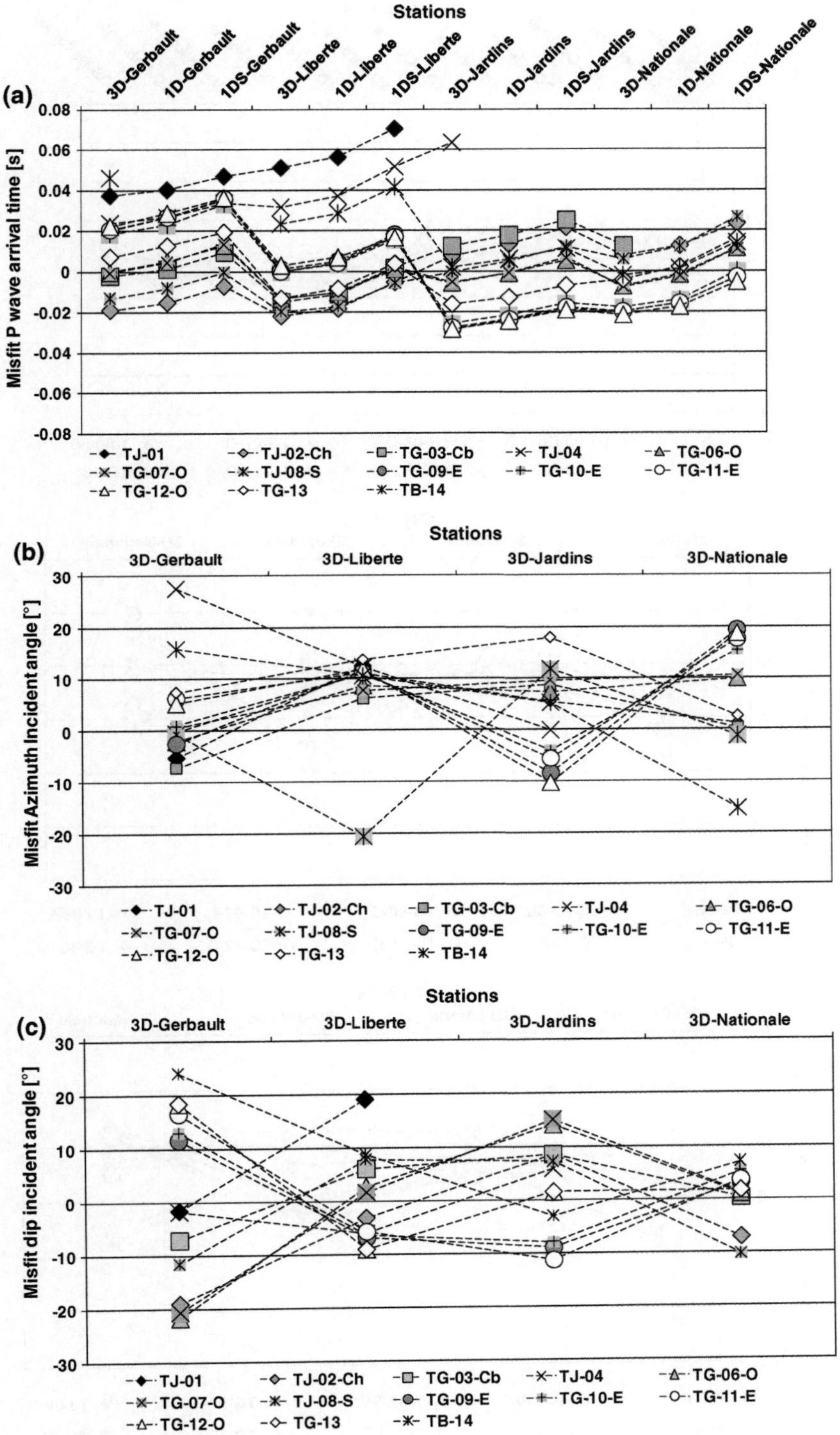

Figure 7
Step 1, inversion of all the data for calculating the velocity model. Differences between calculated and measured values for all the blasts on the Tressange site of **a** the arrival times of P waves, **b** azimuths and **c** dips

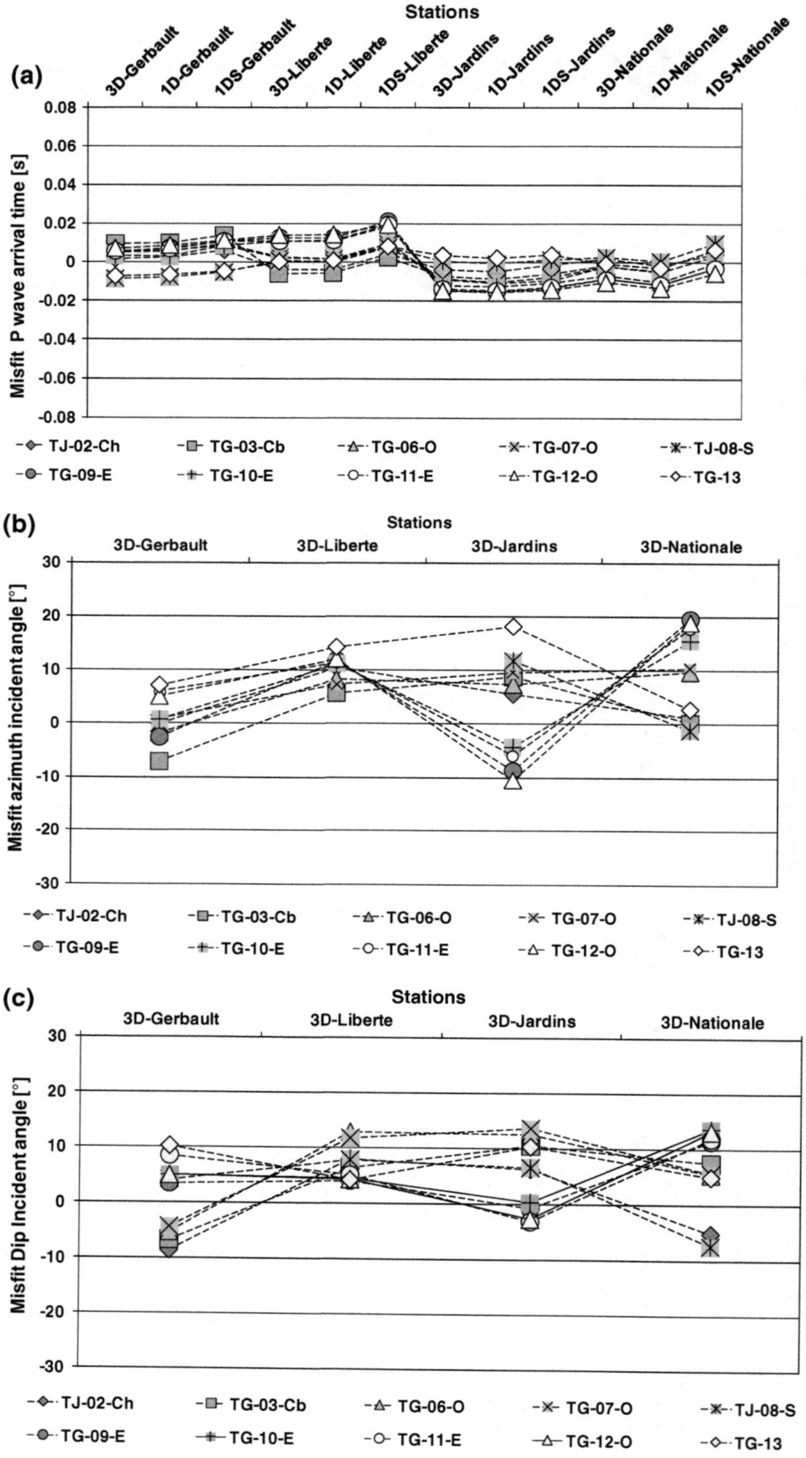

Figure 8
Step 2, inversion without abnormal data for the calculation of the velocity model. Differences between calculated and measured values for all blasts on the Tressange site of **a** the P-wave arrival times, **b** azimuths and **c** dips

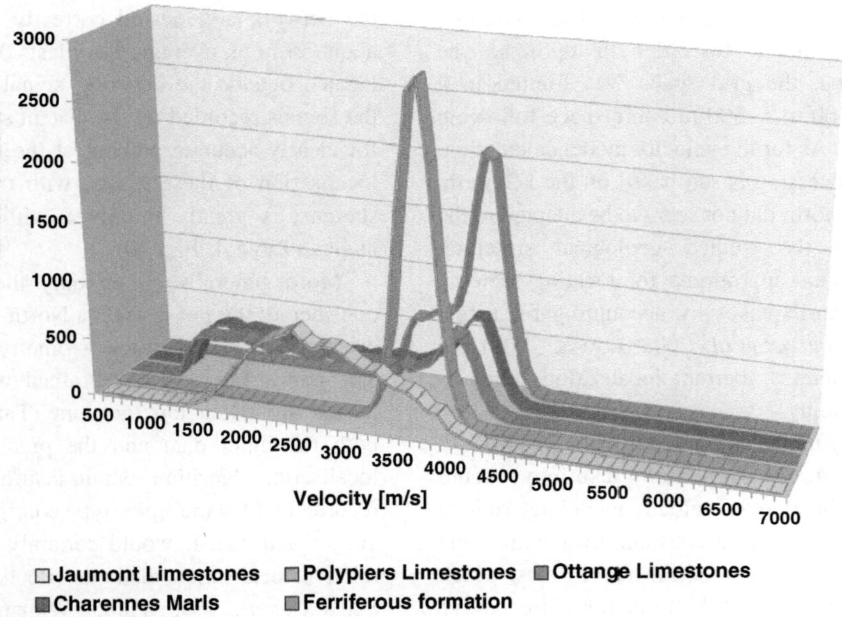

Figure 9
Distribution of the velocities tested by inversion for each of the geological layers considered

Table 3

Best P-wave velocity model obtained by inversion in steps 1 and 2

Formation	Jaumont Limestones	Polypiers Limestones	Haut-Pont/Ottange Limestones	Charennes Marls	Iron-ore formation
Step 1: All data					
Velocity (m/s)	5,530	4,820	4,790	3,235	3,215
Standard deviation (m/s)	430	215	1,280	1,435	1,510
Step 2: Without blasts 1, 4 and 14					
Velocity (m/s)	2,340	3,375	3,855	3,190	3,190
Standard deviation (m/s)	600	165	920	970	305

arrival times, as well as between observed and calculated polarisation angles. The hypocenter with the maximum likelihood is determined by using the Oct-Tree nonlinear method (Lomax and Curtis, 2001), based on a successive division of space into cells depending on the value of the probability calculated for each cell. This approach is used to completely solve the inverse problem and thus provides a representation of the overall pdf of the localisation. The most probable hypocentre corresponds to the pdf maximum. The drawback of this method is that the calculation can be prolonged because the entire solution space is explored for each iteration. To limit time calculations, this method can be combined with the linear C.H.E.A.P method (Tarantola and Valette, 1982) based on the determination of a local maximum by gradient calculation, that is to say of the maximum pdf in the relevant cell. For each event, the pdf, as well as the corresponding reliability ellipse at 68%, is calculated.

4.3.5 Localisation Parameters

Calibration blasts are relocated on the Tressange site with the global Oct-Tree method to avoid the restrictions of the linear C.H.E.A.P method described above. The grid size was set as 10 m minimum and 50 m maximum. In order to optimise the calculation time, the grid space was limited in a 1,500 m × 1,500 m × 500 m cubic space following *XYZ* directions. As for the velocity model calculation, localisation processing is run based on the L2-norm, since the EDT-norm did not seem to be adapted to the dimensions of the studied geological structure. Although it seems appropriate for locating regional or worldwide earthquakes by accounting for actual picking errors (Pinsky *et al.*, 2008; Lomax, 2005), the EDT-norm generates aberrant localisation solutions in the current study.

The velocity model employed is from SYTMISvel software (Sect. 4.3.3). To partially take into account uncertainties related to the velocity model determined previously by inversion, a constant error with value of ±0.002 s is introduced on all travel times, which corresponds to an error of ±70 m/s for a stress wave travelling at 3,500 m/s over a distance of 350 m. Errors on the incidence angles are assessed at ±10° which seems reasonable according to Fig. 8; manual picking error is set at ±0.005 s.

4.3.6 3D Localisation Results

Several relocalisation tests were carried out on the 13 blasts related to the site of Tressange. Firstly, only P-wave arrival times were used; secondly, the polarisation angles (azimuths and dips) were added. The results of these tests are summarised in Table 4 and Fig. 10. When only P-wave arrival times are considered, the mean misfit between the true and calculated positions of the blasts is 260 m with significant errors in the *Z* direction, i.e., depth (210 m on average). When P-wave arrival times and polarisation angles are taken into account in the localisation, the mean misfit between actual positions and calculated positions is reduced to 70 m (Table 4). There is a misfit of 48 m in the *X*–*Y* plane and of 47 m in the *Z* direction (Table 4). The localisation quality is improved in nearly every case, except for blast 08 located directly below the Liberty station. In this specific case, the azimuth measurement has little meaning as it is determined with a very significant error, which affects localisation quality. On the other hand, blasts 09, 10, 11 and 12 located in the centre of the network are located correctly with an error of about 55 m on average. For blasts 01, 04, 13 and 14, located outside the network, signal-to-noise ratio of the signals recorded on the distant stations is too low for clearly accurate picking of the first arrivals. The localisation of these blasts, with only two or three stations, is greatly improved with the polarisation angles (Table 4, Fig. 10).

More generally, when only the arrival time is considered, the pdf reveals a North West–South East orientation, related to the geometry of the network. The pdf is far more constrained when polarisation angles are taken into account (Table 4). However, both the input data and the precision of the 3-D localisation algorithm remain insufficient for the data recorded with a multiplet-type configuration (blasts 9, 10, 11 and 12). It would certainly be useful to use relative localisation methods (Poupinet *et al.*, 1984; Slunga *et al.*, 1995; Abdul-Wahed *et al.*, 2006) to gain precision while relocalizing this kind of data.

4.4. Seismic Energy at the Source

The seismic energy at the source is a key parameter for characterising the intensity of a seismic rupture. It is however necessary to assess the geometrical and anelastic attenuations of the rock mass and the response of the instrumental channel (Urbancic *et al.*, 1993; Boatwright *et al.*, 2002; Yamada *et al.*, 2007) in order to obtain a correct estimate of this energy. Although it is easy to correct the recorded signal of the geometrical attenuation, it is more difficult to perform other corrections without making numerous assumptions. In conventional seismology, the calculation procedures used for characterising the source often involve a double-couple mechanism since fault shearing mechanisms usually generate earthquakes (Brune, 1970; Madariaga, 1976). Based on this approach, the magnitudes of the earthquakes can be assessed, thus allowing the determination of the source energy (Aki and Richards, 1980). However, these techniques cannot be applied directly to assess the source energy of a microeismic event occurring in a mine. Source mechanisms, different from the double-couple, are expected, for instance, traction, implosion or explosion (Gibowicz *et al.*, 1991; McGarr, 1992; Trifu and

Table 4

Summary of the true and recalculated positions of microseismic blasts, hypocentral misfits in the X–Y plane and depth (Z direction), and associated average values

BLAST, charge	TJ01, 4 kg	TJ02-CH, 4 kg	TG03-Cb, 4 kg	TG04, 4 kg	TG06-O, 3 kg	TG07-O, 3 kg	TJ08-S, 4 kg	TG-09-E, 2 kg	TG-10-E, 2.5 kg	TG-11-E, 3 kg	TG-12-O, 3 kg	TG13, 2.9 kg	TB14, 4.5 kg	Average
X, Measured (m)	864,442	864,171	864,171	864,533	863,993	863,993	864,165	863,926	863,923	863,919	863,919	864,088	863,643	
Y, Measured (m)	195,750	195,560	195,560	195,504	195,456	195,449	195,268	195,129	195,126	195,122	195,115	194,914	194,997	
Z, Measured (m)	126	128	114	118	110	110	115	105	105	105	105	96	106	
Blast relocalisation using P-wave arrival times														
X, Calculated (m)	865,031	864,221	864,179	NL	864,010	864,010	864,233	863,858	863,870	863,858	863,846	864,151	863,489	
Y, Calculated (m)	195,717	195,526	195,508	NL	195,409	195,409	195,245	195,022	195,034	195,022	195,022	194,799	194,938	
Z, Calculated (m)	390	−17	54	NL	−64	−71	69	−173	−181	−173	−165	−157	−184	
Hypocentral misfit	646	157	79	NL	181	187	86	305	305	302	295	285	334	263
Misfit at X and Y	590	60	52	NL	51	44	72	127	106	117	118	131	166	136
Misfit at Z	264	145	60	NL	174	181	46	278	286	278	270	253	290	210
Blast relocalisation using P-wave arrival times and polarisation angles														
X, Calculated (m)	864,438	864,198	864,174	864,467	863,993	863,993	864,221	863,893	863,899	863,893	863,882	864,133	863,688	
Y, Calculated (m)	195,778	195,526	195,514	195,491	195,426	195,426	195,245	195,092	195,098	195,092	195,092	194,864	195,028	
Z, Calculated (m)	69	105	63	192	61	46	71	143	92	128	101	8	198	
Hypocentral misfit	63	49	68	100	57	68	76	62	38	45	44	111	107	68
Misfit at X and Y	28	43	46	67	30	23	61	49	36	39	44	68	54	45
Misfit at Z	57	23	51	74	49	64	44	38	13	23	4	88	92	48

Shumila, 2002; Finck *et al.*, 2003; Sileny and Milev, 2006).

Analyses are currently being conducted to determine and quantify the different corrections to be taken into account in the calculation of the source energy in the context of this study. Therefore, this paper only includes a description of the work concerning the empirical estimation of the source energy from calibration blasts. This approach appears to be a good first-order solution since the blasting charges and their positions are known. An empirical relationship between the seismic energy recorded by the sensors and the seismic source energy according to the distance has been defined as follows (Tastet *et al.*, 2007):

$$E_{\mathrm{cpt}} = K \times E_{\mathrm{src}}^{\gamma} \times D^{-\beta}, \tag{1}$$

with E_{src}: source energy linked to the explosive charge (J), see Eq. 2; E_{cpt}: energy recorded by the sensor and calculated from seismograms (J); D: hypocentral distance based on straight ray path assumption (m); K: constant ($\mathrm{J}^{1-\gamma}\,\mathrm{m}^{\beta}$). Note that in such a formulation (1), unknowns are K, β and γ, where β quantifies both geometrical and anelastic attenuations and γ dissipation and nonlinear source size effects.

The seismic source energy depending on the blasting charge can be expressed as follows:

$$E_{\mathrm{src}} = Q \times E_{\mathrm{total}} \times r \tag{2}$$

with: E_{src}, seismic source energy (J); Q, source blasting charge (kg); E_{total}, total energy released by a blasting charge of 1 kg depending on explosive used; r, ratio set empirically at 0.6.

The value of the total energy released (E_{total}) by a blasting charge of 1 kg depends on the blasting agent being used. In this case, it is TITADYN AG 30 which is packaged in cartridges of 50 mm and 38 cm in length. As these calibration blasts have been carried out in boreholes (confined space), i.e. in a "blocked blast" configuration, the maximum detonation velocity of 6,000 m/s has most probably been reached as well as the maximum energy of 4.2×10^{6} J for 1 kg of blasting charge. This energy includes two terms: The first represents the shock energy and the second the gas energy. Only the shock energy can significantly impact the wave transmission in the rock and

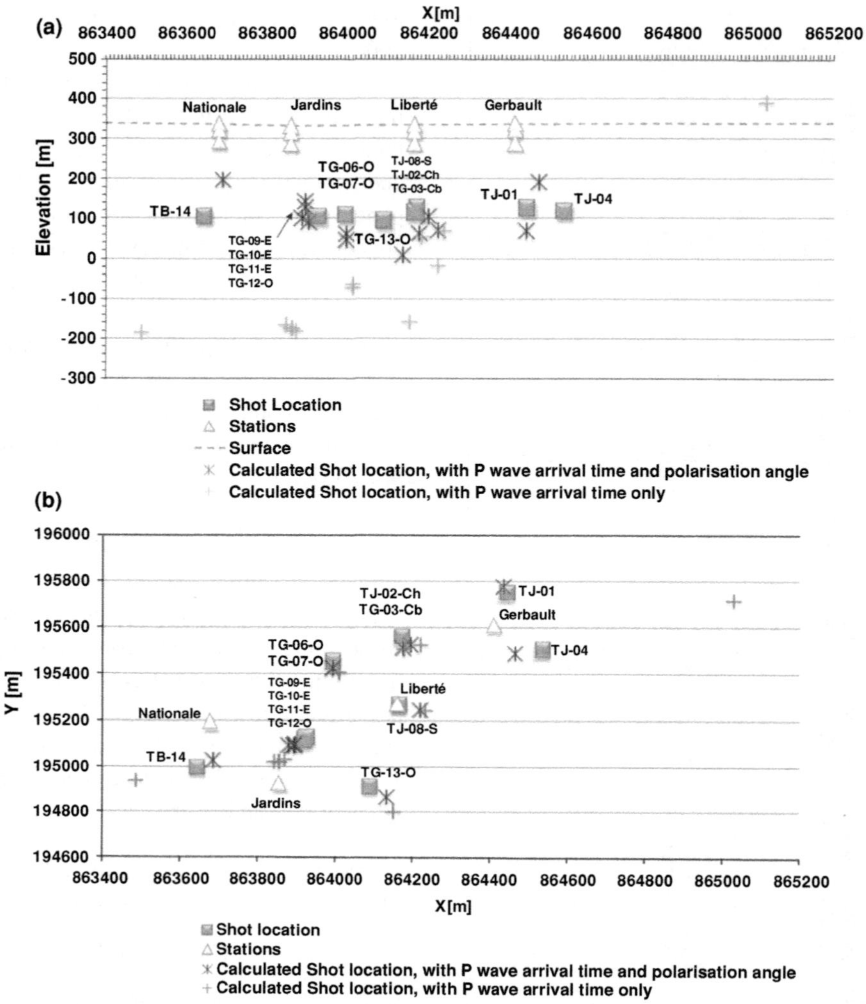

Figure 10
Relocation of Tressange calibration blasts, representation of the actual relocated positions, **a** cross-sectional view and **b** plan view

contribute to the seismic energy radiated into the rock mass with an insignificant participation of detonation gas. Thus, coefficient r with a value of 0.6 has been applied to the total energy in order to account for this phenomenon. Equation (1) leads to the following expression:

$$E^*_{\mathrm{src}} = \left(\frac{E_{\mathrm{cpt}}}{K \times D^{-\beta}}\right)^{1/\gamma}, \quad (3)$$

where E^*_{src}: estimated source energy (J); K, β and γ coefficients are determined, from Eq. (1), for each experimental site. First, β is determined for a given series of blasts of constant source energy E_{src}, defined by the blast charge.

Figure 11 illustrates the correct determination of $[K \times E^{\gamma}_{\mathrm{src}}]$ (considered here as a constant with the blast charge) and power law β.

Secondly, once β has been determined, K and γ coefficients are determined to fit all data for the studied site. Eventually, the robustness of the empirical law (2) is evaluated from the standard misfit between estimated E_{src}* and the true values of E_{src}. The results give a mean error of ~50%, which is very acceptable when one considers all cumulated assumptions. The empirical laws obtained for each site are presented in Table 5.

For each studied site, the dependency of the seismic source energy on the energy recorded by a sensor reveals a good homogeneity with γ^{-1} values close to 0.5 except for the Fontoy site where this value is equal to 0.8 (Table 5). This value illustrates the significant attenuation observed on this site where it was necessary to perform blasts of up to 12 kg. The dependency of the source energy on the hypocentral distance reveals relatively homogeneous exponents between 2.3 and 4. The relative homogeneity observed on sites, which are relatively distant from each other, illustrates the geological homogeneity of the Lorraine iron-basin (Fig. 11a) and contributes to the definition of a source energy estimation law common to all sites. In order to develop this law, the same approach employed for a single site has been adopted (Table 5).

This approach, although empirical, can be used for estimating the source energy of the study sites from the measured sensor energy and the position of

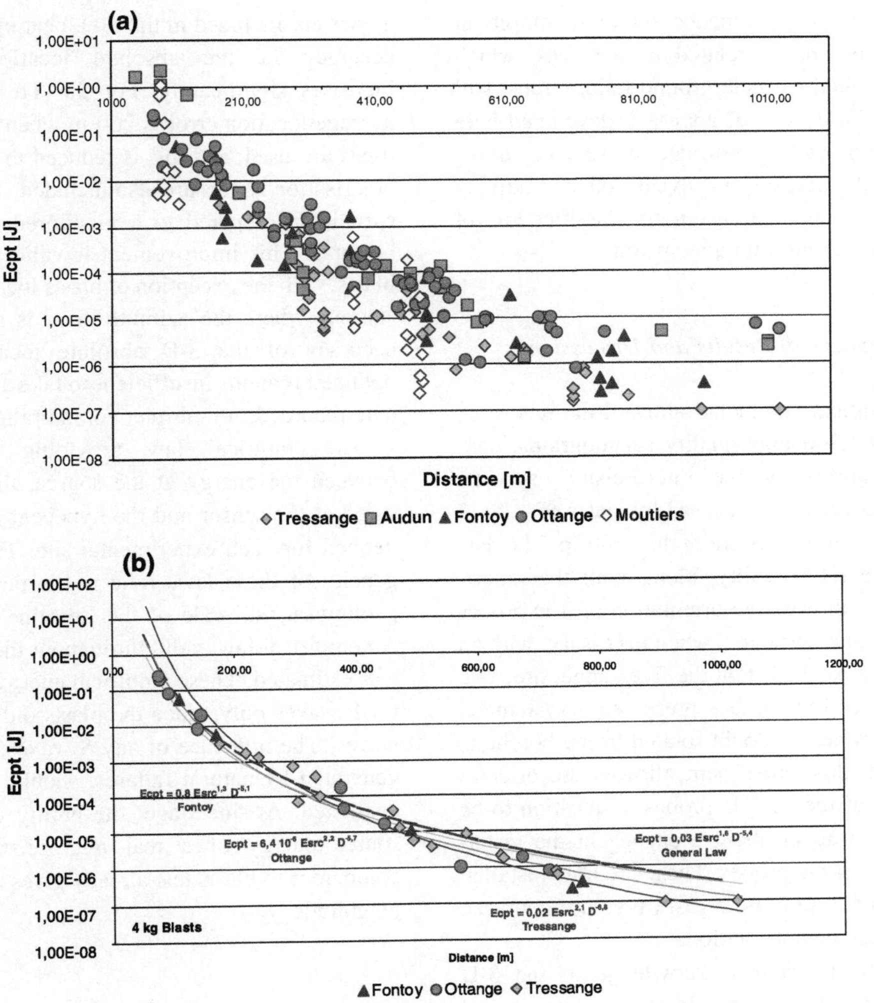

Figure 11

a Figure of the energy recorded at the 3-D sensor as a function of the distance for all experimental sites and all blast explosive charges. **b** Graphic comparison of function for a constant source energy term corresponding to 4 kg of explosive on three different sites (Fontoy, Ottange and Tressange), i.e. 11 stations and 11 blasts

Table 5

Source energy estimation laws from sensor energy for each experimental site

Site	Empirical source energy law
Audun-le-Tiche	$E^*_{src} = 11 \times D^{3.2} \times E^{0.6}_{cpt}$
Fontoy	$E^*_{src} = 1.2 \times D^{4} \times E^{0.8}_{cpt}$
Tressange	$E^*_{src} = 5.2 \times D^{3.2} \times E^{0.5}_{cpt}$
Moutiers Gorcy	$E^*_{src} = 1{,}303 \times D^{2.3} \times E^{0.5}_{cpt}$
Ottange	$E^*_{src} = 230 \times D^{2.6} \times E^{0.5}_{cpt}$
All sites	$E^*_{src} = 8.6 \times D^{3.4} \times E^{0.6}_{cpt}$

the event. However, these relationships are valid only for P waves as the blasts have not generated other types of waves. The influence on these empirical relationships of energy related to S waves, which could be generated by underground failure, must still be assessed. The empirical approach developed here may however be used to estimate the seismic source energy of real microseismic events on a relatively local scale for fast and accurate classification of different events from a unique swarm.

5. *Summary of Results and Discussion*

The experiment conducted allowed the recording of more than 1,200 high-quality seismograms, confirming the ability of the microseismic systems deployed to detect signals caused by blasts of 1 kg of explosives at a distance of more than 300 m. The data signal-to-noise ratio quality, along with the polarisation analysis allowed the orientation of 3-D probes to be checked and corrected, when necessary, with an accuracy close to $\pm 1°$. For the Tressange site, two 3-D probes over four probes presented a systematic error. These probes no doubt rotated in the borehole. On the whole, this experiment allowed the orientations of 7 out of the 22 3-D probes in question to be corrected and brought about a change in the installation protocol of the probes. These are now installed with a centring system, with extra pressure contacts, limiting any accidental rotation.

As mentioned, accurate knowledge of the 3-D sensors orientation is crucial in the processing and analysis of microseismic data. Together with the arrival times measured at the sensors, through the inversion process angles allow a realistic velocity model, in relation to the site's geology, to be calculated. However, for the Tressange site it was shown that the calculation of the velocity model depends on good knowledge of the geological structure of the studied site. It depends as well on a precise analysis of parameters to eliminate any abnormal data. The use of an appropriate norm to minimize any differences between the observed and calculated data is also important. In the current study, the L2-norm is far more accurate than the EDT standard, for both velocity model calculation and blast relocation.

Calculations of azimuth and dip of the incident rays at the 3-D probes is easy and accurate. Once these angles are included in the 3-D location algorithm, the accuracy of the absolute location hypocenters increases significantly. For the Tressange site, the average location error is 260 m when only the arrival times are used, and this is reduced to 70 m when the polarisation angles are also included. This accuracy is sufficient compared to a monitored area of several hectares. This improvement is valid for most of the blasts, with the exception of blasts located right below stations where the azimuth error is significant. The accuracy of the 3-D absolute location algorithm obtained remains insufficient to take advantage of the data recorded in multiplet configurations.

An empirical law providing a relationship between the energy at the source, the energy measured at the sensor and the hypocentral distance was defined for each experimental site. The good homogeneity of these laws reflects the uniformity of the geology at the scale of the Lorraine iron-ore basin. An empirical law, valid throughout the iron-ore field, was estimated. These empirical energy laws are valid for P waves only, since the blasts did not generate S waves. The influence of any S waves, which may be generated by natural failures, should thus be further evaluated. At this stage, the ability to estimate the source energy of a real microseismic event and compare it to blasts test already gives a good order of magnitude.

6. *Conclusion*

The microseismic calibration blasts carried out in the Lorraine iron-ore field form a reference database consisting of 1,200 high-quality seismograms. This unique database allows the validation of the tools developed for microseismic analysis and calibration of the microseismic characteristics of each of the experimental sites. Step-by-step analysis of those data enabled us to: (1) Calculate P-wave velocity models of each site and characterize some geophysical properties of the main geological strata; (2) optimize and validate the 3-D localisation tool; (3) construct empirical source energy and wave propagation laws for the different but geologically similar experimental sites including comparison tests. In the future, we will perform additional research to further characterise the seismic source mechanism. This work will also take into account the relative location method to take advantage of the various blast configurations implemented in this experiment. The results already achieved and the work in progress should allow the characterisation of post-mining instabilities and their early warning systems in order to enhance decision making.

Acknowledgments

Special thanks to the French Ministry of Economy, Finance and Industry for financial support. We fully acknowledge M. E. Rovinalti (former mine manager from ARBED company), M. A. Marnet (DRIRE Lorraine), M. J-P. Josien (former director from GEODERIS) both for authorisations and helpful contributions to the field experiment. Our thanks also go to the ARBED and ARCELOR companies for the technical support and human resources used during the experiment. We are grateful for the two anonymous reviewers for their suggestions that significantly helped to improve this paper. We do not lose sight of Kathe Hooks (School of Mines of Nancy) who fully reading this paper.

References

Abdul-Wahed, M.K., Senfaute, G., Piguet, J.P., *Source location estimation using single station three-component seismic data, Rock Mechanics – A Challenge for Society*, Proc. ISRM Regional Symp. (Eurock 2001, Espoo, Finland, 4 – 7 June 2001).

Abdul-Wahed, M.K., Al Heib, M., Senfaute, G. (2006), *Mining-induced seismicity: Seismic measurement using multiplet approach and numerical modelling*, Int. J. Coal. Geol *66* (1–2), 137–147.

Aki, K. and Richards, P.G., *Quantitative Seismology* (W.H. Freemann, San Francisco 1980).

Bennani, M. and Homand, F. (2004), *Coverage formations at the level of areas in which there is a sudden collapse hazard*, Géodéris R2004/002.

Boatwright, J., Choy, G.L., and Seeking, L.C. (2002), *Regional Estimates of Radiated Seismic Energy*, Bull. Seismol. Soc. Am. *92*, 1241–1255.

Brune, J.N. (1970), *Tectonic stress and the spectra of seismic shear waves from earthquakes*, J. Geophys. Res. *75*(26), 4997–5009.

Contrucci, I., Klein, E., Bigarre, P., Lizeur, A., and Lomax, A., *Early-warning microseismic systems applied to the management of post-mining large-scale ground failures: calibration by a geophysical field experimentof blast swarms* (EGU, April 2008, Vienna, Geophys. Res. Abs., 10, EGU2008-A-09300, 2008).

Couffin, S., Bigarré, P., Bennani, M., Josien, J.P., *Permanent Real Time Microseismic Monitoring of Abandoned Mines for Public Safety. Fields Measurements in Geomechanics* (ed. Myrvoll) (Sweets & Zeitlinger, Lisse 2003) pp. 437–444.

Didier, C. (2008), *The French Experience of Post Mining Management, Post Mining 2008*, Nancy, France.

Driad-Lebeau, L., Lahaie, F., Al Heib, M., Josien, J.P., Bigarré, P., and Noirel, J.F. (2005), *Seismic and geotechnical investigations following a rockburst in a complex French mining district*, Int. J. Coal Geol. *64* (1–2), 66–78.

Finck, F., Kurz, J.H., Grosse, C.U., and Reinhardt, H.W. (2003), *Advances in Moment Tensor Inversion for Civil Engineering, in Non-Destructive Testing in Civil Engineering*, International Symposium (NDT-CE 2003).

Geller, R.J. and Mueller, S. (1980), *Four similar earthquakes in central California*, Geophys. Res. Lett. *7*(10), 821–824.

Gibowicz, S.J., Young, R.P., Talebi, S., and Lawrence, D.J., (1991), *Source Parameters of Seismic Events at the Underground Research Laboratory in Manitoba, Canada: Scaling Relations for Events with Moment Magnitude Smaller than -2*, Bull. Seismol. Soc. Am. 81(4), 1157–1182.

Gibowicz, S-J. (2006), *Seismic doublets and multiplets at Polish coal and copper mines*, Acta Geophysica *54*, 142–157, doi:10.2478/s11600-006-0014-y.

Homand, F. and Dagallier, G. (2004), *Survey of the coverage formations of areas in which there is a non-negligible risk of sudden collapse*, LAEGO, LG.FH.ARB.PSI.RPRE.03.0248.C.

Klein, E., Nadim, C., Bigarre, P., and Dunner, C., *Global monitoring strategy applied to ground failure hazards* (10th Int. Symposium on Landslides & Engineering Slopes - Xi'an, China 2008).

Lees, J.M. (1998), *Multiplet analysis at Coso geothermal*, Bull. Seismol. Soc. Am. *88*, 1127–1143.

Li, T., Cai, M.F., and Cai, M. (2007), *A review of mining-induced seismicity in China*, Internat. J. Rock Mechan. Mining Sci. *44* (8), 1149–1171.

Lomax, A. and Snieder, R. (1995), *Identifying sets of acceptable solutions to non-linear, geophysical inverse problems which have complicated misfit functions*, Nonlinear Processes in Geophys. *2*, 3/4, 222–227.

Lomax, A. (2005), *A reanalysis of the hypocentral location and related observations for the great 1906 California earthquake*, Bull. Seismol. Soc. Am. *91*, 861–877.

Lomax, A. and Curtis, A., *Fast, probabilistic earthquake location in 3-D models using oct-tree importance sampling*, Europ. Geophys. Soc. (Nice 2001).

Lopes, B. (2002), *Etude de l'environnement géologique du site de Tressange*, BRGM/RP-51701-FR.

Madariaga, R. (1976), *Dynamics of an expanding circular fault*, Bull. Seismol. Soc. Am. *66*(3), 639–666.

Magotra, N., Ahmed, N., and Chael, E. (1987), *Seismic event detection and source location using single-station (three-component) data*, Bull. Seismol. Soc. Am. 77(3), 958–971.

McGarr, A. (1992), *An implosive component in the seismic moment tensor of a mining-induced tremor*, Geophys. Res. Lett. *19*(15), 1579–1582.

Maubeuge P.L. (1955), *Geological Observations to the East of the Paris Basin.* Institut National Polytechnique de Lorraine, Vol. 1, pp. 70–300.

Miller, A., Richards, J. A, McCann, D.M., Browit, C.W.A., and Jackson, P.D. (1989), *Microseismic techniques for monitoring incipient hazardous collapse conditions above abandoned mines*, Quart. J. Eng. Geology *22*, 1–18.

Montagne, A., Tincelin, E., Astier, J., and Varoquaux, J.L. (1992), *Les mines de fer de Lorraine*, Chambre syndicale des mines de fer de France, Paris, p. 17; 43; 71-77; 243.

Moriya, H., Fujita, T., Niitsuma, H., Eisenblatter, J., and Manthei, G., (2006), *Analysis of fracture propagation behavior using hydraulically induced acoustic emissions in the Bernburg salt mine, Germany*, Internat. J. Rock Mech. Mining Sci. *43*, 49–57.

Ogasawara, H., Fujimori, K., Koizumi, N., Hirano, N., Fujiwara, S., Otsuka, S., Nakao, S., Nishigami, K., Taniguchi, K., Iio, Y., Nishida, R., Oike, K., and Tanaka, Y. (2002), *Microseismicity induced by heavy rainfall around flooded vertical ore veins*, Pure Appl. Geophys. *159* (1-3), 91–109.

Pinsky, V., Husen, S., Lomax, A. *A comparative study of robust algorithms for rapid, automatic earthquake location* (EGU, Vienna, April 2008) Geophys. Res. Abs. *10*, EGU2008-A-07074, 2008.

Poupinet, G, Ellsworth, W. L., Fréchet J. (1984). *Monitoring velocity variations in the crust using earthquake doublets – an application to the calaveras fault, California*, J. Geophys. Res. 89(NB7) 5719–5731.

Rubin, A.M., Gillard, D., and Got, J.L. (1999), *Streaks of microearthquakes along creeping faults*, Nature *400*, 635–641.

Schaff, D.P., Bokelmann, G.H.R., Beroza, G.C., Waldhauser, F., and Ellsworth, W.L. (2002), *High-resolution image of Calaveras Fault seismicity*, J. Geophys. Res. *107*(B7), 2186, doi: 10.1029/2001JB000633.

Samson, J.C. (1983), *The spectral matrix, eigenvalues, and principal components in the analysis of multichannel geophysical data*, Annales Geophysicae *1*(2), 115–119.

Senfaute, G., Chambon, C., and Bigarré, P. (1997), *Spatial distribution of mining tremors and the relationship to rockburst hazard*, Pure Appl. Geophys. *150* (3-4): 451–459.

Senfaute, G., Abdul-Wahed, M.K., Piguet, J.P., and Josien, J.P., *Qualification of the microseismic monitoring technique applied to the risk of collapse or iron mines* (Eurock 2000, Aachen 2000) pp. 597–602.

Sileny, J., and Milev, A. (2006), *Seismic moment tensor resolution on a local scale: Simulated rockburst and mine-induced seismic events in the Kopanang Gold Mine, South Africa*, Pure Appl. Geophys. *163*, 1495–1513.

Slunga, R., Rognvaldsson, S.T., and Bodvarsson, R. (1995), *Absolute and relative locations of similar events with applisation to microearthquakes in southern Iceland*, Geophys. J. Internati. *123* (2), 409–419.

Tarantola, A. and Valette, B. (1982), *Generalized nonlinear inverse problems solved using the least squares criterion*, Revi. Geophys. Space Phys. *20*(2), 219–232.

Tastet, J., Contrucci, I., Klein, E., Bigarré, P., and Driad-Lebeau, L. (2007), *Large-scale field experiment to calibrate microseismic source parameters applied to real-time monitoring of post-mining instabilities*, Proc. 11th Congress of the International Society for Rock Mechanics, 9-13, July 2007, Lisbon, Portugal. Leiden, The Netherlands (Taylor & Francis, 2007 Vol. *2*, 1147–1150.

Trifu, C. and Shumila, V. (2002), *The use of uniaxial recordings in moment tensor inversions for induced seismic sources*, Tectonophysics *356*, 171–180.

Urbancic, T.I. and Young R.P. (1993), *Space-time variations in source parameters of mining-induced seismic events with* $M < 0$, Bull. Seismol. Soc. Am. *83*(2):378–397.

Volker, O. and Roth, M. (2003), *Automated seismic event location for hydrocarbon reservoir*, Computers and Geosciences *29*, 851–863.

Waldhauser, F. and Ellsworth, W.L. (2000), *A double-difference earthquake location algorithm: Method and application to the Northern Hayward Fault*, Bull. Seismol. Soc. Am. *90*(6), 1353–1368.

Yamada, T., Mori, J.J., Ide, S., Abercrombie, R.E., Kawakata, H., Nakatani, M., Iio, Y., and Ogasawara, H. (2007), *Stress drops and radiated seismic energies of microearthquakes in a South African gold mine*, J. Geophys. Res. *112*, doi:10.1029/2006 JB004553.

(Received October 6, 2008, revised April 22, 2009, accepted May 4, 2009, Published online December 11, 2009)

Pure Appl. Geophys. 167 (2010), 63–75

DOI 10.1007/s00024-009-0001-8

Pure and Applied Geophysics

Coal Mining Induced Seismicity in the Ruhr Area, Germany

MONIKA BISCHOFF,[1] ALPAN CETE,[2] RALF FRITSCHEN,[3] and THOMAS MEIER[1]

Abstract—Over the last 25 years mining-induced seismicity in the Ruhr area has continuously been monitored by the Ruhr-University Bochum. About 1,000 seismic events with local magnitudes between $0.7 \leq M_L \leq 3.3$ are located every year. For example, 1,336 events were located in 2006. General characteristics of induced seismicity in the entire Ruhr area are spatial and temporal correlation with mining activity and a nearly constant energy release per unit time. This suggests that induced stresses are released rapidly by many small events. The magnitude–frequency distribution follows a Gutenberg–Richter relation which is a result from combining distributions of single longwalls that themselves show large variability. A high *b*-value of about 2 was found indicating a lack of large magnitude events. Local analyses of single longwalls indicate that various factors such as local geology and mine layout lead to significant differences in seismicity. Stress redistribution acts very locally since differences on a small scale of some hundreds of meters are observed. A regional relation between seismic moment M_0 and local magnitude M_L was derived. The magnitude–frequency distribution of a single longwall in Hamm was studied in detail and shows a maximum at $M_L = 1.4$ corresponding to an estimated characteristic source area of about 2,200 m^2. Sandstone layers in the hanging or foot wall of the active longwall might fail in these characteristic events. Source mechanisms can mostly be explained by shear failure of two different types above and below the longwall. Fault plane solutions of typical events are consistent with steeply dipping fracture planes parallel to the longwall face and nearly vertical dislocation in direction towards the goaf. We also derive an empirical relation for the decay of ground velocity with epicenter distance and compare maximum observed ground velocity to local magnitude. This is of considerable public interest because about 30 events larger than $M_L \geq 1.2$ are felt each month by people living in the mining regions. Our relations, for example, indicate that an event in Hamm with a peak ground velocity of 6 mm/s which corresponds to a local magnitude M_L between 1.7 and 2.3 is likely to be felt within about 2.3 km radius from the event.

Key words: Mining-induced seismicity, coal mining, magnitude–frequency distribution, *b*-value, energy release, ground velocity.

1. *Introduction*

Early coal mining in the Ruhr area is known since the thirteenth century. In the southern region close to the Ruhr, coal seams reach the surface and small amounts were excavated in open workings. Until the mid-nineteenth century, coal was mined in small mines that reached no more than a few meters depth and in horizontal drifts that reached up to 400 m into the hillsides; mining was confined to the region around the river Ruhr as carboniferous layers dip northwest and lie at greater depths beneath Cretaceous rocks in the northern parts of the area. As the commercial interest in coal increased during the industrial revolution, it became possible to cut through the overlying Cretaceous rock and use of steam engines made it possible to pump water out of the mines. This allowed mining at depths of 400 m and more leading to strongly increased coal production during the first half of the nineteenth century. Beginning in 1957 high excavation costs led to mine closure and the consolidation of the different mining companies. In 2006, six mines were in operation in the Ruhr area producing about 15 million tons of coal annually.

Today mining is confined to the western Ruhr area around Kamp-Lintfort and Rheinberg (Bergwerk West, near station BRHE, Fig. 1), north of Bottrop (Bergwerk Prosper-Haniel, near station ZERL, Fig. 1) and near Dorsten and Marl in the north (Bergwerk Auguste Victoria, near station ZERL, Fig. 1) and in the eastern part in Hamm and Bergkamen (Bergwerk Ost, near station HMES, Fig. 1).

[1] Institute of Geology, Mineralogy and Geophysics, Ruhr-University Bochum, 44780 Bochum, Germany. E-mail: bischoff@geophysik.ruhr-uni-bochum.de

[2] Hustadtring 35, 44801 Bochum, Germany.

[3] DMT GmbH & Co. KG, Am Technologiepark 1, 45307 Essen, Germany.

Coal is mined by longwall mining at depths between 700 and 1,500 m. Longwalls are 300 m wide and 1,000 m long on the average. The thickness of the coal seams is about 2 m. Face advance using modern equipment is about 100 m/month. Excavated areas are not refilled so that surrounding rocks immediately cave. As mining is carried out at different depth levels the mine layout is complex. Mining typically starts at the topmost layer and proceeds to deeper layers. Therefore, each longwall faces different mining conditions. An active longwall can lie in untouched rock mass or at different levels beneath old mine workings in a prestressed area. The tectonic setting of the carboniferous rocks dominated by NE-SW striking folds and thrust faults that are intersected by NW–SE striking normal faults (Drozdzewski *et al.*, 1985) increase complexity.

As early as at the beginning of the twentieth century Mintrop (1909a, b) was interested in induced seismicity caused by coal mining. In 1908 he built the first seismological observatory in Bochum which was in operation until the 1940s when it was destroyed during the Second World War. New seismological stations were installed again by the Westfälische Berggewerkschaftskasse in 1956 and by the Ruhr-University Bochum (Böhmer *et al.*, 1989) which has routinely monitored seismicity of the entire region since 1983. Routine analysis, published in annual reports (Harjes, 1983–2004; Meier, 2005–2006), includes epicenter location using array techniques and determination of local magnitudes. The current seismic network (Fig. 1) consists of a short-period array (BUG-Array) with three vertical-component stations (GS13/S13, 1 Hz) close to the university with an aperture of 2.4 km and the broadband GRSN-station BUG (STS-2, 120 s; Korn, 2002). Local stations (BRHE, ZERL; both MARK L4-3D, 1 Hz and BAHL; STS-2, 120 s) in the mining regions are used for analysis of smaller events.

Local studies using small temporal networks in the mining regions (e.g., Casten and Cete, 1980) obtained a higher resolution of epicenters (±100 m)

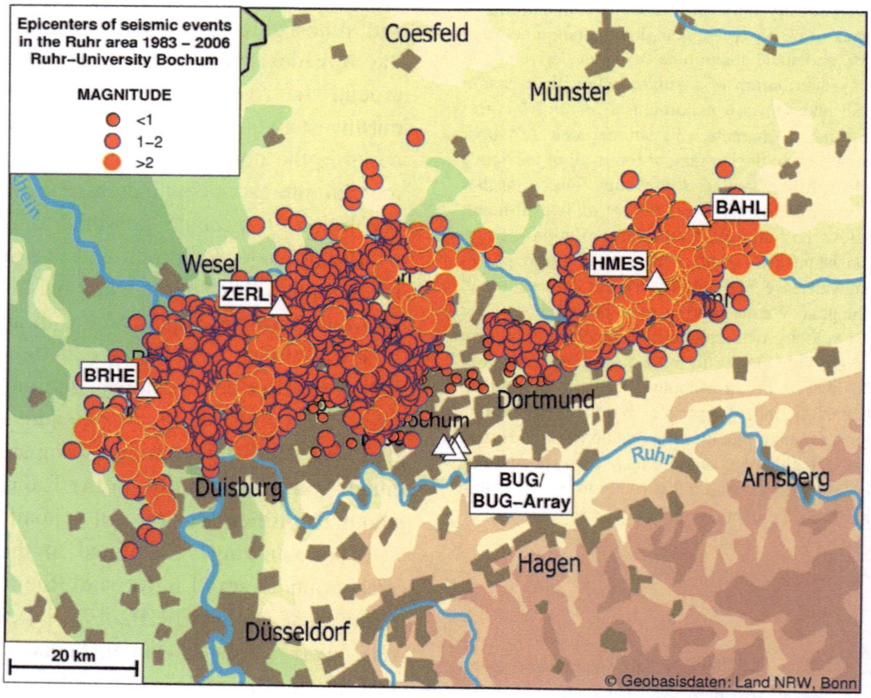

Figure 1

Map of seismic events in the Ruhr area between 1983 and 2006. 13,912 mining-induced seismic events with magnitude $M_L \geq 0.7$ were located. *Triangles* denote the short period BUG-array, broadband GRSN-station BUG and local stations BRHE, ZERL, HMES and BAHL in the mining regions and found a spatio-temporal correlation between seismicity and mining activity in specific longwalls. BAULE and HINZEN (1981) operating a local network in the eastern Ruhr area in Hamm, found a correlation of increased seismicity with remaining pillars. They also related epicenters to the longwall face locating most events within 60 m in front of the face. Source parameters of mining-induced events, such as source radius, stress drop and seismic moment from *P*- and *S*-wave spectra and *P*-wave spectra, respectively, were determined by HINZEN (1982) and GIBOWICZ *et al.* (1990). For seismic events with local magnitude between $1.2 \leq M_L \leq 2.3$ they found typical source radii between 60 and 180 m corresponding to the mining geometry. Their stress drops are small not exceeding a few bars. HINZEN (1982) showed that the corner frequency f_0 does not change with seismic moment M_0 hinting at a fixed source size with seismic moments strongly depending on stress drop. GIBOWICZ *et al.* (1990) found evidence for some non double-couple events. FRITSCHEN *et al.* (1999) determined fault plane solutions assuming pure shear faulting for events from a specific longwall in Hamm and conclude that these seismic events may be caused by failing sandstone strata.

Mining-induced seismicity is a worldwide phenomenon. A review of underground and surface mining regions and related induced seismicity is given by e.g., GIBOWICZ and KIJKO (1994). They conclude that two types of mine tremors are now generally accepted. The first type, directly connected to mining operations, is associated with the formation of fractures at stope faces. The second type, usually with larger events, is associated with the movement on major geologic faults at some distance from the mine faces. These latter events respond to stress changes on the scale of a whole mine. WONG (1993) emphasizes the role of tectonic stresses in the generation of mine seismicity. Comparing focal mechanisms from different regions to data from the World Stress Map Project, he argues that unloading increases the shear stress along preexisting zones of weakness in compressive stress regimes. This could explain the prevalence of observed reverse faulting source mechanisms.

In this article, first we will give an overview of induced seismicity in the Ruhr area. We will discuss temporal variation, cumulative energy release and magnitude–frequency distribution. From observed focal mechanisms a model of the fracture process will be derived. Next we will concentrate on the region around Hamm to analyze seismicity of single longwalls in greater detail. Time-dependent energy release and magnitude–frequency distributions of individual longwalls will be compared. Using a local network we will provide a model for ground velocity decay to estimate the surface effects of induced seismic events.

2. *Regional Induced Seismicity in the Ruhr Area*

Continuous observation of seismicity in the Ruhr area shows that deep coal mining is accompanied by seismic events. No tectonic earthquakes occur in this region. From 1983 to 2006, 13,912 induced events with $0.7 \leq M_L \leq 3.0$ were located (Fig. 1). The epicenter location accuracy is about ±2 km for the region around Hamm and about ±5 km for the other regions. Events with local magnitude $M_L \geq 1.2$ can be felt at the surface. With about 30 such events each month, public interest in mining-induced seismicity is therefore considerable. The largest event with $M_L = 3.3$ occurred near Moers in the western part of the Ruhr region on December 12, 2007 (14:52:31.8 (UT), 51.409°N 6.566°E).

Because network sensitivity increased over time, in recent years about 1,000 seismic events were located annually, for example 1,336 events with $0.7 \leq M_L \leq 3.0$ in 2006. The epicenter map for 2006 (Fig. 2) shows that seismic events cluster in regions known for active mining. Indeed, seismic events can directly be assigned to specific mines. The easternmost cluster, for example, is related to the Bergwerk Ost mine in Hamm. Compared to seismicity of the entire catalog (Fig. 1), epicenters in 2006 concentrate in the northern part consistent with closure of mines in the south.

Temporal correlation between mining and seismic activity is observed. When mining stops fewer seismic events occur. This is illustrated in the average distribution of events per weekday in 2006 (Fig. 3a) which shows a clear minimum during weekends when mining activity is stopped. We find a similar correlation during each day (Fig. 3b) due to hours of

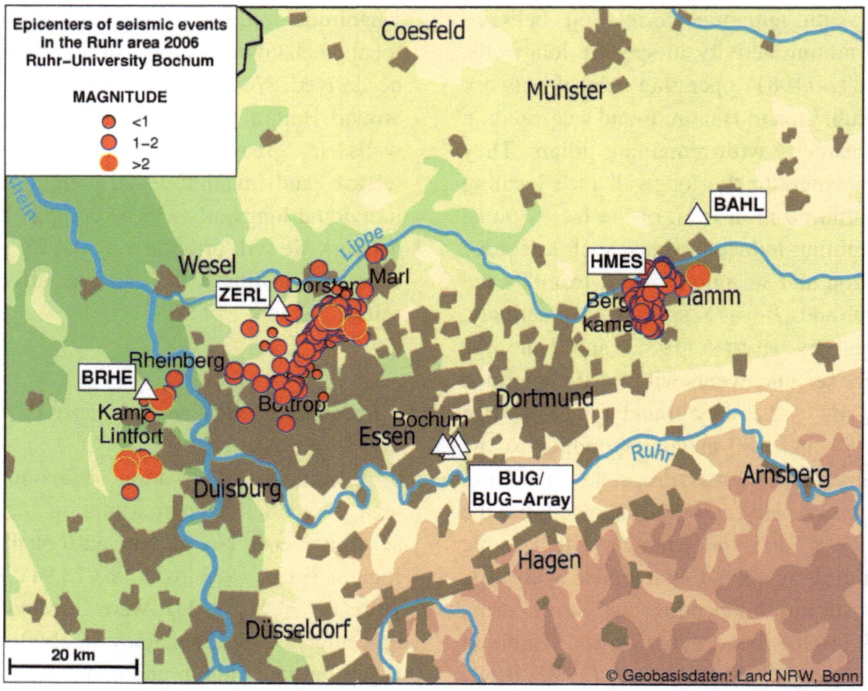

Figure 2
Map of seismic events in the Ruhr area for 2006. 1,336 seismic events with magnitude $M_L \geq 0.7$ were located. Seismic events cluster in the regions of active mining

maintenance. The daily minimum varies between different mines and longwalls resulting in a broad minimum around 7:00 a.m.

From the magnitude time distribution (Fig. 3c) of the entire catalog the cumulative energy release (Fig. 3d) was calculated using the empirical relation by Gutenberg and Richter (1956). According to Hinzen (1982) this is valid for the Ruhr area. Energy is released continuously over time (Fig. 3d). About 100 MJ are released seismically each month. Steps of individual events cannot be distinguished in the cumulative energy release implying that it is not dominated by large but by many small events. The larger step in July 1998 is due to eight $M_L \geq 2$ events that occurred in a specific longwall in Hamm within the short span of 19 days. Immediate energy release corresponds to the observed short-time variability during the week and the day described above (Fig. 3a, b). Thus, seismicity is a direct response to stresses induced by mining.

The seismic region is characterized by the magnitude–frequency distribution (Fig. 4, top). The linear relation between the magnitude of the events and the logarithm of their number, known as the Gutenberg–Richter relation (Gutenberg and Richter 1944), relates the number of small events to the number of large events. The slope of this relation (*b*-value) is typically close to 1 for tectonic earthquakes (e.g., Scholz, 2002). The observed magnitude–frequency distribution (Fig. 4, top) for the Ruhr area follows a Gutenberg–Richter relation. We apply the method by Wiemer and Wyss (2000) to determine the catalog completeness threshold M_C and fit different magnitude intervals $[M_{min}\ 3]$ of the observed distribution to a straight line. The least-squares fit results in the goodness-of-fit as a function of M_{min} (Fig. 4, bottom). Accepting a residual of less than 4% as a good fit, we map the catalog completeness at $M_C = 1.4$ and the associated *b*-value of 2 ± 0.06. The *b*-value as a function of M_{min} (Fig. 4, middle) strongly supports a relatively high *b*-value.

An alternative method to test catalog completeness (Rydelek and Sacks, 1989) assumes that seismic events are uniformly distributed in time. The catalog

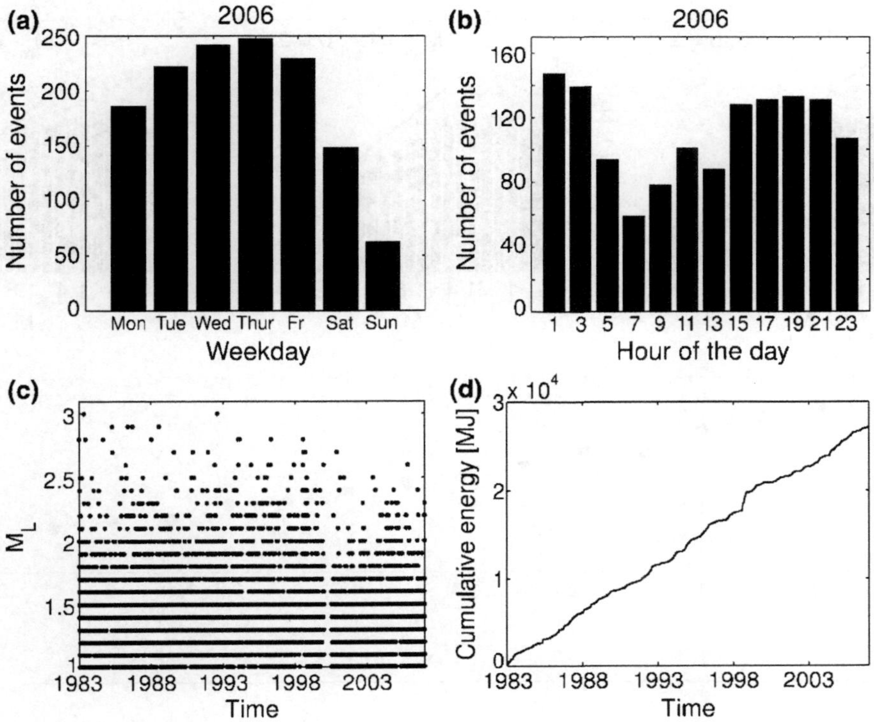

Figure 3
Temporal distributions of seismicity in the Ruhr area. **a** Average weekly variation of seismicity and **b** diurnal variation for 2006. Minima in seismic activity correlate with a stop of mining activity at weekends (**a**) and during hours of maintenance (**b**). **c** Magnitude time distribution (data gap January–June 2000) and **d** cumulative energy release with time for 1983–2006

is regarded as complete if no night-to-day modulation in seismicity is observed. These assumptions are not appropriate for our data set because the occurrence of seismic events is highly dependent on the variability of mining activity.

The high b-value of about 2 indicates a lack of large magnitude events and points to reduced strength of the surrounding rocks. This matches the continuous energy release which is dominated by many small events. However, for $b \geq 1.5$ the effect of very small earthquakes allows the accumulated moment to increase infinitely (MOLNAR, 1979). Thus, a b-value larger than 1.5 is physically not possible for a magnitude range that is extended to minus infinity. That means at smaller magnitudes the actual magnitude–frequency distribution of the Ruhr area should deviate from the Gutenberg–Richter relation with $b = 2$ (Fig. 4, top, black line).

Furthermore, the b-value can be linked to the fractal dimension D (AKI, 1981) by $D = 2b$. For our data this relation cannot be applied because D must be in the range $1 \leq D \leq 3$. These are clear indications that the high observed b-value does not describe the fractal behavior of the entire catalog but is a result of combining distributions of individual longwalls over a time span of 24 years. For each longwall and for short time intervals the magnitude–frequency distribution might differ significantly as will be discussed in Sect. 3.

Local seismic networks can provide an insight into source mechanisms and the origin of seismic events of a certain region. For example, in the eastern part of the Ruhr area (near station HMES, Fig. 1) FRITSCHEN *et al.* (1999) found a strong correlation between the position of longwall mining faces and event locations which indicated that seismic events were directly connected with mining operations. Assuming pure shear failure only two different types of failure were observed. Since the analysis by FRITSCHEN *et al.* (1999) similar observations have been made in the Ruhr area and the Saarland confirming that these two types of failure

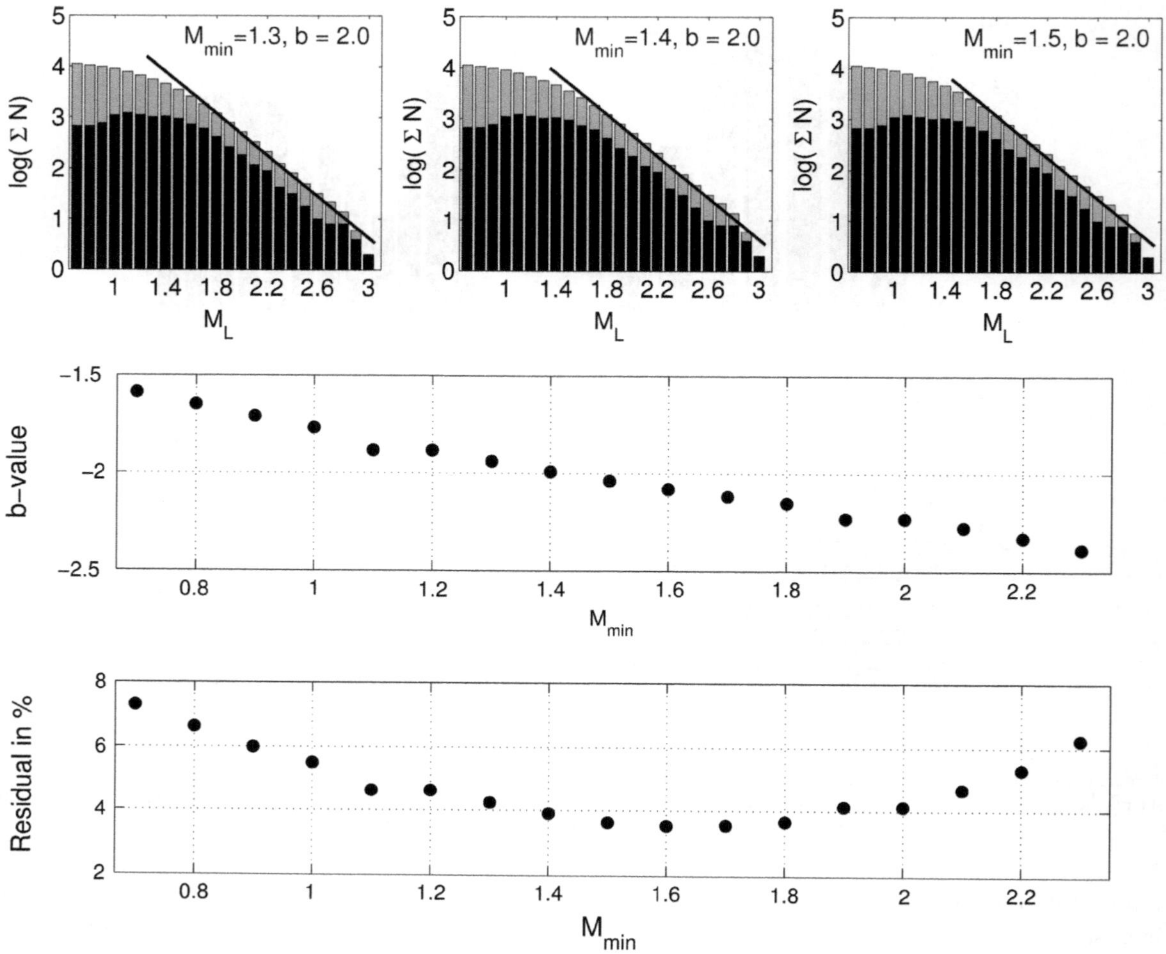

Figure 4
Magnitude–frequency distribution for the Ruhr area from 1983 to 2006 and test of catalog completeness. For different magnitude intervals [$M_{\min}$ 3], *histograms* (*top*) show the absolute (*black*) and cumulative (*gray*) magnitude–frequency distribution and the fit (*black line*) to the Gutenberg–Richter relation. From the residual of the least-squares fit as a function of $M_{\min}$ (*bottom*) we map the minimum magnitude of completeness at $M_C = 1.4$ as the smallest magnitude at which at least 96% (max. 4% residual, *bottom*) of the data are modeled. The *b*-value as a function of $M_{\min}$ (*middle*) supports a high *b*-value of about 2

represent most of the mining-induced events. We show two representatives of typical fault plane solutions by FRITSCHEN *et al.* (1999) in an equal area projection (lower hemisphere, Fig. 5, top and bottom) and from long-time experience we confirm a model (sketch, vertical section, Fig. 5, middle) that can explain the main mechanisms of observed events. For both types the planes interpreted as fracture planes were steeply dipping and parallel to the mining face (Fig. 5), showing almost pure dip-slip movement. Type 1 (Fig. 5, top) resembled the breaking down of the hanging wall with a dislocation towards the goaf. The second type (Fig. 5, bottom) represents shear fractures with dislocations in just the opposite direction. It is very likely that events of Type 2 occurred below the working face. The most often observed events are of Type I. Due to the sparse station distribution (surface stations KIR, HUM, HER, PFE and underground station BSS, Fig. 5) not only *P*-wave polarities but also *S*/*P* amplitude ratios and *S*-polarisation angles had to be used to find the fault plane solutions. Assuming a double-couple mechanism the

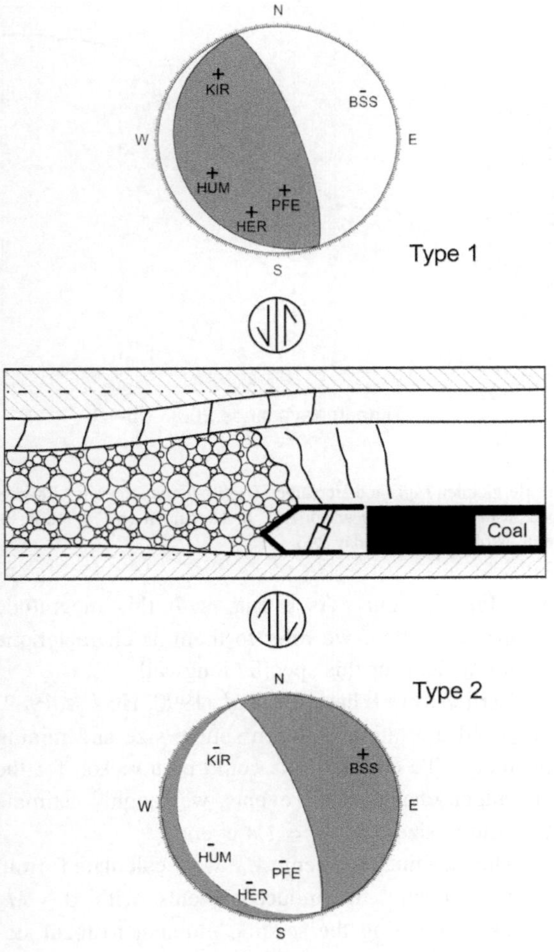

Figure 5
Possible explanation of the observed source mechanisms. The sketch in the middle (*vertical section*) shows steeply dipping fault planes with dislocations in direction towards the goaf. Also shown are representatives of the two types (Type I above and Type II below the longwall) of derived fault plane solutions in an equal area projection (lower hemisphere; FRITSCHEN *et al.*, 1999) together with the station locations of the surface stations (KIR, HUM, HER, PFE) and underground station (BSS) of the local network. In addition to observed P-wave polarities, *S*/*P* amplitude ratios and *S*-wave polarization angles were used to better constrain the fault plane solutions. Assuming a double-couple mechanism the angles of strike, dip and rake could be determined with an uncertainty of approximately 10°

angles of strike, dip and rake could be determined with an uncertainty of approximately 10° (FRITSCHEN *et al.*, 1999). In addition to these two types ALBER *et al.* (2009) found events that can be explained by the reactivation of tectonic faults or by failure of remnant pillars.

3. *Magnitude–Frequency Distribution of Single Longwalls*

Close examination of individual longwalls reveals differences in local seismicity. As an example, we analyzed the seismicity of three longwalls in Hamm and Bergkamen in the eastern part of the Ruhr area (eastern cluster in Fig. 2) for February 2004 to May 2005. In this time interval the longwall S704 was mined in Hamm-Herringen from May 2004 to April 2005 and two adjacent longwalls were mined in Bergkamen beginning in March 2004 and August 2004, respectively. The distance between the two mining regions is about 5 km. Event epicenters and magnitudes were obtained by routine analysis described in Sect. 1. With a location accuracy of about ±2 km, we can distinguish between events in Herringen and Bergkamen.

The cumulative energy release versus time plots (Fig. 6) reveal large variations within and between the two regions. Both in Bergkamen (Fig. 6, left) and Herringen (Fig. 6, right) time intervals with higher energy release (2.5–5.4 MJ/d, marked gray) can be distinguished from times of low seismicity (0.5 MJ/d). In Herringen fading of seismicity due to the cessation of mining is observed after day 296. Because face advance was regular the other variations are not related to changes in mining activity. Therefore, differences in seismicity rate must be caused by local changes in geological conditions or the mine layout. Besides the temporal variation, the total amount of seismic energy varies. In Bergkamen less energy was released than in Herringen although two longwalls were mined there.

Two main differences between mining in Bergkamen and Herringen are known. First, at depths of about 1,500 m mining in Bergkamen is about 400 m deeper than in Herringen. Secondly, in Bergkamen no former mining activity was conducted up to 600 m above the active longwall; in Herringen old workings exist about 60 m above the active longwall and at higher levels. For example, stresses below remnant pillars are increased and when they are undercut by the active mining, this can lead to enhanced seismicity (e.g., ALBER *et al.*, 2009). Although a direct correlation with remnant pillars is not obvious in Herringen, still the mine layout is more complex

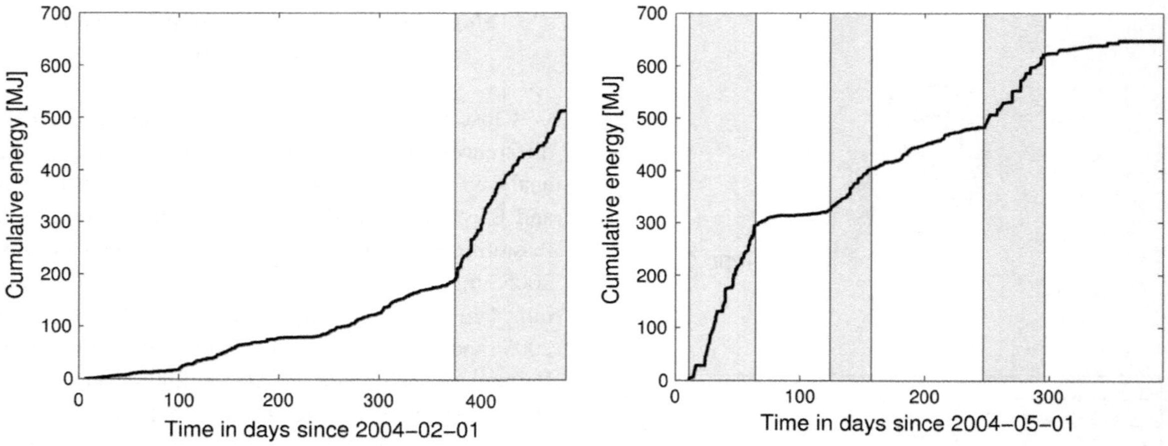

Figure 6
Different cumulative energy release was observed for different longwalls in Bergkamen (*left*) and Herringen (*right*). Time intervals of higher energy release (*marked gray*) of about 2.5–5.4 MJ/d can be distinguished from times of low seismicity. In Herringen (*right*) fading of seismicity due to the cessation of mining is observed after day 296

here. Intricate stress redistribution may have led to the larger energy release and variation in seismicity rate.

These two examples clearly show that each longwall has its own seismic response to mining. Seismicity in Herringen and Bergkamen, 5 km apart, are not correlated. Large variations are also observed between different time spans. A face advance for 100 days corresponds to a change in position of about 300 m. Since observed time intervals of enhanced seismicity may even be smaller (Fig. 6) this means that stress redistribution due to mining is very local.

Magnitude–frequency distributions for Bergkamen (Fig. 7, left) and Herringen (Fig. 7, right) are derived from recordings of station HMES (Trillium 40, 40 s; Fig. 1). This station is directly situated above the active longwall in Herringen and not included in the catalog of the entire region. Therefore, the completeness threshold is significantly lower here with $M_C \approx 0$ for Herringen and $M_C \approx 0.3$ for Bergkamen (Fig. 7).

The magnitude–frequency distributions of the two regions differ strongly. Both distributions cannot be described by the Gutenberg–Richter relation. In Bergkamen larger events still occur less often than smaller ones over the magnitude range down to M_C (Fig. 7, left), but the observed logarithmic distribution is not linear. For Herringen we observe a maximum in the magnitude–frequency distribution around $M_L =$ 1.4 (Fig. 7, right). As events with this magnitude occur more often, we refer to them as characteristic events typical for this specific longwall.

Earlier work (GIBOWICZ *et al.*, 1990; HINZEN, 1982) proposed a relation between source size and mining geometry. To check if this could be a reason for the Herringen characteristic events, we roughly estimate the source size for $M_L = 1.4$ events.

The seismic moment M_0 was calculated from *S*-wave spectra for induced events with $0 \le M_L \le 2.4$. Comparing the seismic moment to local station magnitudes of HMES (Fig. 8) we derive the empirical relation

$$\log M_0 = 1.01\,(\pm 0.02) M_L + 9.70\,(\pm 0.03) \qquad (1)$$

by least-squares fitting to a power law. The model parameters are consistent with observations by others for this magnitude range (e.g., CHEN and CHEN, 1989).

Following ABE (1975) we deduce faulting area from seismic moment by

$$M_0 = 1.23 \times 10^{15} A_r^{3/2}. \qquad (2)$$

For $M_L = 1.4$ we obtain a seismic moment of about 1.3×10^{11} Nm and a rupture area of about $A_r = 2{,}200\ \text{m}^2$. Assuming the length of the fault plane corresponds to the longwall width of 300 m, results in a fault width of about 7 m. As fault planes are possibly very steep this could correspond to the thickness of the layers that failed. Alternatively,

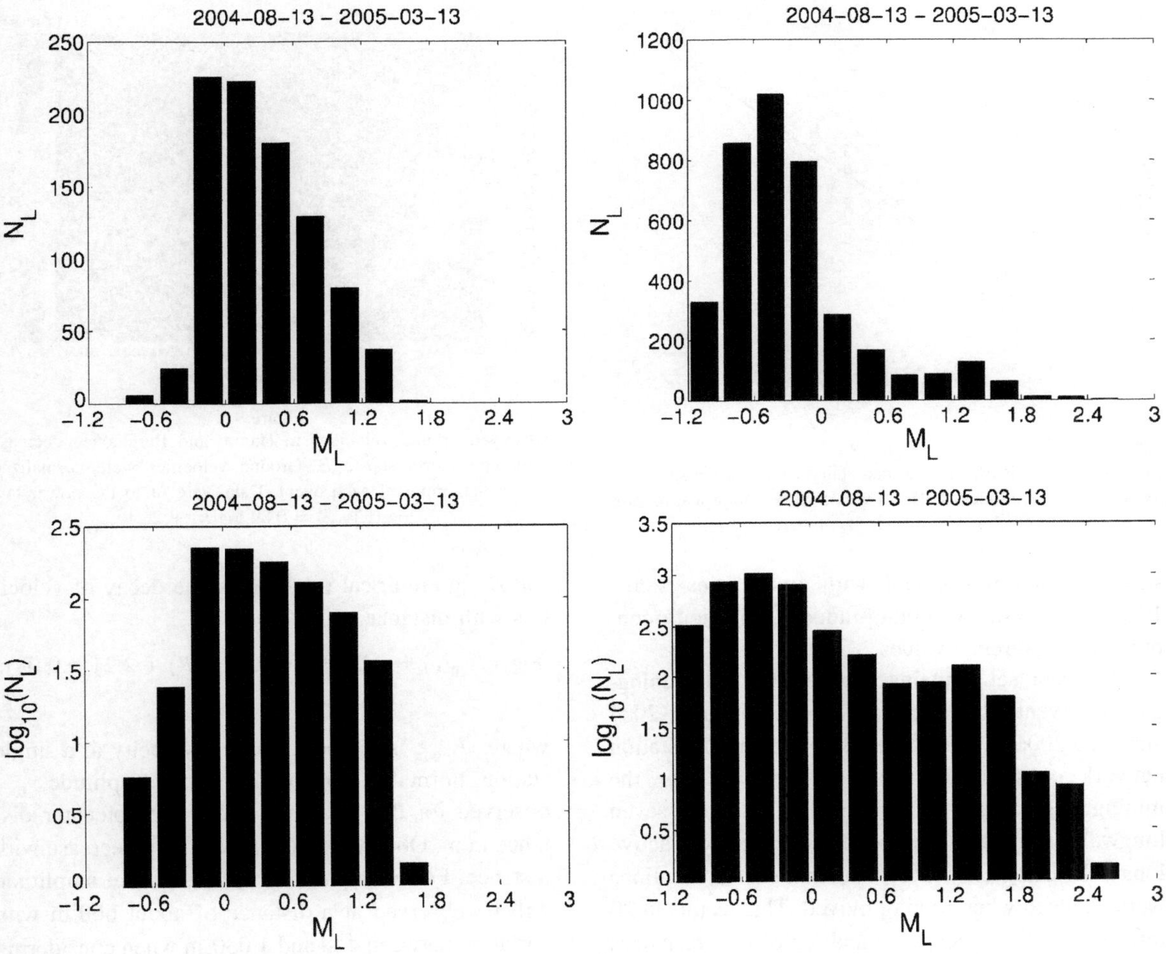

Figure 7
Magnitude–frequency distribution obtained from recordings of station HMES for specific longwalls in Bergkamen (*left*) and Hamm-Herringen (*right*). The number of events was normalized to 300 days (N_L). Absolute (*top*) and logarithmic distributions (*bottom*) show differences between the two regions. The distributions cannot be explained by the Gutenberg–Richter relation, characteristic events with magnitudes of $M_L = 1.4$ are observed in Herringen

assuming a circular fault yields a source radius of about 18 m. These estimates of source dimension generally support the idea that certain geological layers fail in the characteristic events and control their source size.

Knowledge of maximum ground velocities is of interest for the impact of seismic events on their surroundings. Ground-motion amplitude and frequency directly influence people's perception of vibrations and the response of buildings. In contrast to magnitude that characterizes the size of an event, ground velocity can contribute to the measure of instrumental intensity at the station location. For our analysis we use data from a network, which is maintained by DMT GmbH & Co. KG and used for survey reports in Germany. Therefore, ground velocity is measured according to DIN 4150 (Deutsches Institut für Normung e.V., 2001) which demands measurements at the basements of buildings in three orthogonal directions with an instrument having a flat amplitude response in the range between 1 and 80 Hz. DIN 4150 also states, based on experience of the construction industry, that vibrations below certain thresholds for maximum ground velocities induced by industry and construction work are of no harm for buildings (e.g., 5 mm/s for

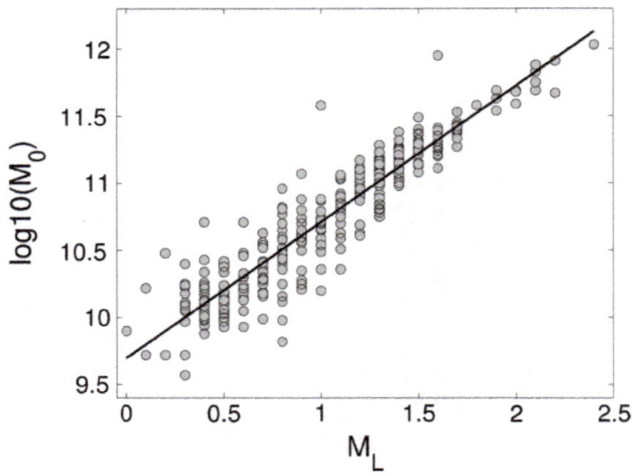

Figure 8
Comparing the seismic moment calculated from S-wave spectra to local station magnitudes of HMES we derive the empirical relation $\log M_0 = 1.01, M_L + 9.70$

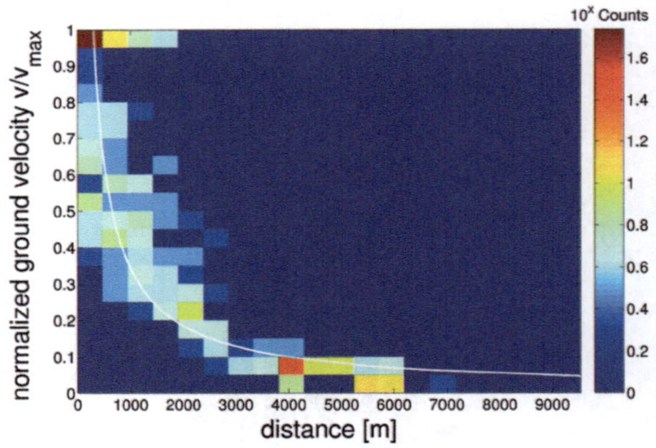

Figure 9
Observed ground velocities in Hamm and Bergkamen decrease with epicenter distance R. Ground velocities were normalized to the maximum of each event. Data were fit to the power law $\log(v/v_{max}) = 0.89 \log(R) + 2.21$

standard buildings exited with frequencies below 10 Hz). Analyses of mining-induced seismicity confirm these reference values.

The data set contains a total of 5,000 mining-induced events that occurred between January 2004 and June 2006 in Hamm and Bergkamen. The station network consists of 10 stations spread over the mining region of 12×5 km^2 and covers seven longwalls. To achieve good coverage of active longwalls using only a few stations, station locations were changed when mining moved. This results in 20 different station locations and epicenter distances ranging from 100 to 9,500 m.

First, we relate observed velocities to epicenter distance to quantify their decay (Fig. 9). Measurements of the smallest events ($v_{max} < 2$ mm/s) were discarded. Because events were not located estimated epicenters are used. We assume that the epicenter for one event coincides with the longwall face which is closest to the station with the earliest P-arrival and shortest S–P time. If more than one longwall is found within the same distance range the event is rejected. A total of 401 velocity seismograms from 69 events was processed with a maximum observed ground velocity of 12.63 mm/s. To include small as well as larger events, amplitudes are normalized to the maximum amplitude measured for each event.

Observations were fit to a power law by least-squares. The frequency distribution of observed amplitude and distance pairs (Fig. 9) results in the following empirical relation for the decay of velocities with distance:

$$\log(v/v_{max}) = -0.89(\pm 0.03)\log(R) + 2.21(\pm 0.08), \tag{3}$$

where v/v_{max} is the peak particle velocity at a single station, normalized to the maximum amplitude v_{max} observed for this event, and R is the epicenter distance in m. Observed velocities rapidly decrease with distance. For example, half the maximum amplitude will be observed at a distance of about 660 m with variation between 440 and 1,030 m when considering the estimated errors.

Secondly, we compare maximum ground velocities from this catalog to the local magnitudes (Fig. 10) which were determined by the BUG-array (Sect. 1). Magnitudes are available for 2,132 events in the range of 0.6–2.3. From the observed relation (Fig. 10), we, for example, estimate the maximum ground velocity for one of the smaller events that might be noticed by people. According to reports of people living in the mining regions mining-induced events with a local magnitude $M_L \geq 1.2$ can be felt at the surface. Due to our observation this mainly corresponds to a maximum ground velocity of about 1 mm/s. Events with higher ground velocities might be felt also at greater distances.

Combining the comparison of ground velocity with magnitude (Fig. 10) and epicenter distance (Fig. 9) we estimate the effect of a relatively large

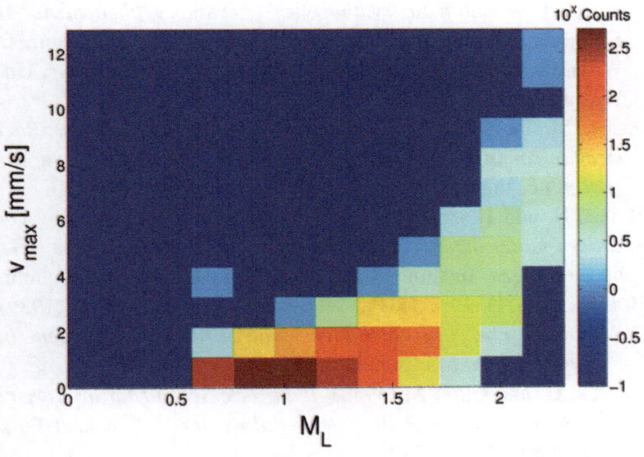

Figure 10
Relation between velocity of ground motion and local magnitude observed for the epicenter region of Hamm and Bergkamen

event with a peak ground velocity of 6 mm/s. This is expected for events of local magnitude M_L between 1.7 and 2.3 (Fig. 10). At a distance of about 2,300 m ± 150 m the amplitude has decayed to one-sixth of the maximum and reaches the threshold of 1 mm/s discussed above. This radius thus defines the area within which the event is likely to be felt. Note that the observed relation can be applied to events in the described region but might differ strongly for other areas.

4. Conclusions

In this article, we continue the analyses of mining-induced seismicity in the Ruhr area. A data set of about 14,000 events with magnitudes up to 3.3 is now available as a result of 24 years of continuous monitoring by the Ruhr University Bochum. This dataset allows us to investigate the general characteristics of mining-induced seismicity in the Ruhr area.

A spatial correlation between mining activity and seismicity is observed. Seismicity is confined to regions of active mining indicating their causal relationship. On a local scale, it is possible to assign seismic events to specific longwalls. Temporal correlation is observed in the decay of seismicity during weekends or maintenance work as well as an increase when mining resumes. This implies that induced stresses are released very quickly. This is supported by a nearly constant energy release per unit time that is dominated by many small events.

The magnitude–frequency distribution for the whole Ruhr area follows a Gutenberg–Richter distribution with a high *b*-value of 2 explaining the lack of larger events and hinting at reduced strength of surrounding rocks. The magnitude–frequency distribution of the entire region results from combining distributions of individual longwalls that show large variabilities.

On a local scale, seismicity of different longwalls may vary significantly even between neighboring panels. Energy release differs by amount, temporal distribution and also magnitude–frequency distribution. These observations reflect the complexity of processes leading to mining-induced seismicity. For example, the depth of excavation and the geometry of old workings probably influence stress conditions at the active longwall and thus seismicity. Stress redistribution is very local and was estimated for a specific longwall to be confined to distances smaller than 300 m. Additionally, local tectonic settings and geological faults may influence the stress field.

For a specific longwall in Herringen a characteristic magnitude of $M_L = 1.4$ was found. An estimation of the fault area from M_0 yields 2,200 m^2 indicating a height of the fault area in the order of 7 m when assuming that the fault length is equal to the width of the longwall (Gibowicz *et al.*, 1990). Characteristic events may be resulting from the failure of certain sandstone layers that lead to similar mechanisms and produce similar fault areas. Alber *et al.* (2009) found evidence for a strong bed of sandstone strata above a longwall in Hamm by numerical modeling and laboratory testing of rock samples. When it is undermined this layer would act as a beam first and then fail if a critical span is exceeded.

Long-time experience confirms that most events can be explained by two different types of failure, that were previously described by Fritschen *et al.* (1999). Polarities of *P* waves, polarization angles of *S* waves and *S*/*P* amplitude ratios of typical events could be fit assuming pure shear failure. Fault plane solutions are interpreted in favor of near vertical fracture planes that are oriented parallel to the longwall face. Seismic events can occur above and below

Reprinted from the journal

the longwall. In either case dislocation towards the goaf is observed. This indicates that new fracture planes are created due to mining-induced stresses.

Regional relations for mining-induced seismic events were derived by regression analysis. The linear relation between the local magnitude M_L and the logarithm of the seismic moment M_0 with a slope of about one fits well to findings by others for this magnitude range (CHEN and CHEN 1989).

Empirical relations between maximum observed ground velocity and local magnitude are of considerable practical benefit. For Hamm we derived the model

$$\log(A/A_{\max}) = -0.89(\pm 0.03)\log(R) + 2.21(\pm 0.08)$$

for the decay of ground motion amplitude with distance. From additional comparison of maximum ground velocities to local magnitude, we estimate, that in the eastern part of the Ruhr area, a relatively large event with a peak ground velocity of 6 mm/s corresponding to a local magnitude between 1.7 and 2.3 is likely to be felt within about 2.3 km.

Acknowledgments

The authors gratefully acknowledge the support of the mine personnel. We wish to thank Bernhard Klotz and Lothar Kühne for operating the stations for many years. Without their technical support this analysis would not have been possible. We further appreciate the routine analysis by Stefanie Erstling. We acknowledge the longtime commitment of Hans-Peter Harjes to studying mining-induced seismicity in the Ruhr area. The authors wish to thank Martin Chapman and an anonymous reviewer who helped to improve the manuscript. This study was partly supported by the German Research Foundation (DFG) within the Collaborative Research Centre 526 'Rheology of the Earth—From the Upper Crust to the Subduction Zone'.

REFERENCES

ABE, K. (1975), *Reliable estimation of the seismic moment of large earthquakes*, J. Phys. Earth *23*(4), 381–390.

AKI, K. *A probabilistic synthesis of precursory phenomena.* In *Earthquake Prediction, an International Review* (Simpson, D. W. and Richards, P. G., eds.), pp. 566–574 (Am. Geophys. Un., Washington, DC 1981).

ALBER, M., FRITSCHEN, R., BISCHOFF, M. and MEIER, T. (2009), *Rock mechanical investigation of seismic events in a deep longwall coal mine*, Internat. J. Rock Mech. Min. Sci. *46*, 408–420.

BAULE, H. and HINZEN, K.-G. (1981), *Erfassung und Deutung von Erderschütterungen, insbesondere zur Ortung von Gebirgsschlägen*, Institut f. Geophysik, Ruhr-Universität Bochum.

BÖHMER, M., HARJES, H.-P., WEBER, V., and WILL, M. (1989), *Seismologische Untersuchungen von Erderschütterungen im Ruhrgebiet*, Das Markscheidewesen *96*(2), 219–226.

CASTEN, U. and CETE, A. (1980), *Induzierte Seismizität im Bereich des Steinkohlenbergbaus des Ruhrreviers*, Glückauf-Forschungshefte *41*(1), 12–16.

CHEN, P. and CHEN, H. (1989), *Scaling law and its applications to earthquake statistical relations*, Tectonophysics *166*, 53–72.

DEUTSCHES INSTITUT FÜR Normung e.V., ed. *DIN 4150: Erschütterungen im Bauwesen*, Beuth Verlag GmbH, Berlin (2001).

DROZDZEWSKI, G., ENGEL, H., WOLF, R., and WREDE, V. (1985), *Beiträge zur Tiefentektonik westdeutscher Steinkohlenlagerstätten.* Geologisches Landesamt Nordrhrein-Westfalen, 1985.

FRITSCHEN, R., RÄKERS, E., PREUßE, A., and BRANDT, K.-H. (1999), *Seismic events due to underground mining activities.* In *18th International Conference on Ground Control in Mining*, pp. 210–215.

GIBOWICZ, S. J., HARJES, H.-P., and SCHÄFER, M. (1990), *Source parameters of seismic events at Heinrich Robert Mine, Ruhr Basin*, Bull. Seismol. Soc. Am. *80*, 88–109.

GIBOWICZ, S. J. and KIJKO, A., *An Introduction to Mining Seismology* (Academic Press. 1994).

GUTENBERG, B. and RICHTER, C. F. (1944), *Frequency of earthquakes in California*, Bull. Seismol. Soc. Am. *34*, 185–188.

GUTENBERG, B. and RICHTER, C. F. (1956), *Earthquake magnitude, intensity, energy, and acceleration*, Bull. Seismol. Soc. Am. *46*, 105–145.

HARJES, H.-P. (1983–2004), *Registrierungen seismischer Ereignisse im Ruhrgebiet*, Institut f. Geologie, Mineralogie und Geophysik, Ruhr-Universität Bochum.

HINZEN, K.-G. (1982), *Source parameters of mine tremors in the eastern part of the Ruhr-district (West-Germany)*, J. Geophys. *51*, 105–112.

KORN, M., ed. (2002), *Ten years of German Regional Seismic Network (GRSN)*, Report 25 of the Senate Comission for Geosciences of the Deutsche Forschungsgemeinschaft, Wiley, Weinheim.

MEIER, T. (2005–2006), *Registrierungen seismischer Ereignisse im Ruhrgebiet.*, Institut f. Geologie, Mineralogie und Geophysik, Ruhr-Universität Bochum.

MINTROP, L. (1909a), *Die Erdbebenstation der Westfälischen Berggewerkschaftskasse in Bochum, Teil 1/2*, Glückauf *45*(11), 357–366.

MINTROP, L. (1909b), Die Erdbebenstation der Westfälischen Berggewerkschaftskasse in Bochum, Teil 2/2, Glückauf *45*(12), 393–403.

MOLNAR, P. (1979), *Earthquake recurrence intervals and plate tectonics*, Bull. Seismol. Soc. Am. *69*(1), 115–133.

RYDELEK, A. and SACKS, S. (1989), *Testing the completeness of earthquake catalogues and the hypothesis of self-similarity*, Nature *337*, 251–253.

SCHOLZ, C. H., *The Mechanics of Earthquakes and Faulting* (Cambridge University Press 2002).

WIEMER, S. and WYSS, M. (2000), *Minimum magnitude of completeness in earthquake catalogs: examples from Alaska, the western United States, and Japan*, Bull. Seismol. Soc. Am. *90*(4), 859–869.

WONG, I. G. W., *Tectonic stresses in mine seismicity: Are they significant*? In *Rockburst and Seismicity in Mines* (ed. Young) (Balkema 1993).

(Received October 30, 2008, revised April 27, 2009, accepted May 17, 2009, Published online December 4, 2009)

Pure Appl. Geophys. 167 (2010), 77–89
© 2009 Birkhäuser Verlag, Basel/Switzerland
DOI 10.1007/s00024-009-0002-7

Pure and Applied Geophysics

Mining-Induced Seismicity in the Saarland, Germany

RALF FRITSCHEN[1]

Abstract—Coal mining in the Saar mine, Germany, is accompanied by mining-induced seismic events. Strong events occur only in certain areas of the mine, other areas exhibit almost no seismicity. Shear events occur simultaneously to non-shear events. The shear events occur in different depths but their epicenters do concentrate in bands. The strike of the bands coincides with the strike of larger regional faults in the area. The seismic events of the Saar mine show some characteristics which distinguish them from seismic events observed in other German coalfields. The Gutenberg–Richter relation, for example, does not hold for these events. Furthermore, radiated seismic energy and extracted coal volume are not correlated. In the Primsmulde field a strong seismic event was observed even before mining in that region started. The event was triggered just by driving roadways into the field. The shear events cannot be explained by the mining process alone. They are presumably induced in certain regions (bands) under tectonic load by an interaction of mining-induced and tectonic stresses. In February 2008, extraction in the Primsmulde field induced a seismic event of magnitude 4, which led to surface vibrations reaching 93 mm/s. After this event, the Primsmulde field had to be abandoned. Future extraction of the Saar mine will be restricted to some small areas not intersected by the event bands found in the Dilsburg Ost and Primsmulde fields. The Saar mine will close in 2011.

Key words: Triggering, Mining induced seismicity, Gutenberg–Richter-Relation.

1. *Introduction*

As in many other regions worldwide, coal mining in Germany is accompanied by mining induced events. Today three areas hold active coal mines; one of them is located in the Saarland, in the far west of Germany (Fig. 1). Although industrial mining in the Saar region started in the middle of the eighteenth century, only a few cases of mining-induced events were reported before the 1990. In 1997, however, several strong seismic events were felt by the population of the villages in the area of the Saar mine. As a result, the Saar mine decided to install a local seismic network in order to determine the peak particle velocities (PPVs) on site and to investigate possible relations between mining and seismicity. Quickly it became clear that the observed seismicity was directly related to mining activities in the Schwalbach seam. Most all seismic events were located in the vicinity of active longwall faces. From 1997 to 2006, these were basically located in the Dilsburg field of the Saar mine. The Dilsburg field consists of an eastern (Dilsburg Ost) and a western (Dilsburg West) part. The parts are separated by an unmined area, protecting the main shaft of the mine (Fig. 2). In 2006, mining of the Schwalbach seam started in the Primsmulde field, planned as the mine's new main production field.

Figure 3 shows the number of seismic events that were recorded since continuous monitoring started in July 1997. Comparison with Table 1, which lists the periods of mining, shows the dependency of seismicity on mining. Particularly mining in the eastern part of the Dilsburg field and in the Primsmulde field led to high seismicity.

Figure 4 depicts the epicenters of seismic events with PPVs > 1 mm/s, which could be located between 1997 and 2008. Clusters of seismic events are located in the northeast and the northwest, corresponding to the location of the Dilsburg Ost and Primsmulde fields. Minimal seismicity was observed in the Dilsburg West field.

Analysis will be focused on the Dilsburg Ost and Primsmulde fields in the following and will show some characteristics of the observed strong seismic events, which particularly distinguish them from most other induced seismic events in that region and especially from those observed in the other German coal regions.

[1] DMT GmbH & Co. KG, Am Technologiepark 1, 45307 Essen, Germany. E-mail: Ralf.Fritschen@dmt.de

 Reprinted from the journal

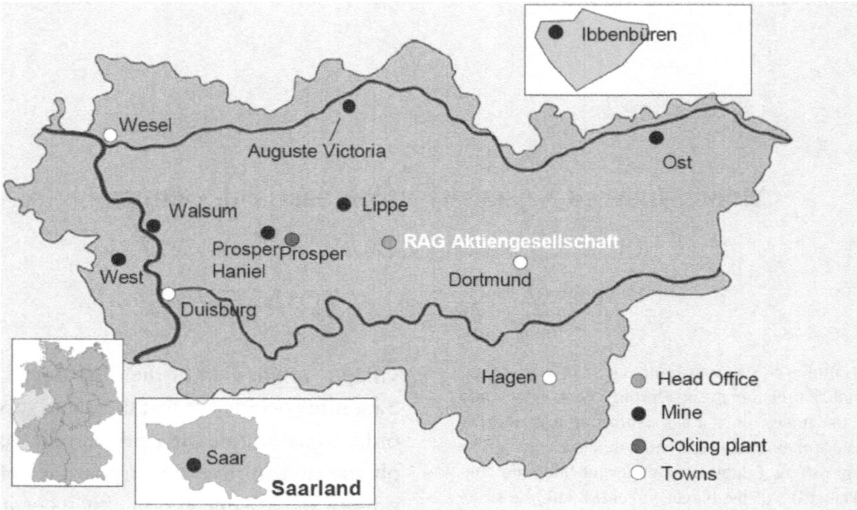

Figure 1
Map of German coal mines (*left inlet* Germany, *right inlet* Ibbenbüren, *middle inlet* Saarland, *big map* Ruhr area). Coal mining is taking place in three German regions: in the Ruhr area and Ibbenbüren, which are both located in the state of North Rhine-Westfalia, and in the Saarland, a small state in the southwest of Germany

2. *Location and Fault Plane Solution*

In 1997, the seismic network of the Saar mine consisted of just four surface stations. The network was not optimized for locating seismic events, but was primarily installed to determine the intensities of the mining induced events. The intensities were measured as PPV values at buildings basements. The recording was done with 3C geophones with flat velocity amplitude responses ranging between 1 and 80 Hz. Over the course of the measurements the network was continuously extended. When the excavation of panels 8.7 o and 8.8 o started in December 1999 the seismic events could be located with an accuracy of 80 m in epicenter position and 150 m in depth. Later, when the network was completed with underground stations, event depths and epicenter positions could be determined with an accuracy of 50 m.

2.1. *Dilsburg Ost Field*

Mining of the Schwalbach seam started in the south of the Dilsburg Ost field in 1979 at a depth of about 500 m. The excavation proceeded to the north. Since the Schwalbach seam is dipping to the north, the working panels reached greater depths from year to year. When mining took place in the panels 8.5 o and 8.6 o at depths of 850 m, seismic events were felt at the surface. In 1998, the extraction of 8.5 and 8.6 Ost was finished and after that no seismicity was observed until mining started again in Dilsburg Ost in January, 2000.

2.1.1 *Mining of Panels 8.7 o and 8.8 o*

Seismicity set in shortly after starting extraction in panels 8.7 o and 8.8 o in January 2000 at a depth of approximately 1,000 m below the surface. The epicenters of the events were located in the middle roadway between the panels, slightly in front of the advancing panel 8.7. They did migrate together with the advancing panels towards the northwest. At first, only relatively small seismic events with PPVs < 3 mm/s occurred, corresponding roughly to local magnitudes $M_l < 2.5$ (Fig. 6). The situation changed in November 2000 with the occurrence of a group of strong seismic events which reached PPVs of about 23 mm/s ($M_l = 3.5$). Fault plane solutions showed that all recorded seismic events could be classified into three different event types corresponding to three different types of rupture: two of them (Types 2 and 3)

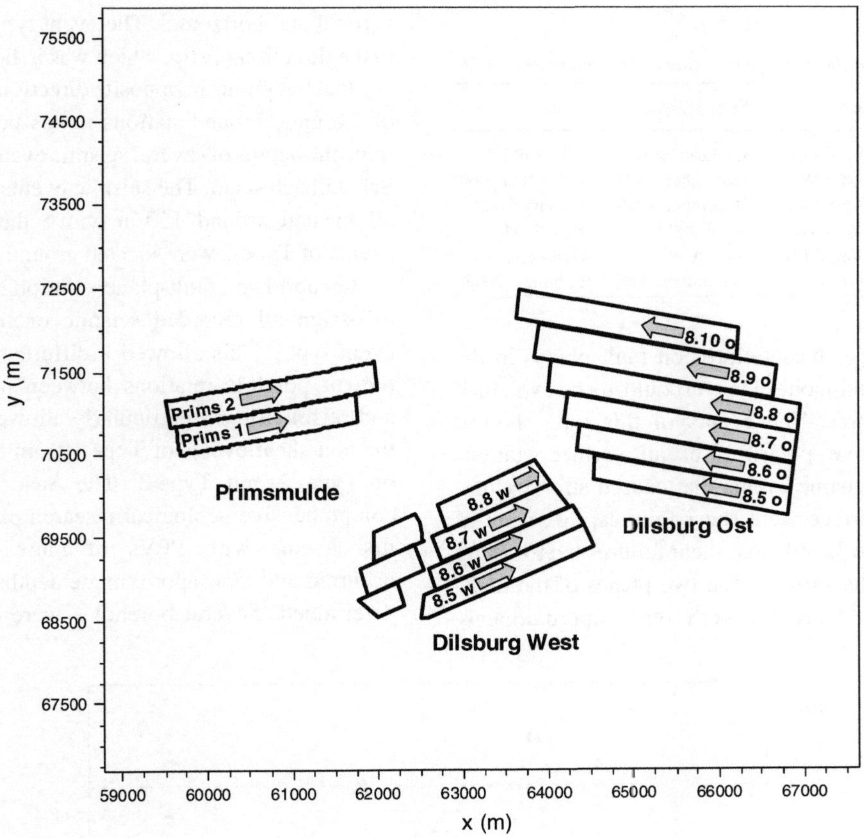

Figure 2
Working panels in the Schwalbach seam at Saar mine since 1997. The *arrows* indicate the direction of mining. Always two working panels were moved together, i.e., panel 8.5 o was directly followed by 8.6 o, 8.7 o was directly followed by 8.8 o aso. The panels are located at a depth from approximately 1,000 m (Dilsburg Ost) to 1,400 m (Primsmulde)

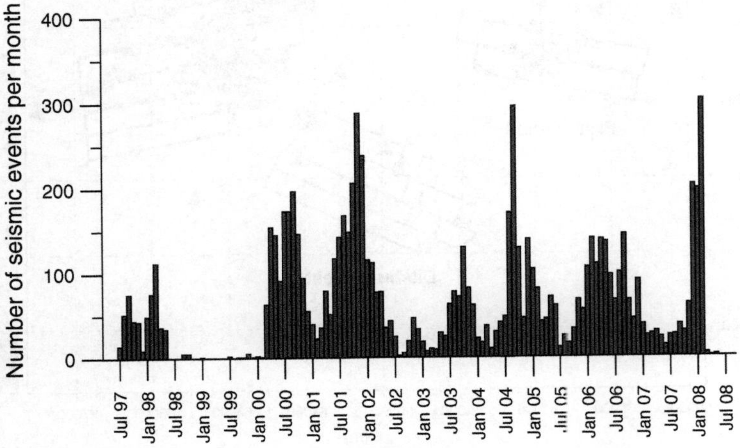

Figure 3
Number of seismic events that were recorded at the surface stations. The highly active times in 2000–2002 are associated with panels 8.7 and 8.8 Ost, the active times in 2004–2005 are associated with panels 8.9 and 8.10 Ost and the activities after December 2007 were caused by Prims 1 and Prims 2

Table 1

Working panels in the Schwalbach seam at Saar mine since 1997

Seam	Panel	Start	End
Schwalbach	8.5/8.6 Ost	January 1996	July 1998
Schwalbach	8.5/8.6 West	December 1997	December 1999
Schwalbach	8.7/8.8 Ost	December 1999	January 2003
Schwalbach	8.7/8.8 West	April 2002	July 2004
Schwalbach	8.9/8.10 Ost	July 2004	October 2007
Schwalbach	Prims 1/2	September 2006	February 2008

were caused by shear failures on fault planes in the rockmass, the third one (Type 1) could not be explained by shear failures. The events of this type showed negative P-wave polarities at all surface stations (Fig. 5). All seismic events that caused strong vibrations at the surface were shear failures, i.e., were of Type 2 or Type 3, and most shear failures led to strong vibrations at the surface. The two planes of the fault-plane solution were for both types approximately vertical and horizontal. The event types differed only in the direction of slip, which was in both cases almost vertical but points in opposite directions. With the help of the underground stations it was possible to determine the depths of several seismic events relative to the Schwalbach seam. The seismic events of Type 2 were all located around 150 m above the seam and the events of Type 3 were located around 20 m below.

Location and fault-plane solution made it possible to assign all recorded seismic events to the three event types. This allowed a differentiated investigation of possible relations between mining, geology and seismicity and particularly allowed separation of the non-shear events of Type 1 from the shear events of Type 2 and Type 3. The Saar mine started a comprehensive geological research program after the first events with PPVs of more than 20 mm/s occurred and their approximate depths of origin were determined. Several boreholes were drilled into the

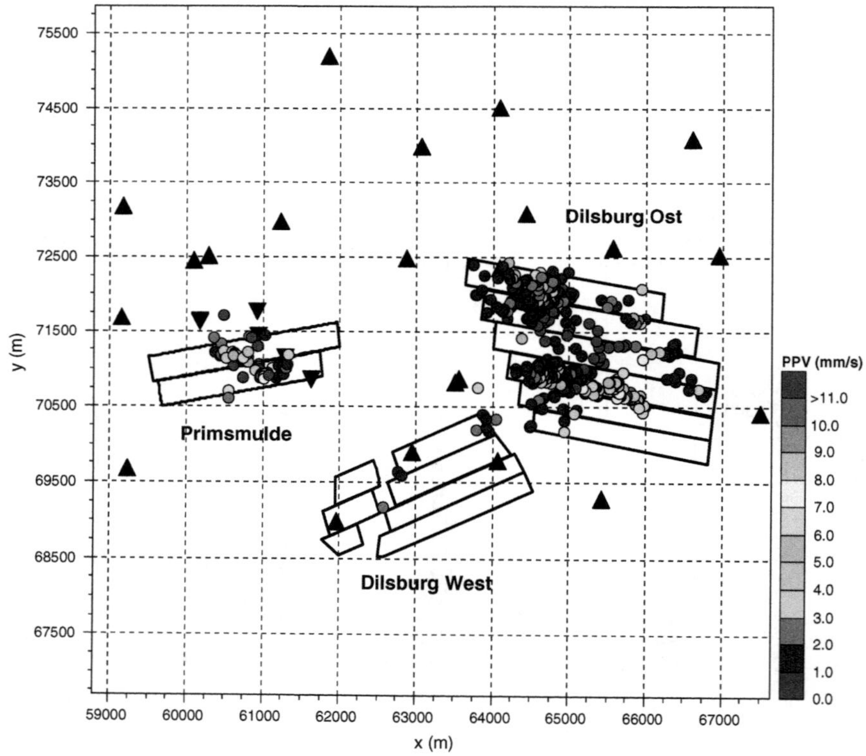

Figure 4

The local seismic network of the Saar mine. The map shows the station locations (*triangles*) and the panel outlines of the Dilsburg West, Dilsburg Ost and Primsmulde fields, where the Schwalbach seam was mined. The epicenters of mining induced seismic events that occurred between 1997 and 2008 with peak particle velocities (PPVs) of more than 1 mm/s are marked as *circles*

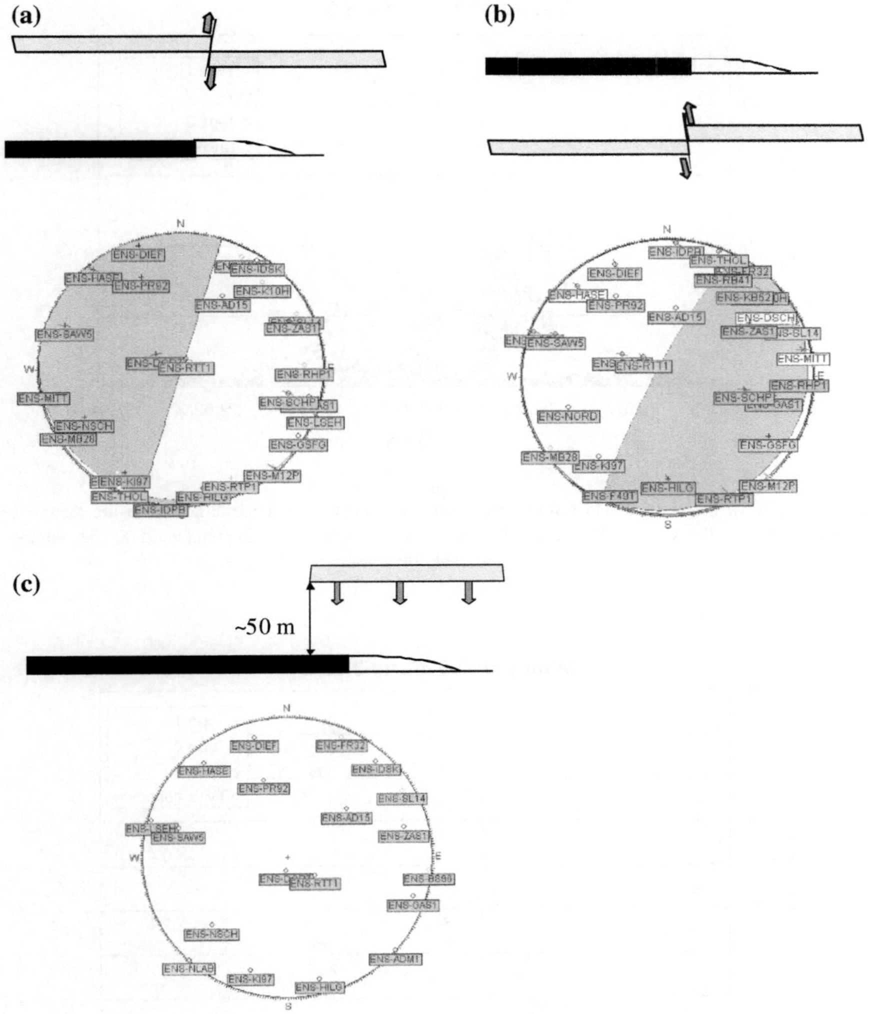

Figure 5

Fault plane solution and the proposed mechanism of typical seismic events: **a** Type 2, approximately 150 m above the Schwalbach seam; **b** Type 3, approximately 30 m below the seam and Type 1, approximately 50 m above the Schwalbach seam. The *upper half* of the focal sphere is shown with possible fault plane orientations for the shear events of Types 2 and 3 that fit to the observed P-wave polarizations and S-wave polarization angles. Type 1 is no shear event and has only negative polarizations at the surface stations

hanging and foot wall to find possible correlations between the seismic events and the rock mass surrounding the Schwalbach seam.

At first, it appeared that the shear events corresponded to two channel sandstones with thicknesses of up to 10 m each, deposited by meandering carboniferous river systems in the footwall about 20–30 m below the Schwalbach seam. The strongest seismic events of Type 3 were recorded when the region, where the two channel sandstones overlap, was overmined by panels 8.7 o and 8.8 o. It was therefore assumed initially that the failure of the channel sandstones was the main cause of the strong seismic events, following the simple concept of a beam under load, which is able to withstand mining-induced stresses up to a certain point until it fails. That competent sandstone layers may be responsible for the generation of strong seismic events was also found by Fritschen *et al.* (1999) for seismic events in another mining area. Interestingly these were also of

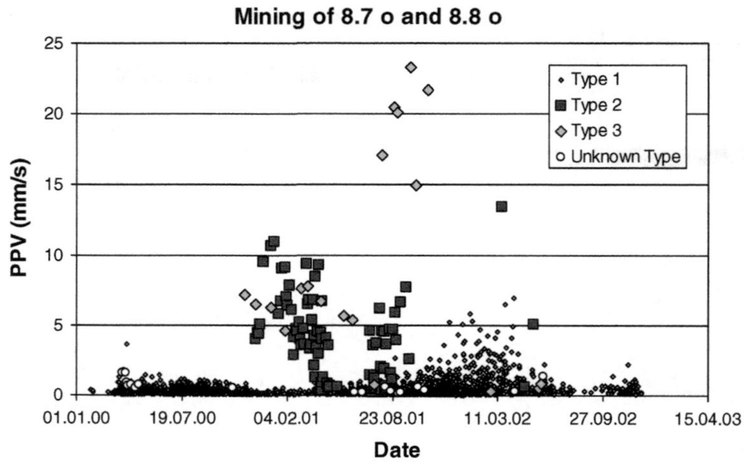

Figure 6
Seismicity during the mining of 8.7 o and 8.8 o. Most of the strong seismic events are of Types 2 and 3, i.e. are shear events, and most of the shear events lead to high PPVs at the surface. When the first shear events occurred in September 2000, the number of Type 1 events considerably decreased

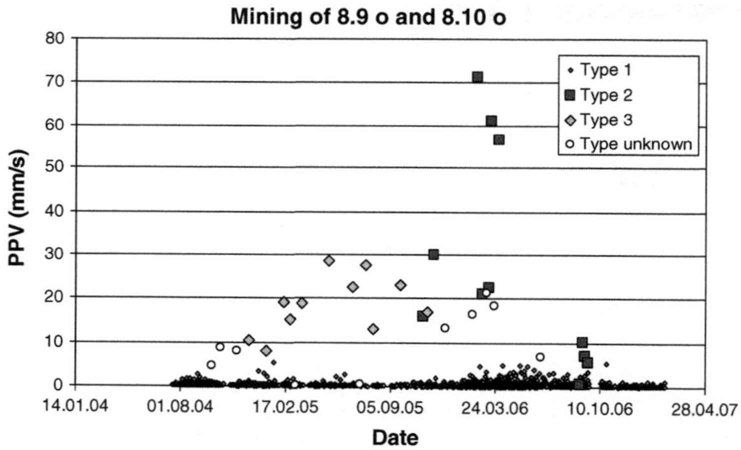

Figure 7
Seismicity during the mining of 8.9 o and 8.10 o. Most of the strong seismic events are of Types 2 and 3, i.e., are shear events, and most of the shear events lead to high PPVs at the surface

Types 2 and 3, i.e., they were shear events with steep fracture planes parallel to the working face with slip directions both towards and away from the gob.

However, the concurrences of competent layers, in this case the channel sandstones, and mining induced stresses cannot explain the occurrence of the strong shear type events alone. Initially, there is the remarkable coincidence in space and time of Type 3 events, about 30 m below the seam, and of Type 2 events, about 150 m above the Schwalbach seam, which suggests, that these events have a similar if not the same cause. It is unlikely that channel sandstones 30 m below the seam have something to do with seismic events 150 m above seam Schwalbach. Secondly, the panels 8.9 o and 8.10 o did cause seismic events which were also of Types 2 and 3 and were closely correlated to the events induced by 8.7 o and 8.8 o. However, not all of these events could be associated with channel sandstones.

2.1.2 *Mining of Panels 8.9 o and 8.10 o*

Mining of the double system 8.9 o and 8.10 o started in July 2004 at a depth of approximately 1,100 m and almost immediately induced seismic events of Type 1. When the face reached the region where Types 2 and 3 events were induced by the preceding panels 8.7 o and 8.8 o, again Type 2 and Type 3 events with considerable intensity occurred (Figs. 7, 10). The Type 2 and Type 3 events of both double systems together jointly formed a pattern, i.e., the epicenters of the events were not evenly distributed around the panels but laid in certain "bands" as depicted in Fig. 8.

2.2. *Primsmulde Field*

In 2004, the Saar Mine started to develop the Schwalbach coal seam in the new Primsmulde coalfield at a depth of approximately 1,400 m. The development of the field started 2 years before coal extraction with the driving of roadways, which later formed the longwall panels. As in the Dilsburg Ost and Dilsburg West fields the extraction of the Schwalbach seam in the Primsmulde was planned as a double panel system. It was not known if this system in the new part of the mine would induce seismic events of considerable intensity because at that time both double panel systems with strong seismic events (Dilsburg Ost) and double panel systems without significant seismicity (Dilsburg West) had been observed. A reliable explanation of the contrary seismicity found in these two fields was not available at the time.

Soon afterwards it was clear that the extraction of coal in the Primsmulde could lead to strong seismic events. In May 2005, after driving the first roadways into the Primsmulde field, a seismic event of $M_l = 3.3$ was induced (Fig. 9). This was the first time that an event, which was considerably felt at surface, was induced in a German coal mine just by developing a new field. No other seismicity was observed after that event in the Primsmulde field, until mining started in October 2006. However, even then seismicity was moderate with events that were hardly detected and not felt at the surface (Fig. 10). The situation changed dramatically in June 2007, approximately 9 months after the start of extraction in the Primsmulde, when a Magnitude 3.6 event with a PPV of 29 mm/s was induced in the seam level, more than 300 m in front of the double seam system.

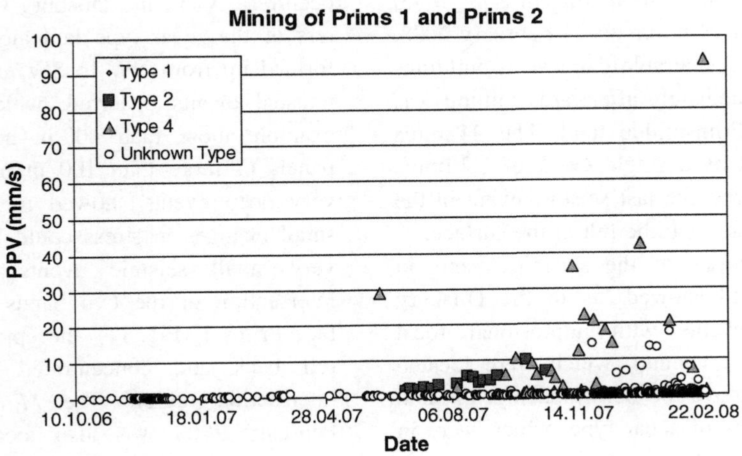

Figure 8
Seismicity during the mining of Prims 1 and Prims 2. Most of the strong seismic events are of Types 2 and 4, i.e. are shear events, and most of the shear events lead to high PPVs at the surface. After the 93 mm/s event of 23 February 2008 mining was abandoned and seismicity stopped the day after

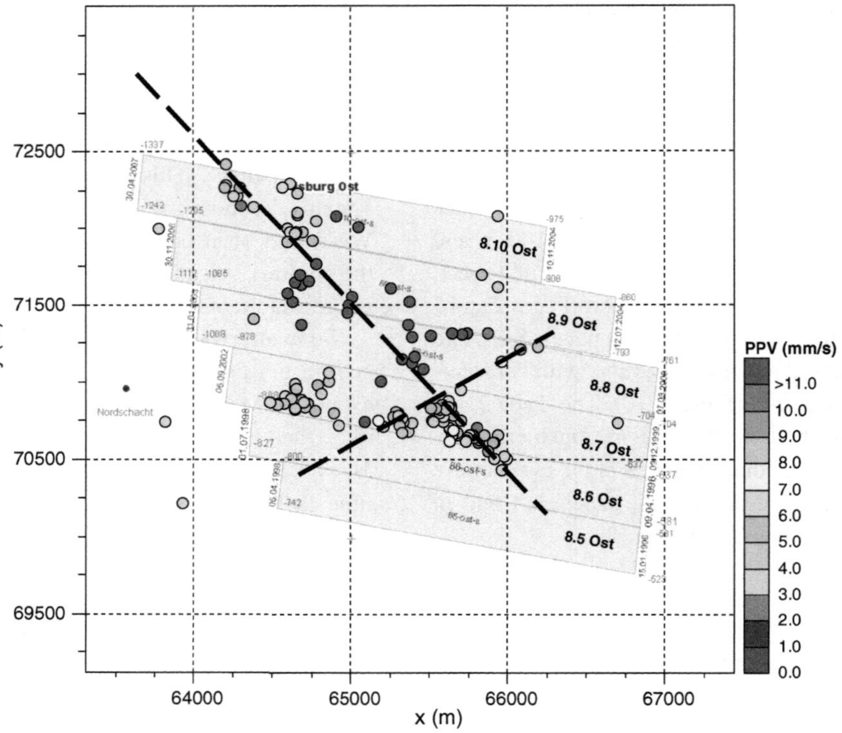

Figure 9
Location of seismic events with PPVs > 3 mm/s. The epicenters of most of the strong shear of the Dilsburg Ost field line up along SE–NW and SW–NE striking bands (*thick dashed lines*)

The event, located approximately. 180 m away from the May 2005 event, marked the beginning of a series of strong seismic events, culminating in a $M_l = 4.0$ event with a PPV of 94 mm/s on 23 February 2008. The event caused considerable damage to buildings on the surface. Immediately afterwards mining was abandoned in the Primsmulde field. The 94 mm/s event was followed by a single event of 1.2 mm/s 1 day after, which was the last seismic event in the Primsmulde field that could be felt at the surface.

Fault-plane solutions of the seismic events in the Primsmulde field showed, as in the Dilsburg Ost field, small events with implosional focal mechanisms (Type 1 events), which were located in the first 100 m above the Schwalbach seam, and strong seismic events of shear type, which were in the Primsmulde located at seam level or approximately 300 m above the seam level. In addition to a NW–SE striking Type 2 event, i.e., with a strike almost parallel to the working face, a new event type (Type 4) with a NE-SW striking steep fault plane (one of the two possible planes of course) occurred. As in the Dilsburg Ost field, the epicenters of the shear type laid along structures, which formed up from NW to SW and SW to NE. More unusual events occurred, which were located for example more than 400 m horizontally from the panels or more than 100 m below the seam. The very deep events showed in particular that even small changes in stress could lead to, in this case, very small seismic events. After passing the intersection of the two bands with the advancing face Prims 1 (Fig. 11), the epicenters of the events fell back and concentrated furthermore at the intersection region. The $M_l = 4.0$ event of 23 February 2008 was also located in that region and, remarkably, was 174 m behind the advancing working face of Prims 1. The event was composed of Type 3 with a hypocenter laying 300 m above the Schwalbach seam.

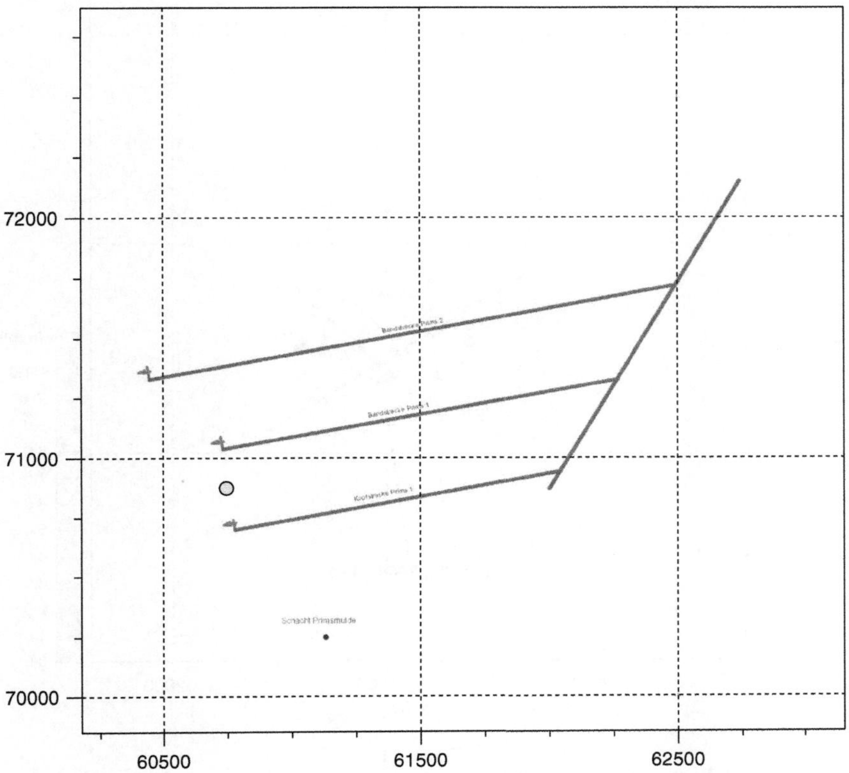

Figure 10
The Primsmulde field was developed by driving roadways into the Schwalbach seam at a depth of approximately 1,400 m. In the middle of the field a magnitude 3.3 event was induced at seam depth (*circle*) in May 2005, before any extraction in the field started

It is notable that the event bands were aligned parallel to the strike of the fault planes. They also coincided with the strike of larger regional faults. The Primsmulde field itself, however, is not intersected by fault systems, as is known from 3-D exploration and driving of underground roadways.

3. *Characteristics of the Seismicity in Dilsburg Ost and Primsmulde*

The strong seismic events that were induced by mining in the fields of Dilsburg Ost and Primsmulde exhibit some characteristics which distinguish them from seismic events observed in other German coalfields. This is at first the occurrence of numerous very strong seismic events, i.e., events with PPVs > 20 mm/s, in a region not influenced by old workings. Also characteristic is the occurrence of strong events in notably a different depth but similar horizontal location, for example in the foot wall and hanging wall in the Dilsburg field or in the seam level and 300 m above in the Primsmulde field. Furthermore the evolution of a band (or region) with strong seismic events that are not only in their horizontal location but also in their depth and fault mechanism distinguishable from the other seismic events, is unusual for German coalmines.

There are two further observations which indicate that the strong seismic events have causes other than the small ones and which also indicate the differences between the observations at the Dilsburg Ost and Primsmulde fields and all other coalfields at Ruhr and Saar: these are on the one hand the correlation between extracted coal volume and seismicity and on the other hand the number of seismic events as a function of seismic energy (or magnitude)—the Gutenberg Richter relation.

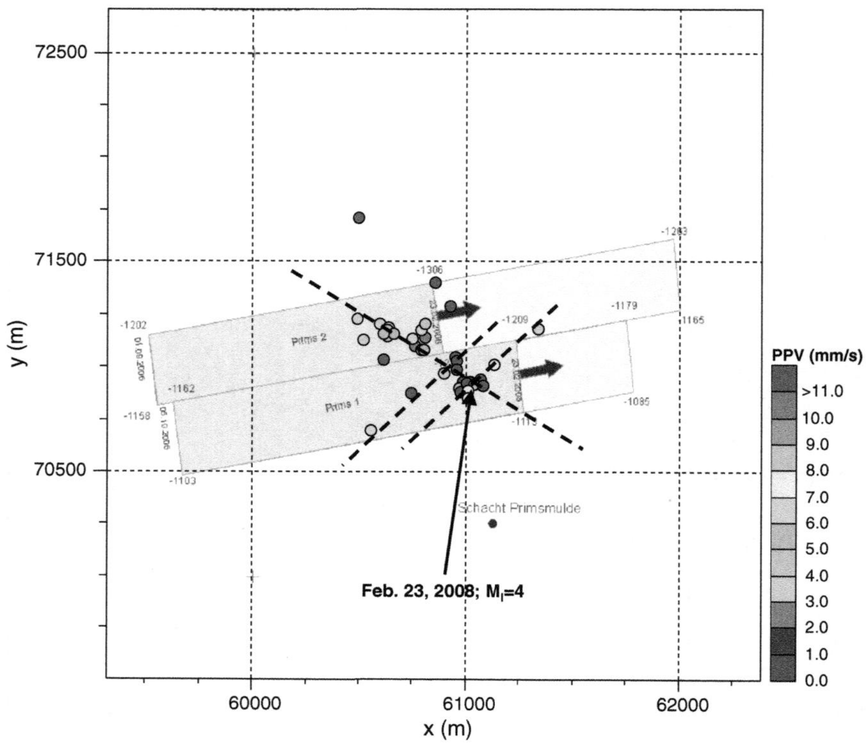

Figure 11
Location of seismic events with PPVs > 3 mm/s. The epicenters of the events in the Primsmulde field are lying, like the epicenters in the Dilsburg Ost field, along the SE–NW and SW–NE striking structures (*thick dashed lines*). The figure shows the face positions of 23 February 2008, when the $M_l = 4$ event occurred after which mining in the Primsmulde field was abandoned

3.1. Correlation between Seismicity and Coal Extraction

The strong dependence of seismicity on the volume of extracted coal is well known and was shown for Czech and Polish mines by Glowacka (1992) and for mines in the German Ruhr area by Fritschen (2002). This dependence follows from the proportionality between the cumulative seismic moment and the induced volume change in a rockmass that was shown by McGarr (1976). Assuming that the volume of extracted coal is proportional to the induced volume change and that the cumulative seismic energy is in average proportional to the cumulative seismic moment, it follows that the volume of extracted coal is proportional to the amount of seismic energy released.

Figure 12(left) shows as an example the monthly values for the volume of extracted coal together with the seismic energy per month for a German coal mine. The seismic energy was calculated as the mean value of the integrals of the squared particle velocities at the seismic stations, which were corrected for attenuation and geometrical spreading according to Boatwright and Fletcher (1984), neglecting the correction for the radiation patterns.

The correlation between seismic energy and volume in Fig. 12(left) is obvious. A similar correlation is found for several other mines in the Ruhr area, particularly when a certain working is not influenced by abandoned workings. The workings in the Dilsburg Ost and Primsmulde fields did not display this relationship, which indicates that the stresses induced by the extraction of coal are not solely responsible for the generation of the observed seismic events.

3.2. Gutenberg–Richter-Relation

The frequency-magnitude or frequency-energy relation, take on a logarithmic scale, the form $\log n = a - bm$, where n is the cumulative number

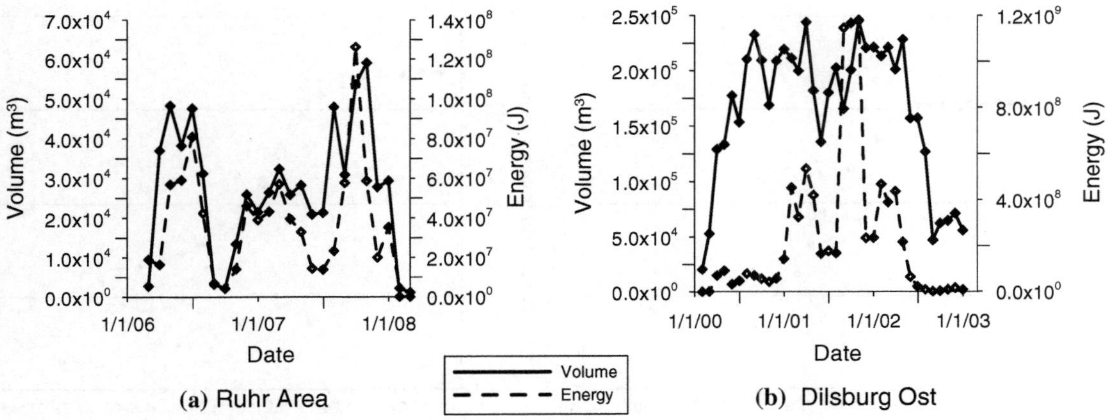

Figure 12
Seismic energy and extracted coal volume per month for two different mining regions. The connection between extracted volume and radiated seismic energy is generally observed in German coal mines (**a**—Ruhr Area). The seismic events observed in Dilsburg Ost and Primsmulde did not follow this rule. There is no direct connection between extracted coal volume and radiated seismic energy (**b**—Dilsburg 8.7 o and 8.8 o). There is even no connection between extracted coal volume and radiated seismic energy when only Type 1 events are taken into account (not displayed)

of seismic events above a magnitude m or above a certain seismic energy log E_s, a and b are parameters. Many observations worldwide indicate that seismic events induced by mining follow this rule in general as do tectonic earthquakes and laboratory experiments. This indicates the self-similarity of observed seismic events, i.e., of the fracture processes involved. Self-similarity means here the absence of a characteristic scale length or order of magnitude determining the occurrence of the seismic events.

Figure 13 shows as an example the relation for events induced by mining in the Ruhr area. The log-linear Gutenberg–Richter Relation does not hold for the seismic events observed in the Dilsburg Ost and Primsmulde fields. Here the events follow a bimodal distribution, i.e., the large events do not follow the same power relation as the small events. However, a range of energies develops, in which the number of events does not decrease with greater energy but stays at the same level. For even greater energies, the cumulative number of seismic events drops again. Similar distributions are described by Gibowicz and Kijko (1994) for mining-induced events in Poland. They provide two possible explanations for the bimodal distribution. On the one hand, a bimodal Gutenberg–Richter Relation can be an indication of a change in geology in the course of the ongoing working face. On the other hand the low energy part of the distribution can be explained by mining-induced relaxation processes around the working and the high energy part by an interaction of mining-induced and tectonic stresses. The interaction of mining-induced stresses with prevailing tectonic stresses or with critically stressed rock may also result, as described by Schneider (2004), in developing strong events at the expense of small ones, which would, however, cause an absence of seismic events in a certain energy range. That the strong shear events of Types 2 and 3 seem to interact with the weaker events of Type 1 can be seen in Fig. 6. After the first shear events in September 2000, the number of recorded Type 1 events considerably dropped. Though when plotting the Gutenberg–Richter-Relation for the Type 1 events alone, it perfectly fits the observations, indicating that altogether no Type 1 events "are missing" (Fig. 13). That only a few shear-type events of small energy exist is remarkable. This is possibly a consequence of a predominant spatial scale, which may be caused by the width of the double panel system influencing a critically stressed region.

Interestingly similar event types, weak implosional events and stronger shear events, were found by Kusznir *et al.* (1980, 1984) for seismic events induced by coal mining in North Staffordshire, UK. Kusznir *et al.* found, however, no violation of the Gutenberg–Richter Relation. Other possible occurrences of

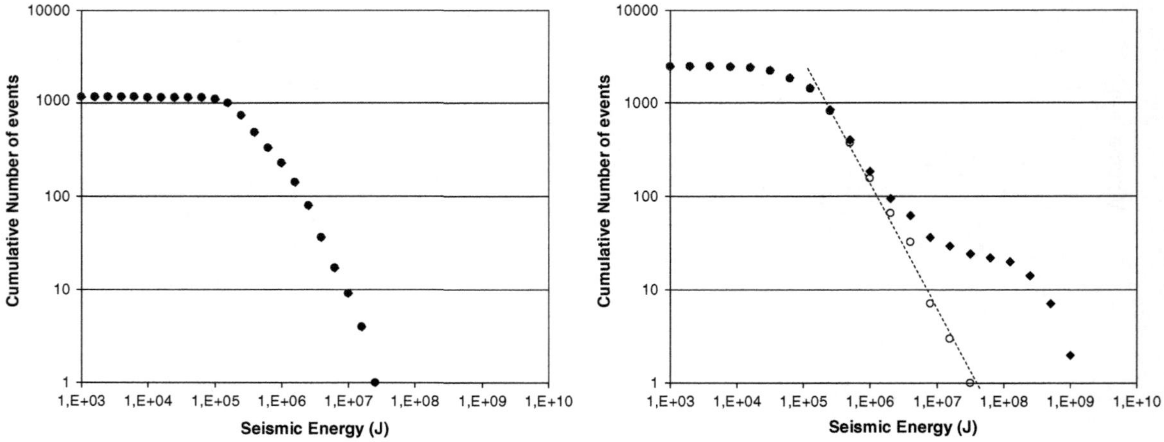

Figure 13
Number of seismic events with energies above a certain value as a function of that value (Gutenberg–Richter-Relation). Generally, a linear trend in the log–log plot is observed for mining-induced seismic events in Germany as is shown in the left figure (eastern part of the Ruhr Area). The right figure shows the relation for Dilsburg Ost (Primsmulde looks, similar). The filled symbols show the relation for all events (Type 1, Type 2 and Type 3). The Type 1 events alone (*open circles*) follow the Gutenberg–Richter Relation remarkably well

implosional events were reviewed by WONG and MCGARR (1990).

3.3. *Triggered versus Induced Seismicity*

Following GIBOWICZ and KIJKO (1994), two broad types of mine tremors are observed almost universally: Those directly connected with mining operations, that is, associated with the formation of fractures at stope faces, and those associated with movement on major geologic discontinuities. In the Ruhr area, where seismic events have been recorded with local networks for several decades, most recorded seismic events are directly connected with the advancing working faces and there is no evidence, that strong seismic events, that are caused by the movement of major geological faults, do exist. The two broad types of mining induced events in the Saarland (the shear-type events and the non-shear events) bring to mind the two types described by GIBOWICZ and KIJKO. In the Dilsburg Ost and Primsmulde fields, however, no geological faults were mapped in the vicinity of the strong seismic events. In the Primsmulde field, even 3-D seismic did not discover any major discontinuities in the entire field.

The observed seismicity fits, however, perfectly into the more general concept of triggered vs. induced seismicity (e.g., MCGARR and SIMPSON 1997 or WANG 1993). Here, induced seismicity is that for which the causative activity can account for either most of the stress change or most of the energy required to produce the seismic events (non-shear events of Type 1 in our case). Triggered events in contrast can be caused by only a small fraction of stress change or energy, associated with the event. Tectonic loading plays here, according to MCGARR and SIMPSON, the primary role (shear events of Types 2 and 3). Recognizing that especially the strong events may be caused by tectonic stress changes has important implications for extraction planning. In the case of Saar mine, the whole mine—not only the work in the affected working panel—was temporarily closed after the magnitude 4 event in February 2008. The state authorities only allowed reopening of the mine after a concept of how to avoid strong seismic events was presented. It is now believed that the strong seismic events were caused by activating (triggering) an area, which was under tectonic load. Since the driving of roadways already suffices to trigger the events, the changing of mining parameters like working speed or panel width would have presumably not helped to change the condition. Consequently the only way to avoid the generation of strong events was to avoid mining areas under tectonic load. Future extractions of the Saar mine will therefore be restricted to only some small areas not intersected by the bands found in the Dilsburg Ost and Primsmulde fields, and the extraction in these areas will be carefully accompanied by seismic monitoring in order to find areas under tectonic load at an early stage. The Saar mine will close in 2011.

4. Conclusions

The extraction of the Schwalbach seam at the Saar mine, Germany, led to the generation of different types of seismic events. All strong seismic events were shear failures, which were restricted to certain areas of the mine. These events occurred at different depths but their epicenters were concentrated in certain regions (bands) in the Dilsburg Ost and Primsmulde fields, which indicated that mining was influenced by tectonic stresses. The strong shear events occurred simultaneously to non-shear events induced by mining. The different nature of the shear type events is illustrated by the Gutenberg–Richter Relation, which is not log-linear when considering all events collectively.

At present, the only approach to avoid strong seismic events is to stay away from areas under tectonic load. In the meantime, numerical studies are performed in order to reproduce the generation of the strong seismic events and to find possible conditions that allow mining in regions under tectonic load that do not lead to the generation of strong seismic events.

References

Boatwright, J. and Fletcher, J.B. (1984), *The partition of radiated energy between P and S waves*, Bull. Seism Soc. Am. *74*, 361–376.

Fritschen, R. Räkers, E. Preusse, A. and Brandt, K.-H. (1999), *Seismic events due to underground mining activities*. In *Proc. 18th Interna. Conf. on Ground Control in Mining, Morgantown*, U.S.A.

Fritschen, R., *Prognose von Erderschütterungen im deutschen Steinkohlenbergbau*, ISBN 3-7739-1503-9 (Verlag Glückauf GmbH, Essen 2002).

Gibowicz, S.J. and Kijko, A. *An Introduction to Mining Seismology* (Academic Press New York 1994).

Glowacka, E. (1992), *Application of the extracted volume of a deposit as a measure of deformation for seismic hazard evaluation in mines*, In Tectonophysics *202*. 285–290.

Kusznir, N.J., Ashwin, D.P., and Bradley, A.G. (1980), *Mining-induced seismicity in the North Staffordshire coalfield, England*. Int. J. Rock. Mech. Sci. and Min. Sci. and Geomech. Abstr., *17*, 45–55.

Kusznir, N.J., AL-Saigh, N.H., and Ashwin, D.P. (1984), *Induced seismicity generated by longwall coal mining in the North Staffordshire coal-field, U.K.* In *Rockbursts and seismicity in mines Johannesburg (South African national group on rock mechanics)* (N.C. Gay and E.H. Wainwright, eds), pp. 153–160.

McGarr, A. (1976), *Seismic moments and volume changes*, J. Geophys. Res. *81*, 1487–1494.

McGarr, A. and Simpson, D., *Keynote lecture: A broad look at induced and triggered seismicity, rock bursts and seismicity in mines*. In Proc. 4th Internat. Symp. on Rockbursts and Seismicity in Mines, Poland (A.A. Balkema 1997), pp. 385–396.

Schneider, G. *Erdbeben – Eine Einführung für Geowissenschaftler und Bauingenieure* (Spektrum Akademischer Verlag, München 2004).

Wong, I.G. and McGarr, A. *Implosional failure in mining-induced seismicity: A critical review*. In *Rockbursts and seismicity in mines Rotterdam/Brookfield* (A. A. Balkema 1990), pp. 45-51.

Wong, I.G. *Tectonic stresses in mine seismicity: Are they significant?* In *Rockbursts and seismicity in mines, Rotterdam/Brookfield* (A. A. Balkema 1993), pp. 273-278.

(Received September 27, 2008, revised May 12, 2009, accepted May 21, 2009, Published online December 8, 2009)

Pure Appl. Geophys. 167 (2010), 91–106

DOI 10.1007/s00024-009-0010-7

Pure and Applied Geophysics

Omori's Law Applied to Mining-Induced Seismicity and Re-entry Protocol Development

J. A. Vallejos[1] and S. D. McKinnon[1]

Abstract—This paper describes a detailed study of the Modified Omori's law $n(t) = K/(c + t)^p$ applied to 163 mining-induced aftershock sequences from four different mine environments in Ontario, Canada. We demonstrate, using a rigorous statistical analysis, that this equation can be adequately used to describe the decay rate of mining-induced aftershock sequences. The parameters K, p and c are estimated using a uniform method that employs the maximum likelihood procedure and the Anderson–Darling statistic. To estimate consistent decay parameters, the method considers only the time interval that satisfies power-law behavior. The p value differs from sequence to sequence, with most (98%) ranging from 0.4 to 1.6. The parameter K can be satisfactorily expressed by: $K = \kappa N_1$, where κ is an activity ratio and N_1 is the measured number of events occurring during the first hour after the principal event. The average κ values are in a well-defined range. Theoretically $\kappa \leq 0.8$, and empirically $\kappa \in [0.3–0.5]$. These two findings enable us to develop a real-time event rate re-entry protocol 1 h after the principal event. Despite the fact that the Omori formula is temporally self-similar, we found a characteristic time T_{MC} at the maximum curvature point, which is a function of Omori's law parameters. For a time sequence obeying an Omori process, T_{MC} marks the transition from highest to lowest event rate change. Using solely the aftershock decay rate, therefore, we recommend T_{MC} as a preliminary estimate of the time at which it may be considered appropriate to re-enter an area affected by a blast or large event. We found that T_{MC} can be estimated without specifying a p value by the expression: $T_{MC} = aN_1^b$, where a and b are two parameters dependent on local conditions. Both parameters presented well-constrained empirical ranges for the sites analyzed: $a \in [0.3–0.5]$ and $b \in [0.5–0.7]$. These findings provide concise and well-justified guidelines for event rate re-entry protocol development.

Key words: Omori's law, Mining-induced seismicity, Aftershock sequences, Mine safety, Ground control, Re-entry protocol.

J. A. Vallejos: On secondment from the Department of Mining Engineering, University of Chile, Santiago, Chile.

[1] Department of Mining Engineering, Queen's University, Goodwin Hall, Kingston, ON K7L 3N6, Canada. E-mail: javier.vallejos@mine.queensu.ca

1. Introduction

In deep, hard-rock mines, seismicity and the associated rockburst problem present hazards to worker safety. Following large seismic events or blasts, there is a short-term increase in levels of seismic activity that, over time, decays to background levels. Restricting access to areas of a mine affected by seismicity for sufficient time to allow this decay of aftershock events is the usual approach in re-entry strategies. In mines, this decay period is generally a matter of hours, however, at the crustal scale, aftershock sequences can occur over months or years after the principal event. Following the principal event, the rate of aftershock occurrence $n(t)$ is considered to decrease with time by (Omori, 1894; Utsu, 1961):

$$n(t) = \frac{K}{(c+t)^p}, \qquad (1)$$

where t is the time measured from the principal event, c is an offset time constant, p is a parameter that determines the speed of decay, and K is an activity parameter related to the number of events in the sequence.

The Modified Omori's law (MOL) has been applied to worldwide earthquake aftershock sequences, where the parameter p differs from sequence to sequence, with a typical range of 0.6–1.6 and a median value of 1.1 (Utsu *et al.*, 1995). This variability has been previously related to the fault system properties and surrounding lithosphere (e.g., Mogi, 1967; Kisslinger, 1996), but it is still not clear which factors are most significant in controlling the p value (Utsu *et al.*, 1995). Using a rate-state dependent fault strength model, Dieterich (1994) indicated that $p > 1$

Reprinted from the journal

may arise when the stresses on the fault surface decrease with time following the principal event.

The majority of published research on aftershock decay refers to crustal as opposed to mining-induced seismicity. McGarr and Green (1978) noted that the cumulative number of aftershocks of two mine tremors could be described by the MOL, with $p = 1$. Scott Phillips *et al.* (1999) found that the production of aftershocks induced by a controlled mine collapse followed the MOL, with $p = 1.3$. Spottiswoode (2000) established, for 11 sequences in 4 mines, that blast aftershocks were in agreement with Omori's law, with p values ranging from 0.54 to 1.10 with an average of 0.82 ± 0.20. These results were obtained through superposition of the time series and by a least-squares fit to the logarithm of $n(t)$. Stacking aftershock sequences permit the study of p values for sequences possessing too few events to allow individual analysis. However, this method may introduce artefacts into the aftershock time sequence and, therefore, influence the estimated parameters (Nyffenegger and Frohlich, 1998).

As noted above, there has been no systematic statistical study of how well Omori's law describes the decay of mining-induced aftershock sequences, and there are no well-defined recommendations of how this equation can be used in practice for re-entry protocol development. Thus, the primary goal of this paper is to address statistically how well Omori's law can describe the decay rate of mining-induced seismicity using a uniform method to fit the equation. Next, we propose practical guidelines for how the MOL can be used for the development of re-entry protocols. These guidelines are applicable for the range of mining conditions that can be found in some mines in Ontario, Canada.

2. *Sequence Identification and Sources of Data*

From seismic event catalogs, aftershock sequences can be identified in various ways, including temporal and spatial clustering. In an evaluation of current re-entry practices (Vallejos and McKinnon, 2008), we found that re-entry protocols were used almost exclusively without formal consideration of the spatial distribution of seismicity. In some mines, the analysis was done for the complete mine, in others the analysis was done level by level, while others tried visually to isolate clusters. To provide guidelines that reflect current re-entry practices that can be used by mine personnel, we relax our analysis to the temporal MOL model without spatial dependence. If space–time clustering schemes for aftershock detection were used, our results would depend to some extent on the assumptions of the clustering method, restricting the applicability of our guidelines.

For a specified zone or volume of interest, aftershock sequences are identified in time mainly by two approaches. The first involves following the extraction sequence provided by the mine. When this identification is not available, the start of a sequence is identified by the ratios method (Frohlich and Davis, 1985). This method evaluates the ratio:

$$r(N_b, N_a) = T_{N_a}/T_{N_b}, \tag{2}$$

where T_{N_a} and T_{N_b} are the time of occurrence of the N_ath and N_bth event following and preceding the principal event, respectively. Subsequent events are identified as aftershocks if the above ratio is smaller than a critical value generated by a random process with a certain probability. For our analysis, we set $N_a = 1$, $N_b = 5$ with a probability of 1%, giving a critical value of $r_c(5,1) = 0.002$. We define the start of the sequence if the ratio $r(5,1)$ is less than the critical value for a group of at least three consecutive events. The first event of the group is identified as the principal event of the sequence regardless of its magnitude. Note that with this definition and assigned probability, 1% of the events will be principal events, even if a Poisson process generated the catalog.

A total of 163 mining-induced aftershock sequences were extracted from the seismic catalogs of the following 4 mines in Ontario:

1. *Copper Cliff North, Sudbury* The study area consists of the upper part of the 100/900 orebodies, between the 2,700 and 3,400 levels, from January 2003 to December 2004. The microseismic monitoring system covering this zone is composed of 13 uniaxial and 2 triaxial accelerometers. In total, 37 aftershock sequences were extracted for the analysis. A description of mining

methods, geology and monitoring system can be found in MALEK and LESLIE (2006).

2. *Craig, Sudbury* Five aftershock sequences were obtained for zones 10 and 11. A description of the mining methods, geology and monitoring system can be found in SIMSER (2006).
3. *Kidd Creek, Timmins* The study region corresponds to the complete mine D, covering a volume of approximately 300 m × 200 m × 500 m, between the 6,800 and 8,800 levels, from August 2004 to December 2007. In this zone, the underground microseismic monitoring system consists of 15 uniaxial and 4 triaxial accelerometers. Blasthole mining with delayed paste backfill is used to extract the ore underground. A general description of the geology of the mine can be found in BOARD *et al.* (2001). A total of 70 aftershock sequences were identified in this catalog.
4. *Macassa, Kirkland Lake* The zone is composed of a set of primary, longitudinal, continuous retreat sublevel longhole stopes with delayed paste backfill at a depth below surface of 1,500 m. The geotechnical domains include very brittle, massive, high-strength tuff rocks in the hangingwall and massive-to-moderately jointed high strength orebody in a series of sub-parallel faults. These faults strike northeast and dip steeply (~70°) to the south. The footwall consists of a medium-strength basic syenite, which is generally quite blocky. The average stope dimensions are approximately 3 m in thickness and 30 m in width. The dominant geological structure of the zone is a single major fault intersecting the stopes, which can be parallel to the hangingwall or footwall contact or be the actual contact itself. Inclusions are infrequent and discontinuous. The underground microseismic monitoring system surrounding the zone consists of a dense array of 66 uniaxial accelerometers. The volume of interest consists of approximately 150 m × 100 m × 100 m. A total of 51 aftershock sequences were identified from December 2004 to May 2007.

3. Methods

3.1. Filtering

Each aftershock sequence was filtered by limiting the source location error-number of triggered sensors (VALLEJOS and MCKINNON, 2008), and by a lower limit of moment magnitude, selected at the magnitude bin with the highest frequency of events in a non-cumulative frequency-magnitude distribution (WIEMER and WYSS, 2000; WOESSNER and WIEMER, 2005). The effect of these two filters was to remove poorly located seismic events from the analysis and to provide some degree of uniformity to the data, respectively.

To evaluate the effect of the first filter quantitatively, we compare the maximum likelihood estimate of the MOL parameters with and without error-filtering for 50 mining-induced aftershock sequences. We found that, in 75% of the cases, the error-filtering effect did not change the MOL parameters by more than 10%. For the other 25%, we found an improvement in the fit of the MOL equation to the data (decrease in the Anderson–Darling statistic). We conclude that this pre-filter does not significantly affect the MOL parameters and, in general, improves the fit to the data.

3.2. Maximum Likelihood Estimate of the MOL Parameters

If the occurrence times t_i ($i = 1, \ldots, N$) of the individual events are available at a time interval $[T_a, T_b]$, then the parameters K, p, c and their uncertainties σK, σp, σc can be estimated by the maximum likelihood method (OGATA, 1983). Assuming that all the events in an aftershock sequence are conditionally independent and distributed according to a non-stationary Poisson process with intensity function given by Eq. 1, the following log-likelihood function is obtained:

$$\ln L(K,p,c,T_a,T_b) = N \ln K - p \sum_{i=1}^{N} \ln(t_i + c) - KA(p,c,T_a,T_b), \quad (3)$$

where

$$A(p,c,T_a,T_b) = \begin{cases} \ln(T_b+c) - \ln(T_a+c) & \text{for } p=1, \\ \left[(T_b+c)^{1-p} - (T_a+c)^{1-p}\right]\Big/(1-p) & \text{for } p \neq 1. \end{cases} \tag{4}$$

The maximum likelihood estimates (MLE) of the parameters K, p, c and are those that maximize Eq. 3.

3.3. Statistical Criteria

One of the applications of the MLE is to the statistical model selection on the basis of the Akaike Information Criterion (AIC), given by (Akaike, 1974):

$$\text{AIC} = -2\max\{\ln L(K,p,c,T_a,T_b)\} + 2n, \tag{5}$$

where n is the number of parameters of the model and $\ln L(K, p, c, T_a, T_b)$ is the log-likelihood function. A model with a smaller AIC is considered to be a better fit. We will make use of this criterion later to compare different rate models. It should be noted that the AIC only measures the relative fit of competing models and cannot be used as an absolute measure of suitability of fit (Nyffenegger and Frohlich, 1998). To evaluate whether the MOL-fitted parameters adequately describe the time sequence, therefore, we make use of the non-parametric Anderson–Darling statistic A^2 (Anderson and Darling, 1954; Lewis, 1961), defined by:

$$A^2 = -N - \sum_{i=1}^{N} \frac{(2i-1)}{N}[\ln(u_i) + \ln(1-u_{N+1-i})], \tag{6}$$

where

$$u_i = \begin{cases} \frac{\ln(t_i+c)-\ln(T_a+c)}{\ln(T_b+c)-\ln(T_a+c)} & \text{for } p=1, \\ \frac{(t_i+c)^{1-p}-(T_a+c)^{1-p}}{(T_b+c)^{1-p}-(T_a+c)^{1-p}} & \text{for } p\neq 1. \end{cases} \tag{7}$$

$A^2 = 0$ indicates a perfect fit to the data. Sequences with $A^2 \leq 2.0$ have been considered to follow the Omori distribution, and those with $A^2 \leq 1.0$ have been considered to fit well (Nyffenegger, 1998; Nyffenegger and Frohlich, 1998, 2000).

3.4. Power-law MOL Parameters

From a preliminary analysis of mining-induced aftershock sequences, it was determined that the MLE of the MOL parameters were extremely sensitive to the time interval $[T_a, T_b]$ used to fit the equation. An example is presented in Fig. 1 for a blast-related aftershock sequence at the Kidd Creek mine. To represent the seismicity rate, we followed the method proposed by Utsu (1962), where the data are divided into logarithmic intervals. In this figure, the reported parameters, their uncertainties and goodness of fit, have been estimated by the maximum likelihood method using the complete duration of the sequence, i.e., $[T_a, T_b] = [t_0, t_N]$, where t_0 and t_N are the occurrence times of the principal and last events detected by the ratios method, respectively. Inspection of Fig. 1 provides the following:

1. There is high uncertainty in the MLE of the MOL parameters.
2. The p value seems high compared to the usual range reported in the crustal literature (0.6–1.6).

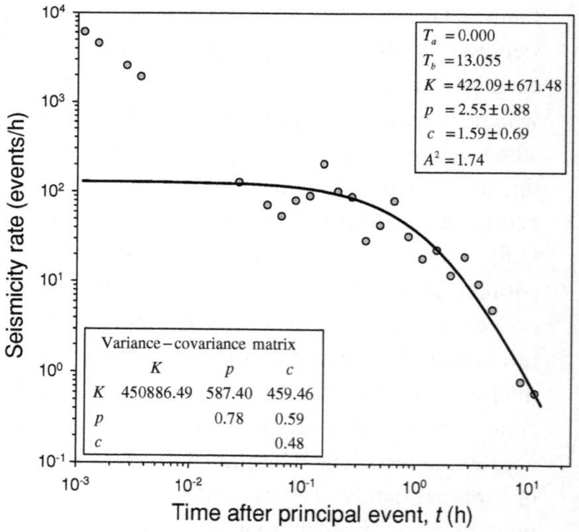

Figure 1
Seismicity rate and maximum likelihood estimate of the MOL parameters for the complete time interval $[t_0, t_N]$ for a blast-related aftershock sequence at the Kidd Creek site.

3. Based on the Anderson–Darling goodness of fit, the sequence can be considered to follow an MOL distribution ($A^2 \leq 2.0$).
4. The fitted MOL does not adequately describe the aftershocks at very short occurrence times ($< \sim 0.1$ h).

This parameter sensitivity, also recognized for some crustal sequences, motivated us to investigate in depth the effects of the time interval on the MLE of the MOL parameters. We conclude that, to estimate consistent decay parameters, it is necessary to exclude some aftershocks from the start and end of the sequence and to consider only the time interval that satisfies power-law behavior. We propose, therefore, a uniform algorithm based on formal properties of aftershock time sequences and numerical simulations. The method is described as follows:

1. Calculation of the inter-event median times relative to the principal event: $\bar{t}_i = (t_i + t_{i-1})/2$. Based on an examination of the early portion of the cumulative-event data, define time intervals of possible start and end time of power-law decay as: $\Delta T_S = [\bar{t}_1, \bar{t}_i]$, $\Delta T_E = [\bar{t}_{i+1}, t_N]$. Generally, it is sufficient to set the division between ΔT_s and ΔT_E at the event closest to one time unit, i.e., $t_i \approx 1$.
2. For each possible combination of inter-event median times T_a and T_b contained in ΔT_S and ΔT_E, respectively, the MLE of the MOL parameters with their uncertainties is obtained, corresponding goodness of fit A^2, and maximum curvature point given by:

$$T_{\mathrm{MC}} = \left[Kp\sqrt{\frac{1+2p}{2+p}} \right]^{\frac{1}{1+p}} - c. \qquad (8)$$

3. From the solutions determined in the previous steps, consider only inter-event median times that conform to a power-law behavior: $A^2 \leq 1$ and $c = 0$.
4. Next, consider all the inter-event median times that include the maximum curvature point: $T_a \leq T_{\mathrm{MC}} \leq T_b$. This constraint was identified from numerical simulations. Essentially, we generated 100,000 MOL time sequences with a normally distributed p value, $c = 0$ and N events in a time interval $[T_a, T_b]$. The variables N, T_a and T_b were varied uniformly over a certain range, ΔN, ΔT_a and, ΔT_b, respectively. We chose reasonable values for the above ranges using, as a reference, crustal aftershock sequences: $p = 1.1 \pm 0.2$, $\Delta N = [10, 1{,}000]$, $\Delta T_a = [0.0001, 1.0]$, $\Delta T_b = [2{,}100]$ (KISSLINGER and JONES, 1991; NYFFENEGGER and FROHLICH, 1998). We found that the condition $T_a \leq T_{\mathrm{MC}}$ is always satisfied and that there is a small likelihood (9.4%) of $T_{\mathrm{MC}} > T_b$ occurring. The last situation takes place when the power-law decay is forced to end promptly. If, after applying $T_a \leq T_{\mathrm{MC}} \leq T_b$, no feasible solution is found, then the condition $T_{\mathrm{MC}} \leq T_b$ is relaxed.
5. From the remaining solutions, select as a reference the time interval $[T_{ar}, T_{br}]$ which maximizes $T_{\mathrm{MC}} - T_a$. This also fixes the reference decay constant $p_r \pm \sigma p_r$. This criterion was established by considering that, for a power-law decay, the maximum change in rate is always found between T_a and T_{MC}.
6. The optimal solution is selected as the one that maximizes the time interval relative to the reference solution, and is referred to as start T_s and end T_E times of power-law decay. This condition is defined as the inter-event median time interval that maximizes the ratio T_b/T_a and satisfies $T_a \leq T_{ar}$, $T_b \geq T_{br}$. To avoid selecting an optimal solution too different from the reference one, we constrain the optimum p value within a 95% confidence limit of p_r, i.e., $|p - p_r| \leq 1.96\ \sigma p_r$.

Figure 2 shows the same example presented in Fig. 1, but with the MOL parameters estimated for the time interval of power-law decay determined with the above method. It is seen that the events occurring at very short ($< \sim 0.1$ h) and long ($> \sim 5$ h) delay times have been automatically excluded. Also, the uncertainties of the MOL parameters have been considerably reduced and the p value seems to be in the usual reported range for crustal sequences. The data segment used to fit the equation conform to MOL decay much better than the entire data set used in Fig. 1 (lower A^2).

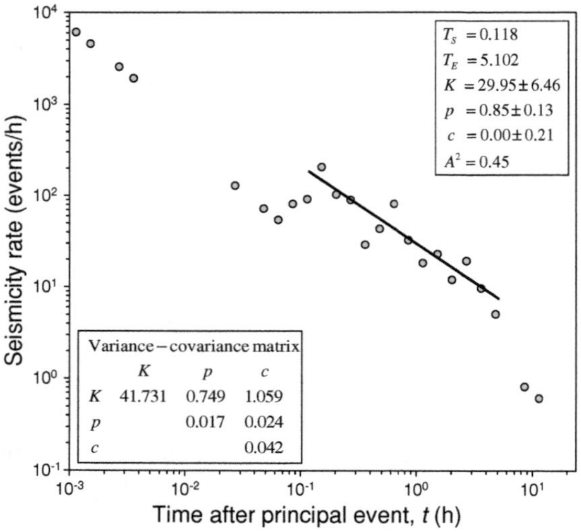

Figure 2
Seismicity rate and maximum likelihood estimate of the MOL parameters for the determined time interval of power-law decay [T_s, T_E] for a blast-related aftershock sequence at the Kidd Creek site.

4. *Justification and Interpretation of the Time Interval of Power-law Decay*

By using the AIC criterion, we will statistically verify that the start and end times of power-law decay determined by the proposed method represent justifiable conditions for the onset and cessation of power-law behavior. To interpret the results, we make use of the Akaike weights (w_i). If AIC_i is the AIC value for model i, and $\min_{m \in M}\{\text{AIC}_m\}$ is the minimum AIC value of a set of M candidate models, then the Akaike weights are expressed by:

$$w_i = \frac{\exp(-\Delta_i/2)}{\sum_{m=1}^{M}\exp(-\Delta_m/2)}, \tag{9}$$

where Δ_i is a measure of each model relative to the best model of the set of candidate models:

$$\Delta_i = \text{AIC}_i - \min_{m \in M}\{\text{AIC}_m\}. \tag{10}$$

The w_i's are interpreted as the probability that the model is the best among the set of candidate models.

4.1. *End Time of Power-law Decay T_E*

We want to demonstrate that there is a justifiable reason to split the sequence at T_E independently of the selected start time (t_0 or T_s). We compare an MOL adjusted for the complete time interval [t_0, t_N] with a fixed MOL for the time interval [t_0, T_E] followed by a different rate model from the following six combinations:

$$n(t)=\begin{cases} \dfrac{K_1}{(c_1+t)^{p_1}} & \text{for } t_0 \le t \le T_E \\ \left[\begin{array}{ll} \dfrac{K_1}{(c_1+t)^{p_1}}+\dfrac{K_2}{(c_2+t-T_E)^{p_2}} & (11)\\ \dfrac{K_1}{(c_1+t)^{p_1}}+K_2e^{-\alpha(t-T_E)} & (12)\\ \dfrac{K_1}{(c_1+t)^{p_1}}+K_2 & (13)\\ \dfrac{K_2}{(c_2+t)^{p_2}} & (14)\\ K_2e^{-\alpha t} & (15)\\ K_2 & (16) \end{array}\right] & \text{for } T_E < t \le t_N \end{cases}$$

Equations 1 and 11–16 give a total of three different functions and seven combinations of rate models to compare. Equations 11–13 are associated with two superimposed mechanisms after T_E while Eqs. 14–16 are representative of a transition between two independent systems. The three types of functions are well known in the literature. Equations 11 and 15 have been proposed by OGATA (1983) to confirm significant secondary aftershocks and a possible transition from a power-law to an exponential decay. A permanently present exponential and power-law (Eq. 12) was suggested by NARTEAU *et al.* (2002). Finally, we considered, as an alternative rate model, a transition to background levels (Eqs. 13 and 16). The above comparison also will enable us to identify whether there is any alternative consistent rate model after T_E. This result will indicate whether or not a systematic new physical process commences after the initial power-law decay.

To ensure stable solutions, we only consider sequences with at least 10 events in the time interval [T_E, t_N]. As a result, only 42 of the sequences were retained for the analysis. This low number of retained cases is a direct consequence of the mining process. Before the power-law ends, a new sequence is generally induced (by blasting). We found that the rate formulas (Eqs. 11–16) gave better results than a single MOL (Eq. 1), in 39 of the 42 cases with an average probability of 95%. The implication is that a single MOL does not hold after T_E, and a new process becomes established, which was correctly identified by the proposed method. Similar results were obtained if the comparison was made using the start time of power-law decay T_s instead of t_0. From this analysis we conclude the following:

1. The most repeated rate models that give the best fit after T_E, in descending order, are: Eqs. 16, 13, 11, 15, 12, and 14, i.e., the models that include a transition to background levels.
2. We found a low number of cases (only six) for which Eq. 11 was found to be the best model indicating that the ratios method performs well in extracting main sequences from the catalogs.
3. We found that Eqs. 14 and 15 are the most effective alternative rate models for justifying the division of the sequence at T_E. Equation 15 identifies a transition from a power-law to an exponential decay, which has been interpreted in the literature as a "correlation time", after which aftershocks cease to occur and healing dominates (NARTEAU *et al.*, 2002).

4.2. *Start Time of Power-law Decay T_s*

For the start time of power-law decay, we compare the AIC obtained for the time interval [t_0, T_E] with the composite of AIC obtained for a MOL in the time interval [T_s, T_E] preceded by the following alternative rate models:

$$n(t)=\begin{cases}\left[\begin{array}{l}\dfrac{K_2}{(c_2+t)^{p_2}} \quad (17)\\ K_2e^{-\alpha t} \quad (18)\\ K_2+K_3t \quad (19)\\ K_2 \quad (20)\end{array}\right] & \text{for } t_0 < t \le T_S\\ \dfrac{K_1}{(c_1+t)^{p_1}} & \text{for } T_S \le t \le T_E\end{cases}$$

Equation 17 is equivalent to the procedure outlined by OGATA (1983). We have incorporated into the analysis three additional alternative models. The exponential and constant rate models were already mentioned in the previous section, while a linear decay function, Eq. 19, was proposed by NARTEAU *et al.* (2002). Again, we only considered the cases with at least 10 events in the time interval [t_0, T_s]. A total of 37 sequences were retained for the analysis. We found that the combined rate formulas (Eqs. 17–20), provide better results than a single MOL (Eq. 1) for only 65% of the cases. This result suggests that, based on the AIC criterion, there is no strong motivation for dividing the sequence at T_s, and that there is no consistent best-fit rate model in the time interval [t_0, T_s]. We leave open as a possibility, that there is another equation that can fit the data better than those considered here. However, in our method, T_s has a fundamental statistical and physical meaning. For a given T_E, T_s is always the first inter-event median time that satisfies a power-law process, i.e., the solution constraints of $c = 0$ and $A^2 \le 1$. In the context of developing guidelines for re-entry protocols, there is a practical advantage to determining power-law constants with these constraints, and being able to compare them for different mining environments.

5. *Results*

5.1. *Start T_s and End T_E Times of Power-law Decay*

Figure 3a presents the population distributions of the start and end times of power-law decay for the Copper Cliff North, Kidd Creek and Macassa mines. Most of the analyzed sequences start the power-law decay less than 1 h after the principal event, indicating that, for mining-induced seismicity, aftershock sequences display non-power-law behavior only briefly (<1 h). The only sequence detected with $T_s > 1$ presented a nearly constant rate, followed by a build-up during the first hour before the onset of power-law decay. The distribution of T_s for Copper Cliff North and Macassa mines suggests that these monitoring systems are more sensitive in detecting events above their corresponding lower limit of magnitude closer to the principal event than the Kidd Creek array. A more useful interpretation is derived

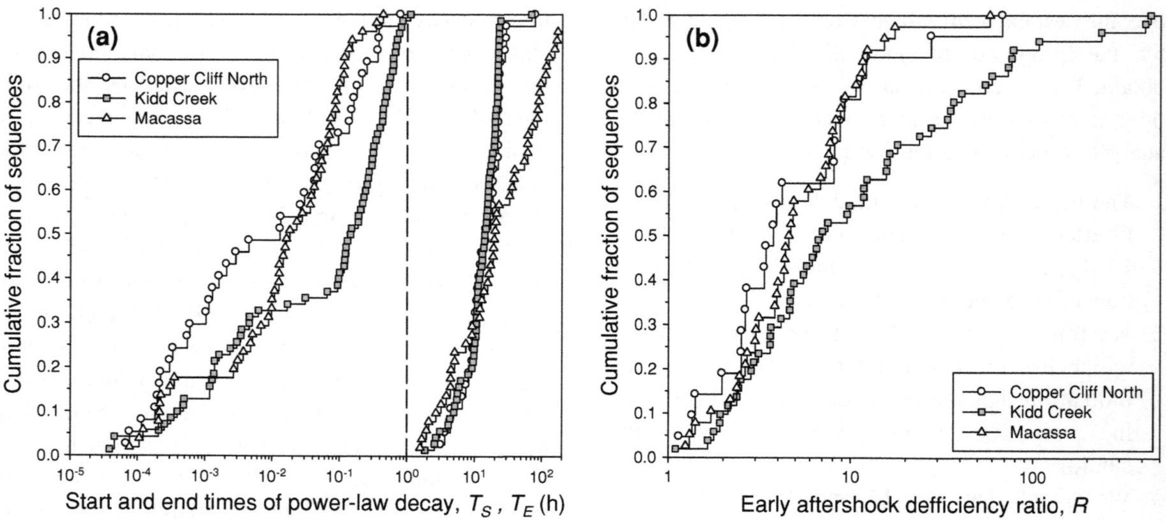

Figure 3
Normalized cumulative population distributions of the start and end times of power-law decay (frame **a**) and the early aftershock deficiency ratio (frame **b**) for the aftershock sequences at the Copper Cliff North, Kidd Creek and Macassa sites.

by using the concept of Early Aftershock Deficiency (EAD), introduced by PENG *et al.* (2006). We quantify the EAD by extrapolating the determined power-law MOL curve toward the principal event and calculating the number of events between the first measured event $\bar{t}_1 = (t_1 - t_0)/2$ and the start time of power-law decay T_s:

$$N^{\text{calc}}_{\bar{t}_1 - T_S} = \frac{K}{1-p}\left(T_S^{1-p} - \bar{t}_1^{1-p}\right). \tag{21}$$

To evaluate the degree of EAD, we define the ratio:

$$R = \frac{N^{\text{calc}}_{\bar{t}_1 - T_S}}{N^{\text{meas}}_{\bar{t}_1 - T_S}}, \tag{22}$$

where $N^{\text{meas}}_{\bar{t}_1 - T_S}$ is the measured number of events between $\bar{t}_1$ and T_s. Figure 3b confirms that both the Copper Cliff North and the Macassa data present lower EAD ratios than the Kidd Creek catalog.

5.2. *p Values*

For each sequence, the MOL parameters, with their uncertainties and the corresponding goodness of fit A^2 were estimated for the following three time intervals:

1. The complete duration of the sequence detected by the ratios method $[t_0, t_N]$.
2. The determined time interval of power-law decay $[T_s, T_E]$.
3. From the principal event until the end time of the power-law decay $[t_0, T_E]$.

We compare these three solutions to draw more conclusions regarding the MLE parameter sensitivity, and also to emphasize the importance of selecting an adequate time interval to estimate the MOL parameters, focusing on the p value. First, we evaluate the effect of selecting an adequate end time by comparing $[t_0, t_N]$ with $[t_0, T_E]$. We observe that in, 66% of the cases, the p value determined for $[t_0, t_N]$ is lower than that obtained for $[t_0, T_E]$. This underestimation of the p values is probably due to the inclusion of background seismicity on the estimated p for $[t_0, t_N]$. Next, we evaluate the effect of selecting an adequate start time. OGATA (1983) suggested that the interplay between p and c in the MOL, combined with the difficulty of detecting early aftershocks following the principal event, may bias p toward higher values. By combining the time intervals $[t_0, T_E]$ and $[T_S, T_E]$, we established the following empirical relationships between p and c:

$$\begin{aligned} p_{[t_0,T_E]} &= -0.18 \pm 0.17 c_{[t_0,T_E]} + 1.02 \pm 0.02 p_{[T_S,T_E]} \quad & R^2 = 0.915 \quad & \text{for Copper Cliff North,} \\ p_{[t_0,T_E]} &= 0.61 \pm 0.06 c_{[t_0,T_E]} + 0.95 \pm 0.04 p_{[T_S,T_E]} \quad & R^2 = 0.830 \quad & \text{for Kidd Creek,} \\ p_{[t_0,T_E]} &= 0.93 \pm 0.18 c_{[t_0,T_E]} + 0.99 \pm 0.01 p_{[T_S,T_E]} \quad & R^2 = 0.875 \quad & \text{for Macassa.} \end{aligned} \tag{23}$$

From Eq. 23, it is clear that the effect of the MLE of c on p is a bias toward higher values relative to the power-law MOL solution. The only exception is Copper Cliff North. We evaluated the hypothesis that the regression coefficients are equal to zero using the Student t statistic. We found that this hypothesis can be confidently rejected at a 5% significance level in all cases, except for the coefficient $c_{[t_0,T_E]}$ for Copper Cliff North. In this particular case, the effect of the MLE of c on p is negligible.

Figure 4 presents the cumulative distributions of the p values estimated for the time interval of power-law decay. We found that p can, in general, range from 0.4 to 1.6, with average values of 0.81 ± 0.20, 0.92 ± 0.22, 1.04 ± 0.33 and 0.98 ± 0.18 for Copper Cliff North, Craig, Kidd Creek and Macassa, respectively. By using the Anderson–Darling test, the hypothesis that p is normally distributed for all the mines cannot be rejected at the 20% significance level.

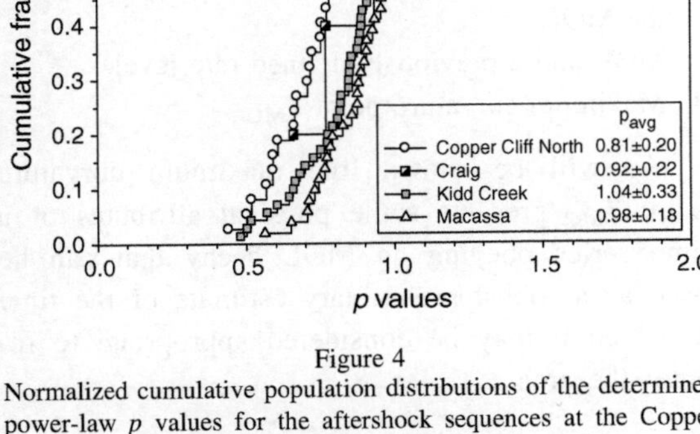

Figure 4
Normalized cumulative population distributions of the determined power-law p values for the aftershock sequences at the Copper Cliff North, Craig, Kidd Creek and Macassa sites.

5.3. K Values

By definition, for a sequence conforming to power-law decay, the parameter K corresponds to the rate of activity one time unit after the principal event. However, rate estimates can present high fluctuations during the first time unit and any evaluation will depend on the widths of the bins used to represent the data. This can bias estimates of K. Also, we found no significant correlation between K and the magnitude of the principal event, as suggested by HELMSTETTER and SORNETTE (2002). Based on these observations, we investigate an alternative criterion to estimate K, based on the number of events. For a power-law MOL, the calculated number of events at time t since $\bar{t}_1$ is given by:

$$N(t) = \int_{\bar{t}_1}^{t} n(t)\mathrm{d}t = \begin{cases} K \ln\left(\frac{t}{\bar{t}_1}\right) & \text{for } p = 1, \\ \frac{K}{1-p}\left\{ t^{1-p} - \bar{t}_1^{1-p} \right\} & \text{for } p \neq 1. \end{cases} \tag{24}$$

By assuming $t = 1$ in Eq. 24 the following activity ratio can be defined:

$$\kappa = \frac{K}{N(t=1)} = \begin{cases} -\frac{1}{\ln \bar{t}_1} & \text{for } p = 1, \\ \frac{1-p}{1-\bar{t}_1^{1-p}} & \text{for } p \neq 1. \end{cases} \tag{25}$$

It is seen from Eq. 25 that κ depends on the time occurrence of the first event $\bar{t}_1$ and the p value of the aftershock sequence. For the four different mines being studied, we found that $\bar{t}_1$ can vary between 0.00001 and 0.1 h. Figure 5 presents κ as a function of p for two different values of $\bar{t}_1$. We observed that, for the same $\bar{t}_1$, a higher p value implies a lower κ, and that, in theory, the only way for $\kappa \geq 1$ is when $p = 0$ in the first time unit of data. We have included in Fig. 5 actual data from mining-induced aftershock sequences using the estimated MOL power-law parameters and the measured number of events during the first hour after the principal event N_1. Most of the cases (77%) are within the boundary

Reprinted from the journal

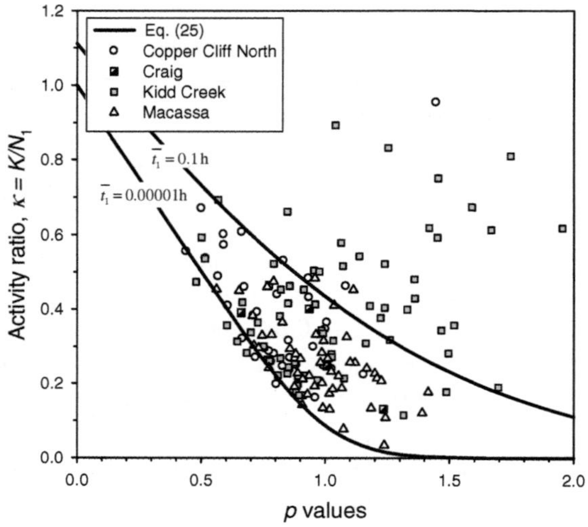

Figure 5
Activity ratio κ as a function of p for different values of $\bar{t}_1$.

values of Eq. 25. A large number of the occurrences (18%) with high κ values also correspond to the Kidd Creek site. These cases are a direct consequence of the natural variations in the inter-event times, limited by the sensitivity of the monitoring system in the early portion of the first hour after the principal event. We found average κ values of 0.39 ± 0.16, 0.29 ± 0.10, 0.41 ± 0.19 and 0.26 ± 0.11, for the Copper Cliff North, Craig, Kidd Creek and Macassa sites, respectively. By using the Anderson–Darling test, the hypothesis that κ is normally distributed cannot be rejected at the 20% level of significance for all the sites under study. This analysis provides the following conclusions:

1. The parameter K does not correspond to the number of events occurring during the first time unit, as considered by Liu (1984), Hudyma *et al.* (2003) and Heal *et al.* (2005), and previously commented on by Page (1986), unless $p = 0$ in the first time unit of data.
2. The normal distribution of κ suggests that some average correlation may exist between K and N_1. Figure 6 presents this correlation for the four sites under study. Note that K can be satisfactorily expressed as:

$$K = \kappa N_1. \tag{26}$$

For the usual range of p and $\bar{t}_1$ values of mining-induced aftershock sequences we found that, theoretically, $\kappa \leq 0.8$. Similarly, the data analysis of four different mining environments suggests that $\kappa \in [0.3–0.5]$.

5.4. *Guidelines for Re-entry Protocol Development*

We have demonstrated that the MOL can be adequately used to describe the aftershock decay rate of mining-induced aftershock sequences. In terms of the MLE of the MOL parameters, we basically have two parameters to specify: p and K. Both can be estimated based on the analysis of locally observed historical data. The applicability of these findings for developing a real time re-entry protocol is as follows:

1. The most probable p value can be assumed or, alternatively, can be used as a normally distributed variable.
2. Following the principal event and after recording the number of seismic events during the first hour N_1, the value of K can be established with an associated level of confidence.

This indicates that we have provided a rigorous approach to estimating the parameters of the MOL for re-entry protocol development and to delineating an expected decay rate. However, the task remains of developing a criterion for how the MOL can be used in practice to make a preliminary estimate of the time at which it may be considered appropriate to re-enter the area. The following three re-entry criteria will be presented and discussed:

1. Preset level in the cumulative density function of the MOL.
2. MOL and a previously defined rate level.
3. Maximum curvature point T_{MC}.

As will be shown, the maximum curvature point T_{MC} presents some physical attributes of a time series obeying an MOL decay that can be used as a robust preliminary estimate of the time at which it may be considered appropriate to re-enter the area.

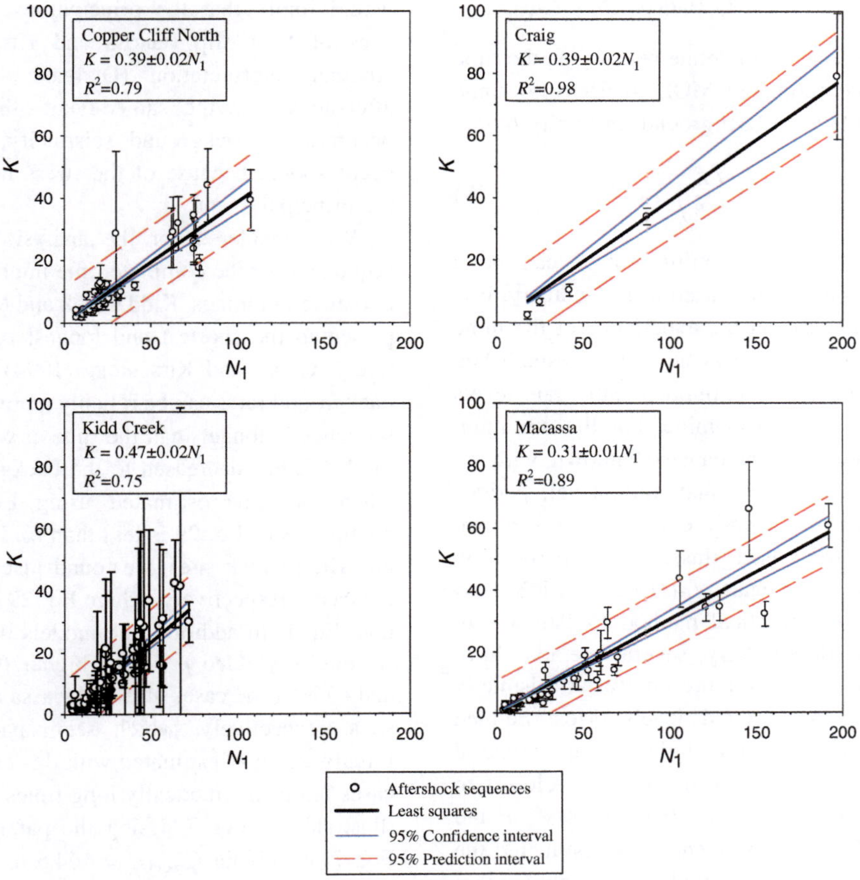

Figure 6

Correlation between the parameter K and the measured number of events occurring during the first hour after the principal event N_1.

5.5. *Preset Level in the Cumulative Density Function of the MOL*

For a power-law MOL, the cumulative density function at time t since $T_a \neq 0$ is given by Eq. 7 with $c = 0$. This criterion consists of presetting some level D in Eq. 7, which corresponds to a percentage of the total number of events occurring in a given time window $[T_a, T_b]$ after the principal event. The re-entry time is obtained directly from Eq. 7:

$$t_{\text{re-entry}} = \begin{cases} T_a \exp\left\{ D \ln\left(\frac{T_b}{T_a}\right)\right\} & \text{for } p = 1, \\ \left\{ D\left(T_b^{1-p} - T_a^{1-p}\right) + T_a^{1-p}\right\}^{\frac{1}{1-p}} & \text{for } p \neq 1. \end{cases} \tag{27}$$

It is clear from Eq. 27 that the re-entry time estimated by this criterion does not depend on the activity parameter K, which seems unrealistic. For example, assuming reasonable values for the parameters: $D = 0.9$, $[T_a, T_b] = [0.001, 24]$ hours, and $p = 0.8$, 1.0 and 1.2, we obtain $t_{\text{re-entry}} = 15.3$, 8.8, and 2.0 h, respectively, which is an impractically wide range for re-entry purposes.

A similar criterion has been used for re-entry purposes after blasts at the Laronde mine (HEAL *et al.*, 2005). They defined the re-entry time at 90% of the total measured cumulative percentage of the number of events or seismic energy released in a time window of 24 h. The disadvantages of this method are that it can only be used for back analysis (not for real time re-entry decision making), and, by definition, small blasts that do not produce a well-defined decay sequence can cause large re-entry times.

5.6. MOL and a Previously Defined Rate Level

It seems reasonable to define re-entry as the time required for the estimated MOL to decay to some predefined rate level or background seismicity B, i.e.:

$$t_{\text{re-entry}} = \left(\frac{K}{B}\right)^{\frac{1}{p}} - c. \tag{28}$$

We found that the definition of B can be a challenging process. When used in the industry, it is mainly defined based on the experience of the mine personnel. Malek and Leslie (2006) established typical minimum and maximum re-entry rate levels at the Copper Cliff North mine. For the minimum level, they used a maintenance shutdown with no mining activities. For the maximum level, periods when no anomalous seismic events or large crown blasts occurred were used. Based on a superposition of seismicity in a 24-h chart, Vallejos and McKinnon (2008) proposed a practical method to estimate an upper level for B. Similarly, several methods have been identified in the literature on crustal seismicity for recognizing background levels. Most defined background levels as the independent temporal component that yields the time sequence closest to a Poisson process (McNally, 1976; Habermann and Wyss, 1984). For some sequences, we estimated the B value using the above methods and concluded that using a previously defined B value and the power-law MOL parameters can lead to long re-entry times. For example, using the parameters listed in Fig. 2 and $B = 2$ events/h produces $t_{\text{re-entry}} = 24.1$ h, which is longer than the actual measured duration of the sequence.

Next, we evaluated a more rigorous method, proposed by Utsu *et al.* (1995), which also makes use of background seismicity. We compared the AIC values of Eq. 1 with the following alternative rate model:

$$n(t) = \frac{K}{(c+t)^p} + B. \tag{29}$$

If $B > 0$ and the AIC of Eq. 29 is smaller than that for Eq. 1, then the duration of the aftershock sequence can be defined by replacing the estimated parameters of Eq. 29 in Eq. 28. The continuation of background seismicity during a decay sequence would imply that the principal event creates new sites of later slip (Gross and Kisslinger, 1994). Another interpretation (Dieterich, 1986) is that aftershocks are the same events that would have occurred as background seismicity, however they occur sooner because of the stress load imposed by the principal event.

We considered for the analysis the aftershock sequences for the complete time interval $[t_0, t_N]$ from two different mines, Kidd Creek and Macassa, as they presented the shortest and longest t_N times, respectively. Gross and Kisslinger (1994) concluded that background rates can be reliably recovered only if the sequence is longer than the time at which the rate of the MOL has decreased to the background rate, i.e., when the time estimated using Eq. 28 with the parameters of Eq. 29 is less than t_N. For the Macassa and Kidd Creek sites, we found just 59 and 34% of the cases, respectively, where Eq. 29 gave a better fit than Eq. 1. In addition, the models with background seismicity yielded p values higher than 2.0, for 16 and 47% of the cases at the Macassa and Kidd Creek sites, respectively, which seem particularly high. Finally, re-entry estimated with this criterion in some cases led to impractically long times. An example is illustrated in Fig. 7. Using the parameters listed in Fig. 7, we obtain $t_{\text{re-entry}} = 164.5$ h. Based on these observations we did not investigate this criterion further.

5.7. Maximum Curvature Point

The power-law form of Eq. 1 indicates that there is no characteristic timescale and, for large time ($t \gg c$), the equation is temporally self-similar (Ito and Matsuzaki, 1990). However, if the curvature of the MOL is traced in time, given by:

$$\rho(t) = \frac{\ddot{n}(t)}{\left(1 + (\dot{n}(t))^2\right)^{3/2}}, \tag{30}$$

then a characteristic point emerges at the time of maximum curvature T_{MC}, given by Eq. 8, which depends on the estimated MOL parameters. We will demonstrate that T_{MC} has a physical meaning suitable for developing a defendable re-entry protocol. Taking the first derivative of Eq. 1:

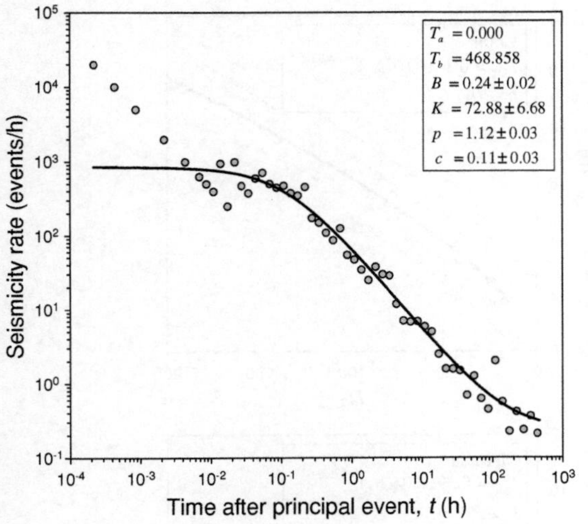

Figure 7
Seismicity rate and maximum likelihood estimate of the MOL parameters with background seismicity for a blast-related aftershock sequence at the Macassa site.

$$\dot{n}(t) = \frac{dn(t)}{dt} = -p\frac{K}{(c+t)^{1+p}}. \tag{31}$$

The negative sign of Eq. 31 reflects the decrease in the change of event rate as time increases. Replacing Eq. 8 in Eq. 31:

$$\dot{n}(t = T_{MC}) = -\sqrt{\frac{2+p}{1+2p}}. \tag{32}$$

Using the typical range of p values between 0.4 and 1.6 we find that $\dot{n}(t = T_{MC})$ can range from -1.15 to -0.93, respectively. The implication is that, after T_{MC}, the event rate starts to decrease at a rate approximately less than one, indicating a transition from high-to-low event rate change. In the case of a power-law equation, the maximum change in rate is always found between T_s and T_{MC}. For example, using the parameters listed in Fig. 7, we find that this sequence decayed to a rate of $n(T_{MC}) = 7.0$ events/h during the first $T_{MC} = 7.9$ h, while it took over 16 additional hours to decay to two events per hour.

The general form of Eq. 8 indicates that T_{MC} is a scale-invariant function of K. By replacing Eq. 26 in Eq. 8 for a power-law process, the following expression is obtained:

$$T_{MC} = aN_1^b, \tag{33}$$

where

$$a = \left[\kappa p\sqrt{\frac{2-b}{1+b}}\right]^b \qquad b = \frac{1}{1+p} \tag{34}$$

Using the estimated power-law parameters and performing a non-linear regression between T_{MC} and N_1 we find the following preliminary correlations (see Fig. 8):

$$\begin{aligned} T_{MC} &= 0.47 \pm 0.09(N_1)^{0.58\pm0.05} & R^2 &= 0.82 \quad \text{for Copper Cliff North} \\ T_{MC} &= 0.30 \pm 0.07(N_1)^{0.66\pm0.05} & R^2 &= 0.99 \quad \text{for Craig} \\ T_{MC} &= 0.40 \pm 0.07(N_1)^{0.65\pm0.05} & R^2 &= 0.80 \quad \text{for Kidd Creek} \\ T_{MC} &= 0.39 \pm 0.04(N_1)^{0.57\pm0.03} & R^2 &= 0.92 \quad \text{for Macassa} \end{aligned} \tag{35}$$

The main advantage of the above correlations is that T_{MC} can be estimated directly based only on N_1 without the necessity of specifying a p value.

6. Conclusions and Recommendations

We have demonstrated that the event decay rate of mining-induced aftershock time sequences can be satisfactorily described by the Modified Omori's law (MOL). However, for estimating consistent decay parameters, it is necessary to exclude aftershocks from the start and end of the sequence and to consider the time interval that satisfies power-law decay. For all the considered aftershock sequences, the proposed method performed well in estimating a proper power-law time interval, which was statistically confirmed by means of the Akaike Information Criterion. We found that an exponential transition systematically commenced after the determined end time of power-

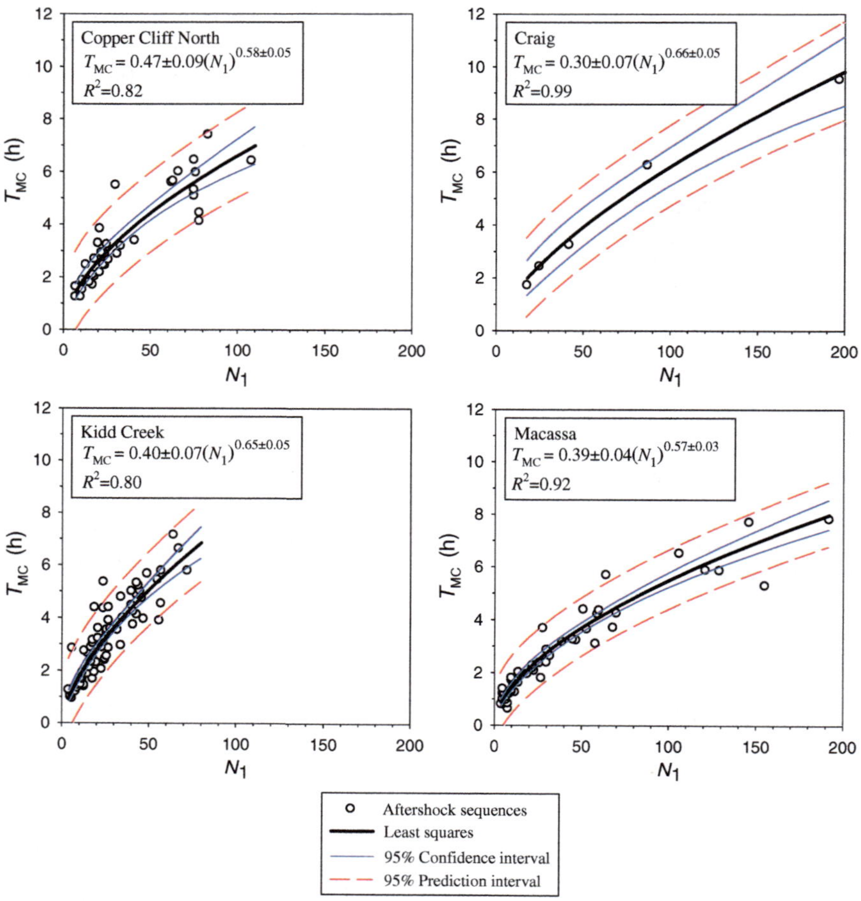

Figure 8
Correlation between the MOL maximum curvature time T_{MC} and the measured number of events occurring during the first hour after the principal event N_1.

law decay. This time has been previously interpreted in the literature as a "correlation time", after which aftershocks cease to occur and healing dominates (NARTEAU *et al.*, 2002). Similarly, we found that mining-induced aftershock sequences display a non-power-law decay only for short times (<1 h) and that there is no consistent rate model in the time interval $[t_0, T_s]$. This suggests that the parameter c does not represent any physical properties of aftershock sequences.

The p value varies from sequence to sequence, with most (98%) being within a typical range of 0.4–1.6. The sequences from mine sites analyzed presented p values that are normally distributed, with average values of 0.81 ± 0.20, 0.92 ± 0.22, 1.04 ± 0.33 and 0.98 ± 0.18 for the Copper Cliff North, Craig, Kidd Creek, and Macassa sites, respectively. Each distribution seems to represent specific conditions of the local environment, with higher decay values for the Macassa and Kidd Creek sites than the Sudbury sites (Copper Cliff North and Craig). This variability can be related to the properties of the fault system, as suggested by KISSLINGER (1996) for crustal earthquakes, but it is still not possible to draw conclusions of the significant factors that control the p value in mining-induced aftershock sequences. The parameter K can be adequately expressed as: $K = \kappa N_1$, where κ is an activity ratio and N_1 is the measured number of events occurring during the first hour after the principal event. We found that the κ values fall in a well-constrained range. Theoretically, $\kappa \leq 0.8$, and, empirically,

$\kappa \in [0.3–0.5]$. Based on the analysis of locally observed historical seismicity, both the p and κ values can be estimated.

Using the MOL formula, it is possible to make a preliminary estimate of the time at which it may be considered appropriate to re-enter the area. We presented and discussed three different criteria. The maximum curvature point T_{MC} resulted in a robust criterion for developing a defendable re-entry protocol. This point seems to apply to all power-law distributions. In the case of a time series obeying a MOL process, this point defines the transition between the highest to lowest event rate change. Based solely on the aftershock decay rate, therefore, we recommend using T_{MC} as a preliminary estimate of the time at which it may be considered appropriate to re-enter the area. We found that T_{MC} can be estimated, without the need of specifying a p value, by the expression: $T_{MC} = aN_1^b$, in which a and b are two parameters dependent on local conditions. Both parameters presented remarkably well-defined average empirical ranges for the sites analyzed: $a \in [0.3–0.5]$ and $b \in [0.5–0.7]$. However, this criterion does not incorporate the strength or magnitude of the event into the re-entry protocol. Based on an analysis and evaluation of current re-entry practices, VALLEJOS and MCKINNON (2008) recommended monitoring the event rate and the associated magnitude/strength release of events for the determination of the re-entry time. Generally, the industry practice for re-entry was based on the requirement for the monitored parameter to return to a previously defined level of seismicity for a specified time window. We recommend incorporating a 2-h time window and a magnitude threshold for resetting after the estimated T_{MC}. This significantly increases the reliability and level of confidence in the re-entry protocol.

Acknowledgments

We would like to thank two anonymous referees for their critical reviews, which led to improvements in the manuscript. This research was supported by a grant from the Workplace Safety and Insurance Board (WSIB) of Ontario. The authors wish to acknowledge the permission of the four mines studied to publish this work.

REFERENCES

AKAIKE, H. (1974), *A new look at the statistical model identification*, IEEE Trans. Autom. Control AC. *19*, 716–723.

ANDERSON, T.W. and DARLING, D.A. (1954), *A test of goodness of fit*, J. Am. Stat. Assoc. *49*, 765–769.

BOARD, M., BRUMMER, R., and SELDOM, S. (2001), *Use of numerical modeling for mine design and evaluation*, In Underground Mining Methods: Engineering Fundamentals and International Case Studies (eds. HUSTRULID, W.A., and BULLOCK, R.L.), pp. 483–491.

DIETERICH, J. (1986), *A model for the nucleation of earthquake slip*, In Earthquake Source Mechanics (eds. DAS, S., BOATWRIGHT, and SCHOLZ, C.), pp. 37–47.

DIETERICH, J. (1994), *A constitutive law for rate of earthquake production and its application to earthquake clustering*, J. Geophys. Res. *99*, 2601–2618.

FROHLICH, C. and DAVIS, S. (1985), *Identification of aftershocks of deep earthquakes by a new ratios method*, Geophys. Res. Lett. *12*, 713–716.

GROSS, S.J. and KISSLINGER, C. (1994), *Test of models of aftershock rate decay*, Bull. Seismol. Soc. Am. *84*, 1571–1579.

HABERMANN, R.E. and WYSS, M. (1984), *Background seismicity rates and precursory seismic quiescence: Imperial Valley, California*, Bull. Seismol. Soc. Am. *74*, 1743–1755.

HEAL, D., HUDYMA, M., and VEZINA, F. (2005), *Seismic hazard at Agnico-Eagle's Laronde mine using MS-RAP*, CIM Maintenance Engineering and Mine Operators Conference, Sudbury, Canada, Canadian Institute of Mining.

HELMSTETTER, A. and SORNETTE, D. (2002), *Subcritical and supercritical regimes in epidemic models of earthquakes aftershocks*, J. Geophys. Res. *107*(B10), 2237, doi:10.1029/2001JB001580.

HUDYMA, M., HEAL, D., and MIKULA, P. (2003), *Seismic monitoring in mines—old technology—new applications*, Proc. 1 Australian Ground Control in Mining Conf., Sydney, pp. 201–218.

ITO, K. and MATSUZAKI, M. (1990), *Earthquakes as self-organized critical phenomena*, J. Geophys. Res. *95*, 6853–6860.

KISSLINGER, C. (1996), *Aftershocks and fault zone properties*, Adv. in Geophys. *38*, 1–36.

KISSLINGER, C. and JONES, L.M. (1991), *Properties of aftershock sequences in Southern California*, J. Geophys. Res. *96*, 11947–11958.

LEWIS, P.A.W. (1961), *The distribution of the Anderson-Darling statistic*, Ann. Math. Stat. *32*, 1118–1124.

LIU, Z.-R. (1984), *Earthquake frequency and prediction*, Bull. Seismol. Soc. Am. *74*, 255–265.

MALEK, F. and LESLIE, I.S. (2006), *Using seismic data for rockburst re-entry protocol at Inco's Copper Cliff North mine*, Proc. of the 41 US Rock Mech. Symp., Golden, Colorado, ARMA/USRMS 06-1163.

MCGARR, A. and GREEN, R.W.E. (1978), *Microtremor sequences and tilting in a deep mine*, Bull. Seismol. Soc. Am. *66*, 1679–1697.

MCNALLY, K. C. (1976), *Spatial, temporal and mechanic character in earthquake occurrence, a segment of the San Andreas Fault in central California*, PhD. dissertation, University of California, Berkeley, California.

MOGI, K. (1967), *Earthquakes and fractures*, Tectonophysics *5*, 35–55.

NARTEAU, C., SHEBALIN, P., and HOLSCHNEIDER, M. (2002), *Temporal limits of the power law aftershock decay rate*, J. Geophys. Res. *107*(B12), doi:10.1029/2002JB001868.

NYFFENEGGER, P. (1998), *Aftershock occurrence rate decay for individual sequences and catalogs*, PhD. dissertation, University of Texas at Austin.

NYFFENEGGER, P. and FROHLICH, C. (1998), *Recommendations for determining p values for aftershock sequences and catalogs*, Bull. Seismol. Soc. Am. *88*, 1144–1154.

NYFFENEGGER, P. and FROHLICH, C. (2000), *Aftershock occurrence rate decay properties for intermediate and deep earthquake sequences*, Geophys. Res. Lett. *27*, 1215-1218.

OGATA, Y. (1983), *Estimation of the parameters in the modified Omori formula for aftershock frequencies by the maximum likelihood procedure*, J. Phys. Earth. *31*, 115–124.

OMORI, F. (1894), *On the Aftershocks of earthquakes*, J. Coll. Sci. Imp. U. Tokyo. *7*, 111–200.

PAGE, Z. (1986), *Comments on "Earthquake frequency and prediction"*, by Z.-R. Liu, Bull. Seismol. Soc. Am. *76*, 1491–1496.

PENG, Z., VIDALE, J.E., and HOUSTON, H, (2006), *Anomalous early aftershock decay rate of the 2004 M_w6.0 Parkfield, California, Earthquake*, Geophys. Res. Lett. *33*, L17307.

SCOTT PHILLIPS, D., PEARSON, CRAIG, YANG, X. and STUMP, B. W. (1999), *Aftershocks of an explosively induced mine collapse at White Pine, Michigan*, Bull. Seismol. Soc. Am. *89*, 1575–1590.

SIMSER, B.P. (2006), Mine wide seismic monitoring at Falconbridge's Craig Mine, Proc. of the 41st US Rock Mech. Symp., Golden, Colorado, ARMA/USRMS 06–1024.

SPOTTISWOODE, S. (2000), *Aftershocks and foreshocks of mine seismic events*, 3 Int. Workshop on the *Application of Geophysics to Rock and Soil Engin.*, Melbourne, Australia. (prior to GeoEng 2000).

UTSU, T. (1961), *A statistical study of the occurrence of aftershocks*, Geophys. Mag. *30*, 521–605.

UTSU, T. (1962), *On the nature of three alaskan aftershock sequences of 1957 and 1958*, Bull. Seismol. Soc. Am. *52*, 279–297.

UTSU, T., OGATA, Y., and MATSU'URA, R. (1995), *The centenary of the Omori formula for a decay law of aftershock activity*, J. Phys. Earth. *43*, 1–33.

VALLEJOS, J. and MCKINNON, S. (2008), *Guidelines for development of re-entry protocols in seismically active mines*, Proc. of the 42 US Rock Mech. Symp., San Francisco, California, ARMA/US-RMS 08-97.

WIEMER, S, and WYSS, M. (2000), *Minimum magnitude of complete reporting in earthquake catalogs: Examples from Alaska, the Western United States, and Japan*, Bull. Seismol. Soc. Am. *90*, 859–869.

WOESSNER, J. and WIEMER, S. (2005), *Assessing the quality of earthquake catalogues: estimating the magnitude of completeness and its uncertainty*, Bull. Seismol. Soc. Am. *95*, 684–698.

(Received September 1, 2008, revised March 30, 2009, accepted April 7, 2009, Published online December 22, 2009)

Pure Appl. Geophys. 167 (2010), 107–117

DOI 10.1007/s00024-009-0004-5

Pure and Applied Geophysics

Statistical Properties of Mine Tremor Aftershocks

T. E. Kgarume,[1,2] S. M. Spottiswoode,[1] and R. J. Durrheim[1,2]

Abstract—Mine tremors and their aftershocks pose a risk to mine workers in the deep gold mines of South Africa. The statistical properties of mine-tremor aftershocks were investigated as part of an endeavour to assess the hazard and manage the risk. Data from two gold mines in the Carletonville mining district were used in the analysis. Main shocks were aligned in space and time and the aftershock sequences stacked and analysed. The aftershocks were found to satisfy Gutenberg–Richter scaling, with a *b* value close to 1. Aftershock activity diminished with time in accordance with the modified Omori law, with *p* values close to 1. However, the relationship between the main shock and its biggest aftershock violated Båths law, with $\Delta M_L \approx 1.9$ for main shocks with $M_L < 3$ and increasing for main shocks with $M_L > 3$. The aftershock density was found to fall-off with distance as $r^{-1.3}$, suggesting triggering by dynamic stress.

Key words: Mine seismic events, aftershock event rate, aftershock density decay, omori's law.

1. *Introduction*

In October 2006 an incident occurred in a deep South African gold mine that motivated a renewed effort to understand the properties of mine tremor aftershocks. A $M_L = 2.4$ seismic event caused severe damage to a stope, fatally injuring several mine workers. In the investigation it was noted that an $M_L = 2.0$ event had occurred nearby 25 min prior to the $M_L = 2.4$ event, but had not caused damage to any active working places so they were not evacuated. In reviewing the incident, it was suggested that the $M_L = 2.4$ event might have been triggered by quasi-static redistribution of stress following the $M_L = 2.0$ event. The Acting Chief Inspector of Mines directed members of the "Minimising the rockburst risk" research team to investigate changes in seismic hazard following the occurrence of large seismic events, and, if warranted by the findings, to formulate guidelines regarding the evacuation of workers following such events. We report on the first phase of the investigation here—a study of the statistical properties of mine tremor aftershocks.

Local magnitude M_L is used throughout this study as it is widely used to describe the size of seismic events in South African mines. Most events recorded by mine seismic networks have local magnitudes between -1.3 and 4.0. Mine seismic networks typically use 4.5 Hz geophones, which cannot record the low frequencies radiated by the larger events and this could lead to underestimates of the magnitudes of events with $M_L > 3$. However, the catalogue of the South African National Seismological Network (SANSN), which has a threshold sensitivity of about $M_L = 2.0$, is used to calibrate mine networks. Various methods are used to calculate M_L. On the mines considered in this study, a combination of energy and moment is used.

2. *Properties of Mine Tremor Aftershocks*

Mining excavations induce stresses within the surrounding rockmass, causing elastic and inelastic deformation. These stress changes can alter the stability of pre-existing zones of weakness, such as faults or the contact between dykes and the host rock. Slip may be triggered, resulting in so-called 'shear-type' events (Jager and Ryder 1999). The induced stresses are sometimes so extreme that intact rock fails, creating new shear ruptures. Seismic events

[1] CSIR, PO Box 91230, Auckland Park, Johannesburg 2006, South Africa. E-mail: rdurrhei@csir.co.za
[2] University of the Witwatersrand, Private Bag 3, Wits 2050, South Africa.

may cause rock to be ejected or shaken loose from the hanging- and sidewalls, posing a danger to mine workers, damaging infrastructure, and hindering production. Seismic events are often followed by aftershocks, which are generally considered to be triggered by quasi-static stress transfer (SHCHERBAKOV *et al.*, 2005). These aftershocks may also pose a hazard to mining personnel and underground infrastructure, as past experience has shown that events as small as $M_L = 0.5$ can cause injuries.

The four empirical relations used in this study to analyse the properties of aftershocks are briefly described below.

2.1. Gutenberg–Richter Frequency–Magnitude Scaling

$$\log N = a - bM, \tag{1}$$

where N is the cumulative number of events greater than or equal to magnitude M, and a and b are constants. The constant a is a measure of the level of seismicity and gives the logarithm of the number of events with magnitudes greater than zero in the specified region and time window. The "b value" is an indication of the ratio of large to small events. For global earthquakes, it is generally in the range $0.8 < b < 1.2$ (FROHLICH and DAVIS 1993).

In the case of mining-related tremors, the b value may vary from $0.5 < b < 1.5$, depending on the mining geometry. It has been suggested that the b value is a function of the type, or "fractal dimension" ($D = 2b$) of the source region (TURCOTTE *et al.*, 2000). The dimensionality can be interpreted in terms of the shape and dimensions of the mining excavations, interactions with nearby faults and dykes that may be the locus of seismic activity, and the spacing of shear zones that may limit the size of the blocks in which strain energy is stored (LEGGE and SPOTTISWOODE, 1987).

1. If $b = 0.5, D = 1$ and planar seismic source zones with L(length) $\gg W$(width) are indicated. This situation arises when there is extensive mining parallel to a fault or dyke. The excess shear stress lobes produced by the excavations extend for hundreds of metres in the strike direction of the fault or dyke (L), but only for tens of metres in the dip-direction (W).
2. If $b = 1, D = 2$ and planar seismic source zones with $L \approx W$ are indicated. This situation arises where the source dimension parallel to the mining face (L) is limited to tens of metres by the leads and lags between adjacent panels.
3. If $b = 1.5, D = 3$ and volumetrically distributed sources are indicated with $L \approx W \approx S$, where S is the shear-zone spacing. In practical terms, this means that the volume of rock contributing to each event is further limited by the presence of shear zones.

Frequency–magnitude statistics for mine tremors are often bi-modal, indicating that more than one population of events is involved. EBRAHIM-TROLLOPE (2001) reports that most events with $M_L < 0$ are a result of the fracturing of the solid rock mass after blasting. This population has a much higher b value than the events generated by local structural failure, e.g., slip along faults and dykes. RICHARDSON and JORDAN (2002) identified two main classes of events: "Type A" events that are tightly clustered in time and space and generally within 100 m of an active mining face, associated with tensile fracturing and/or blasting close to the mine openings, and with an upper moment-magnitude cut-off at $M \approx 1.0$; and "Type B" events that are distributed in space and time throughout the mining region, have a distinct lower moment-magnitude cut-off at $M \approx 0.0$, and are interpreted to be due to "friction-dominated" rupture occurring on existing faults or other weak geological structures at near-lithostatic normal stresses. SPOTTISWOODE and LINZER (2004) contested this interpretation, arguing that most, if not all, "Type A" events are actually development blasts and that the lower magnitude cut-off of "Type B" events can be explained by the triggering logic of the seismic system.

2.2. Modified Omori Law

$$n(t) = \frac{K}{(t+c)^p}, \tag{2}$$

where t is the time after the main shock, $n(t)$ is the number of events occurring at time t, K and c are parameters and p is a rate constant of aftershock decay (NANJO *et al.*, 1998). SPOTTISWOODE (2000) studied the aftershock and foreshock time sequences of blasts and seismic events from four mines, and found them to follow Omori's law with $0.6 < p < 1.03$.

2.3. *Båth's Law*

$$\Delta M = M_L - M_{as}^{max} \approx \text{constant}, \tag{3}$$

where M_L is the magnitude of the main shock and M_{as}^{max} is the magnitude of the largest aftershock (BÅTH 1965). Båth's law states that the difference in magnitude between the main shock and its largest aftershock (ΔM) is a constant independent of the magnitude of the main shock. For earthquakes ΔM typically has a value of about 1.2 (BÅTH 1965; SHCHERBAKOV *et al.*, 2006).

2.4. *Aftershock Density*

$$\rho(r) = \frac{A}{r^n}, \tag{4}$$

where r is the distance from the main shock epicentre, $\rho(r)$ is the number of events occurring at distance r, A is a constant that varies with the number of aftershocks, and n is a rate constant of aftershock decay (FELZER and BRODSKY 2006). The rate of decay of the aftershock density with distance from the main shock is an indicator of the stress transfer mechanism: dynamic stresses induced by the passing seismic wave fall-off as $r^{-1.3}$, while quasi-static stresses fall-off as r^{-3}.

3. *Statistical Analysis of Mine Tremor Aftershocks*

The aftershock sequences were analysed following 342 $M_L \geq 2$ main shocks at two gold mines in the Carletonville mining district exploiting the Ventersdorp Contact Reef and Carbon Leader Reef at depths of 2,500–3,500 m (note that the term "reef" denotes a quartz pebble conglomerate). Two difficulties had first to be addressed: (1) the influence of blasting during the normal mining production cycle on seismicity, and (2) the small number of aftershocks recorded after any single main shock.

3.1. *Influence of Production Blasting on Seismicity*

Mines are evacuated during blasting time and for several hours thereafter to avoid exposing workers to the hazards posed by the fumes and dust produced by the explosives and the seismicity triggered by stress redistribution following the advance of the mining face. Adjacent mines usually coordinate blasting to prevent seismicity triggered by blasting on one mine exposing workers in the adjacent mine to undue risk.

Figure 1 shows the distribution of aftershocks by hour of day for the two mines considered in this study.

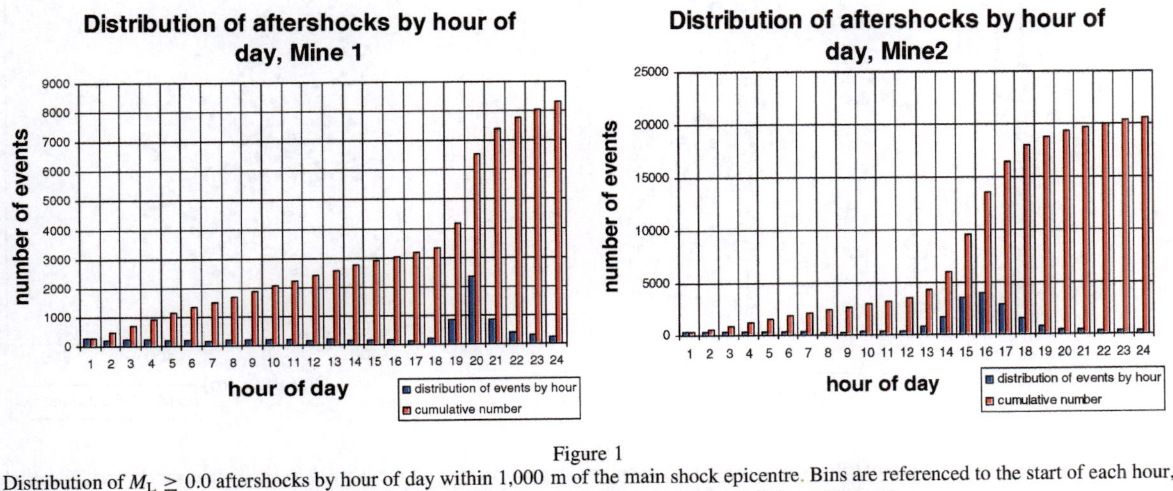

Figure 1
Distribution of $M_L \geq 0.0$ aftershocks by hour of day within 1,000 m of the main shock epicentre. Bins are referenced to the start of each hour, e.g., bin 1 contains the number of events between 00:00:00 and 01:00:00

The number of events classified as aftershocks increases significantly at the onset of blasting time (18h00 at Mine 1 and at 12h00 at Mine 2). Smaller events ($M_L < 0.5$) occur more commonly during blasting time, while the larger events are distributed throughout the 24-h cycle. Main shocks occurring during the blasting period or in the period 4 h prior to the onset of blasting were excluded from the analysis to avoid contamination of the aftershock sequence by the flurry of seismicity triggered by normal production blasting.

3.2. Stacking of Aftershock Sequences

Although larger seismic events in deep-level mines are often followed by an increased rate of seismicity, the number of aftershocks following any single main shock is quite small. Typically there are fewer than five aftershocks of $M_L > 0$ for each $M_L > 2.5$ main shock, which are too few for reliable statistical analysis of event rate decay or frequency–magnitude relationships. This difficulty was overcome by stacking many main-shock–aftershock sequences, using the origin time and epicentre of the main shock as a reference (Fig. 2).

Before routinely implementing the stacking procedure, we first checked whether the aftershock generation process had any dependency on mining conditions. We investigated this by classifying the main shocks into categories based on the calculated strain rate and stress, and the proximity to dykes and faults. We then compared the aftershock productivity in areas with contrasting characteristics: (1) high and low strain rates, (2) high and low stresses, and (3) near to, or distant from, faults and dykes. No significant variation in aftershock productivity was found.

Bin sizes were adjusted to ensure that each bin contained approximately the same number of events. Consequently the bin size grew as the time after the main shock increased, or the distance from the epicentre increased. The aftershock rate was then calculated as a function of time after the main shock and as a function of the distance from the main shock epicentre.

3.3. Analysis Procedure

The following steps were followed to analyse the data:

1. The catalogue consisted of almost 31,000 events (see Table 1). All events with $M_L \geq 2$ were classified as "main shocks" with the exception of those that occurred within the blasting period or

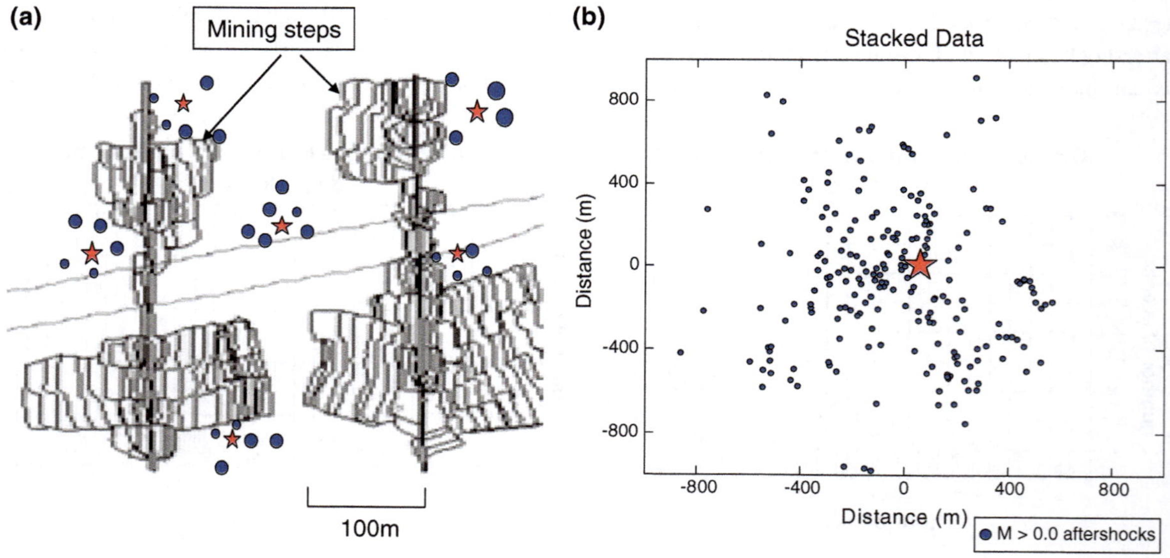

Figure 2
Stacking of different main-shock–aftershock sequences. **a** Cartoon showing a plan view of a mine with several main-shocks and aftershock clusters. The *bands* represent the position of the stope face at monthly intervals. **b** Data stacked at the origin in space

Table 1

Mine seismicity data sets

	M_L range	Mine 1		Mine 2	
Orebody (Reef)		Ventersdorp Contact		Carbon Leader	
Start catalogue:		2003/01/01		1998/01/02	
End catalogue:		2007/11/09		2007/01/08	
Blasting time:		17h00–21h00		12h00–18h00	
No. events in catalogue	$-1 \leq M_L < 4$	10,203		20,611	
Main shocks	$1 \leq M_L < 2$	693		1,184	
	$2 \leq M_L < 3$	72		247	
	$3 \leq M_L < 4$	8		15	
Including/excluding blasting time		Including	Excluding	Including	Excluding
Aftershock analysis 1					
Main shocks	$2 \leq M_L$	80		262	
Aftershocks: occurring prior to next	$0 \leq M_L < 1$	449	300	1,349	498
main shock or within 24 h	$1 \leq M_L < 2$	110	63	443	171
	$2 \leq M_L < 3$	70	5	63	34
	$3 \leq M_L$	3	3	8	4
Aftershock analysis 2					
Main shocks	$1 \leq M_L < 2$	693		1,184	
Aftershocks: occurring prior to next $1 \leq M_L$ main shock or within 4 h	$0 \leq M_L < M_{ms}$	449	70	1,349	498
Aftershock analysis 3					
Main shocks	$2 \leq M_L < 3$	72		247	
Aftershocks: occurring prior to next $2 \leq M_L$ main shock or within 5 h	$0 \leq M_L < M_{ms}$	163	143	350	206
Aftershock analysis 4					
Main shocks	$3 \leq M_L < 4$	8		15	
Aftershocks: occurring prior to next $3 \leq M_L$ main shock or within 11 h	$0 \leq M_L < M_{ms}$	61	51	103	62

All main shocks occurring during blasting time were excluded. Aftershocks were defined as any $M_L \geq 0$ event that occurred within 1,000 m of the main-shock epicentre event during the specified time window

in the four hours prior to the blasting period. These were excluded from the analysis to avoid contaminating the aftershock sequences with the many small events that are triggered by blasting.

2. Events with magnitude $0 \leq M_L < M_{ms}$ occurring within 1,000 m of the main-shock epicenter and prior to the next main shock or within 24 h of the main shock were classified as "aftershocks." Although there undoubtedly are aftershocks with $M_L < 0$, only aftershocks with $M_L \geq 0$ were considered in this study, as smaller events do not pose a significant risk. The aftershock sequences were aligned in space and time.
3. Aftershock sequences were binned and stacked to determine the rate of decay of aftershocks as a function of time and distance. Gutenberg–Richter, modified Omori, Båth's law and aftershock density analyses were conducted for various main-shock and aftershock thresholds. The details of each analysis are described in the respective sections below.

4. Findings

As the results for Mine 1 (Ventersdorp Contact Reef) and Mine 2 (Carbon Leader Reef) are similar, in most cases only one example is shown.

4.1. Gutenberg–Richter Frequency–Magnitude Scaling

The frequency–magnitude statistics for $M_L \geq 0$ events that occurred within 1 h and 1,000 m of $M_L \geq 2$ main shocks were compared with a data set comprised of nearly 3,000 $M_L \geq 0$ seismic events that occurred more than 6 h after an $M_L \geq 2$ main shock. The latter data set were considered unlikely to

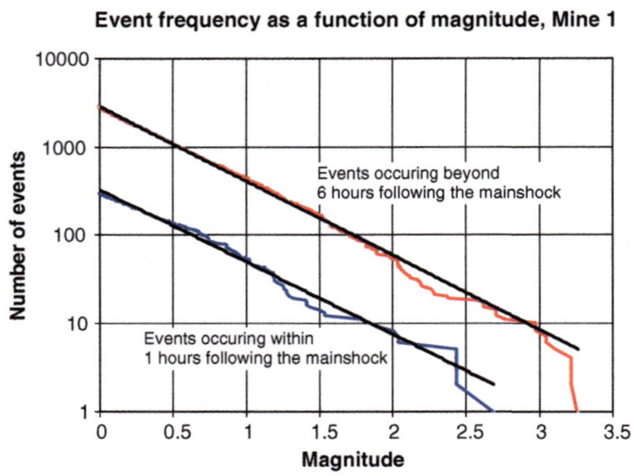

Figure 3
Gutenberg–Richter frequency–magnitude relation for potential aftershocks and background seismicity

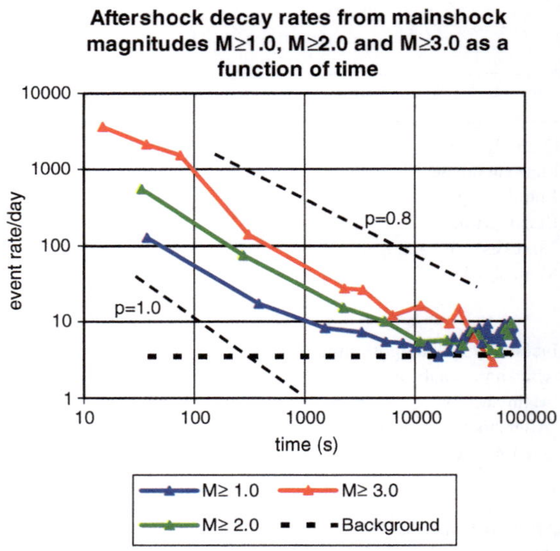

Figure 4
Investigation of the modified Omori law, showing aftershock activity as a function of main shock magnitude. The p value was estimated within the interval 15 s $\leq t \leq$ 83365 s

be aftershocks and taken to represent background seismicity (Fig. 3). Both the aftershock and background frequency–magnitude data are uni-modal and the b values are virtually identical with $b = 0.8$.

4.2. *Modified Omori Analysis*

The decay in the rate of aftershocks with time for three main shock thresholds ($M_L = 1$, 2 and 3) is shown in Fig. 4. The following procedure was used to analyse the data:

1. All events with magnitudes greater than the particular main shock threshold M_L were selected and designated as main shocks. The aftershock sequences corresponding to each main shock threshold were selected. The lower threshold for aftershocks was $M_L = 0$.
2. Bin sizes were adjusted to ensure that each bin contained the same number of events (20 events per bin for Mine 1, 40 events per bin for Mine 2).
3. The event rate for each bin was calculated (expressed in events/day) and plotted at the average time of the events within the bin. The bin size has no effect on the event rate as it is taken into account in the calculation.
4. A curve was fitted to the data to determine the parameters of the modified Omori model.

While larger main shocks were more productive in terms of the number of seismic events, all subsets yielded p values in the range $0.8 < p < 1.0$ (Fig. 4). The lower aftershock threshold magnitude (Fig. 5) and the radius of the 'zone of influence' (Fig. 6) had a negligible effect on the calculated p values. Few aftershocks were recorded at distances greater than 300 m from the main shock.

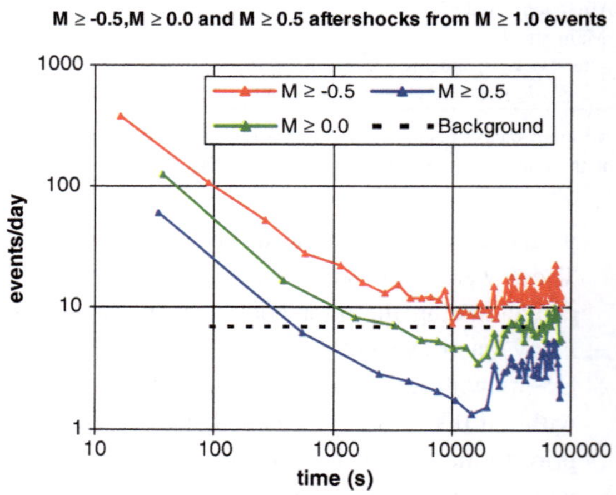

Figure 5
Investigation of the modified Omori law, showing aftershock activity for various lower aftershock threshold magnitudes. The p value was estimated within the interval 16 s $\leq t \leq$ 84407 s

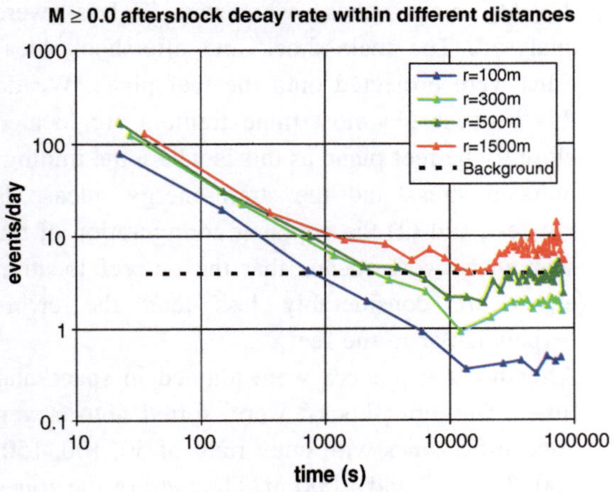

Figure 6
Investigation of the modified Omori law, showing aftershock activity as a function of distance from the main-shock epicentre

The 'time offset' parameter was in the range 1 s < *c* < 20 s (Fig. 7). The *c* value was found to increase with increasing main shock magnitude, as has been found for tectonic earthquakes (SHCHERBAKOV *et al.*, 2004). It was inversely related to the maximum aftershock magnitude. The *c* value is interpreted to be an artefact of the system-triggering logic. It is likely that some aftershocks (especially the smaller ones) that occur soon after a main shock are lost owing to automatic adjustments in the data-adaptive triggering algorithm.

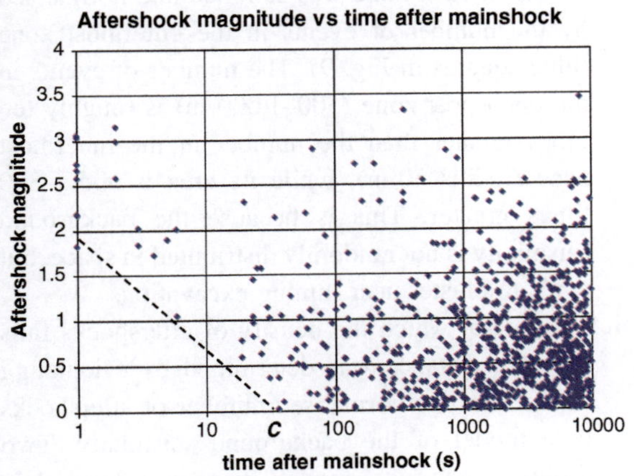

Figure 7
Investigation of the modified Omori law, showing the time offset parameter

4.3. *Båth's Law*

The magnitude of the biggest mine tremor aftershock was found to be, on average, 1.9 magnitude units less than the main shock magnitude (Fig. 8). This difference increases as the main-shock magnitude increases above 3.0. We believe that this is due to the fact that the maximum magnitude of mining-related tremors is controlled by the mining span. The largest events occur when the entire zone subjected to excess shear stress (ESS) experiences slip, leaving relatively little residual ESS to be released in aftershocks.

4.4. *Aftershock Density*

The aftershock density fall-off rate is a parameter that may be used to distinguish between different aftershock triggering mechanisms. The procedure

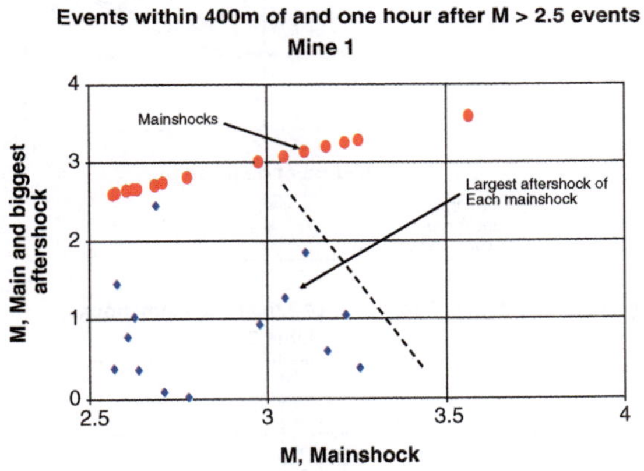

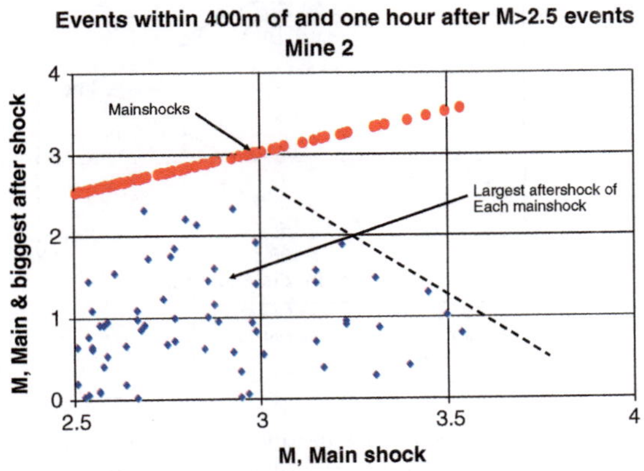

Figure 8
Investigation of Båth's law

used to calculate the fall-off rate is illustrated by means of the example shown in Fig. 9. It should be noted that the data are normalised to the number of events in the outermost zone to facilitate the display. Our primary interest is in the change of density with distance (i.e., the fall off rate) and not the absolute value of the aftershock density.

1. Aftershocks following 246 main shocks and $2 \leq M_L < 3$ main shocks on the Carbon Leader Reef mine occurring within 2 h or prior to the next $2 \leq M_L$ main shock, whichever is less, were analysed. The main-shock and aftershock locations were projected onto the reef plane. We do this because (1) most mine tremors are located close to the reef plane as this is where the mining-induced stress and the strain-energy release is greatest, and (2) the coplanar configuration of the seismic network means that the on-reef location errors are considerably less than the errors perpendicular to the reef.

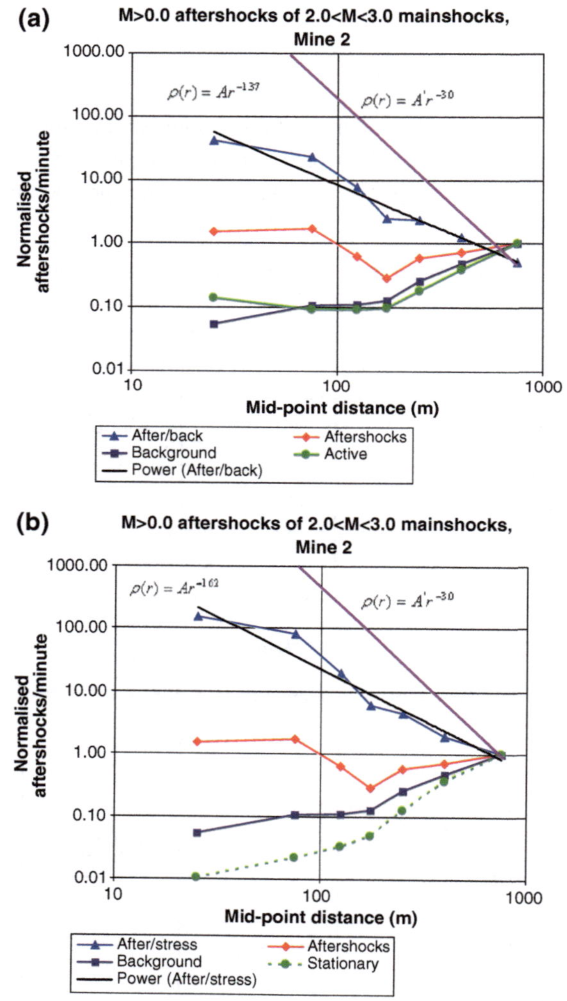

Figure 9
Investigation of the aftershock density as a function of distance from the main shock. The background seismicity is modelled as a function of (**a**) energy release rate or "active mining" and (**b**) stress or "stationary mining." Details of the analysis are given in the text

2. Aftershock sequences were aligned in space and time. The aftershocks were sorted into seven concentric zones with outer radii of 50, 100, 150, 200, 300, 500 and 1,000 m. The radii of the zones were chosen to give approximately equal numbers of aftershocks in each zone. The number of aftershocks in each zone was summed and normalised by the number of events occurring in the outermost zone (500–1,000 m). While the number of events in each zone (red diamonds in Fig. 9) is roughly the same, it should be noted that the density of aftershocks decreases with distance, as the area of each zone increases with distance from the epicentre.
3. The background seismic activity was calculated by considering the number of events within 1,000 m of the actual location of each main shock during a 7 day period of normal mining activity. The background events were aligned in space and time and sorted into zones with the same radii as were used to bin the aftershocks. The number of events in each zone was summed and normalised by the number of events in the outermost zone (blue squares in Fig. 9). The number of events in the outermost zone (500–1,000 m) is roughly ten times greater than the number in the innermost zone (radius 50 m), while its area is about 300 times greater. This is because the background seismicity is not randomly distributed in space, but tends to cluster near mining excavations.
4. The rate at which the density of aftershocks falls off with distance was determined by calculating the ratio of the normalised number of aftershocks to a model of the background seismicity. Two different numerical models were used to explain the background seismicity (Spottiswoode *et al.*, 2008).

(a) The "active" model considers the strain energy released by active mining to be the principal driver of both main- and aftershocks. An area is defined as "active" if mining took place in the month prior to the main shock. The normalised number of events in each zone predicted by this model is remarkably similar to the observed background seismicity (green circles and blue squares in Fig. 9a, respectively). A curve (black line in Fig. 9a) was fitted to the aftershock/background ratio (blue triangles in Fig. 9a), showing a constant fall-off as $r^{-1.37}$. Also shown in Fig. 9a is a line showing the fall-off of aftershock density as r^{-3}, which would have been followed if the triggering was due to quasi-static stresses.

(b) The "stationary" model considers the possibility that aftershocks may also be triggered in the stressed ground around old mining faces. On-reef stress in excess of 300 MPa was considered to be sufficiently high for the rock to be in a state of incipient failure. The normalised number of events in each zone predicted by this model is considerably smaller than the observed background seismicity (green circles and blue squares in Fig. 9b, respectively). A curve (black line in Fig. 9b) was fitted to the aftershock/"stationary" ratio (blue triangles in Fig. 9b). A fit to the data shows a constant fall-off as $r^{-1.62}$.

Felzer and Brodsky (2006) found that that the density of earthquake aftershocks falls off with the distance from the main shock as $r^{-1.3}$, which is comparable to the decay of the maximum seismic wave amplitude (a proxy for dynamic stresses). We obtain a similar fall-off rate for mine tremors. Therefore our analysis, like that of Felzer and Brodsky (2006), supports the view that most aftershocks are triggered by dynamic stresses (which is proportional to the amplitude of radiated seismic wave) rather than by quasi-static stress redistribution, even though many aftershocks do not occur at the time when the shaking occurs. Furthermore, it supports the conclusion by Spottiswoode *et al.*, (2008) that seismicity is driven primarily by the strain energy changes due to active mining, rather than by high stresses, which are found in both areas of active and past mining.

5. *Hazard Associated with Aftershocks*

The ultimate objective of this investigation is to provide guidelines governing: (1) the evacuation of workers following a main shock during a working shift, and (2) the entry of workers following a main shock during blasting time. The procedures described above can be used to determine objectively and quantitatively the period following a large seismic event during which people should be excluded from working areas.

Our results and those of Spottiswoode (2000) demonstrate that the modified Omori model is applicable to mine seismic data and can be used to estimate the hazard posed by aftershocks. It is evident that the probability of a seismic event with the potential to cause injury increases significantly immediately after a moderate-sized event, say $M_L > 2$. The number of aftershocks falls off with time, reaching background levels after 4–10 h, depending on the main-shock magnitude.

Guidelines for the evacuation of and/or re-entry to working places are summarized in Table 2. The time periods for which the aftershock event rate has reached 10× and 3× the background rate were determined from the Omori decay curve of

Table 2

Time period for the aftershock event rate to reach 10× and 3× the background rate

$M_{main\ shock}$	10× (mins)		3× (mins)	
	0–200 m	200–400 m	0–200 m	200–400 m
Mine 1: VCR mine				
$1 \le M_L < 2$	5	2	24	30
$2 \le M_L < 3$	16	5	180	162
$3 \le M_L < 4$	48	60	300	(480)
Mine 2: CLR mine				
$1 \le M_L < 2$	3	2	16	(78)
$2 \le M_L < 3$	10	3	48	15
$3 \le M_L < 4$	60	24	120	(120)

Time is given in minutes and distance in meters from the main-shock epicentre. The times given in brackets are subject to considerable uncertainty owing to a small number of events

Reprinted from the journal

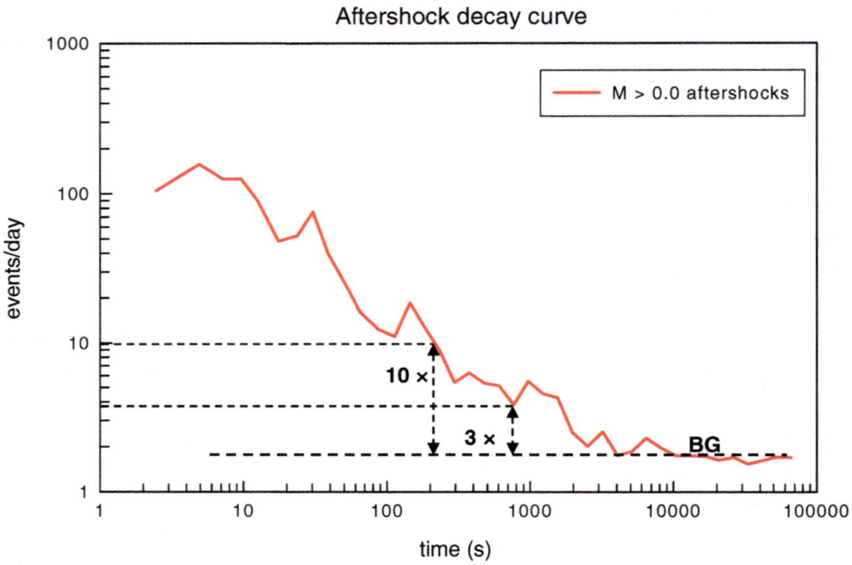

Figure 10
Use of the Omori decay curve to determine the time taken for seismic hazard to decrease to acceptable levels. 10× and 3× represent the time taken for the aftershock rate to decrease to ten times and three times the background rate

aftershocks (Fig. 10). As an example, in case of an occurrence of a $3.0 \leq M_L < 4.0$ main shock at the VCR mine, it takes about 48 min for aftershock activity to decreased to 10× the background level.

6. Conclusions

Aftershocks of tremors occurring in two deep gold mines in the Carletonville area were found to satisfy Gutenberg–Richter scaling with a b value close to 1. Aftershock activity diminished with time in accord with the modified Omori law, with p values close to 1. However, the difference between the magnitude of the main shock and its biggest aftershock was not in accordance with Båths law for earthquakes. The difference was to average 1.9 and increasing for main shocks with $M_L > 3$. The aftershock density was found to fall-off with distance as $r^{-1.3}$, suggesting triggering by dynamic stresses and not simply quasi-static stress changes.

The procedures described here can be applied to different mines, and used to determine the period for which people should be excluded from working areas should a large seismic event occur. The period for which exposure should be reduced depends on the level of risk that is considered to be tolerable. The definition of "acceptable risk" is a necessary and significant consideration in any mining project or operation and is inextricably linked to ethics. This requires input from executive management, government, labour, and civil society

Acknowledgments

This study forms part of project SIM 050302 "Minimising the rockburst risk" sponsored by the Mine Health and Safety Council (MHSC). Mine management is thanked for providing the seismic data.

References

Båth, M. (1965), *Lateral inhomogeneities in the upper mantle*, Tectonophysics *2*, 483–514.

Ebrahim-Trollope, R. (2001), *Gutenberg–Richter relationship and mine-induced seismicity as observed at the African Rainbow Minerals mines—Klerksdorp*, *Proc. Fifth Int. Symp. Rockbursts and Seismicity in Mines* (Van Aswegen, G., Durrheim, R.J., and Ortlepp, W.D., eds), The South African Institute of Mining and Metallurgy, Symposium Series *S27*, pp. 501–508.

Felzer, K.R. and Brodsky, E.E. (2006), *Decay of aftershock density with distance indicates triggering by dynamic stresses*, Nature *441*(8), 735–738.

FROHLICH, C. and DAVIS, S.D. (1993), *Teleseismic b values—or, much ado about 1.0*, J. Geophys. Res. *98*. B1, 631–644.

JAGER, A.J. and RYDER, J.A., *A Handbook on Rock Engineering Practice for Tabular Hard Rock Mines* (SIMRAC, Johannesburg, 1999).

LEGGE, N. and SPOTTISWOODE, S.M. (1987), *Fracturing and microseismicity ahead of a deep gold mine stope in the pre-remnant and remnant stages of mining. Proc. Sixth Congress of the Int. Soc. Rock Mechanics* (G. Herget and S. Vongpaisal, eds) (A. A. Balkema, Rotterdam, The Netherlands, 1987) pp. 1071–1077.

NANJO, K., NAGAHAMA, H., and SATOMURA, M. (1998), *Rates of aftershock decay and the fractal structure of active fault systems*, Tectonophysics *287*, 173–186.

RICHARDSON, E. and JORDAN, T.H. (2002), *Seismicity in deep gold mines of South Africa: implications for tectonic earthquakes*. Bull. Seismol. Soc. Am. *92*, 1766–1782.

SHCHERBAKOV, R., TURCOTTE, D.L., and RUNDLE, J.B. (2005), *Aftershock statistics*, Pure Appl. Geophys. *165*, 1051–1076.

SHCHERBAKOV, R., TURCOTTE, D.L., and RUNDLE, J.B. (2004), *A generalized Omori's law for earthquake aftershock decay*, Geophys. Res. Lett. *31*, L11613, doi:10.1029/2004GL019808.

SHCHERBAKOV, R., TURCOTTE, D.L., and RUNDLE, J.B. (2006), *Scaling properties of the Parkfield aftershock sequence*, Bull. Seismol. Soc. Am. *96*, S376–S384.

SPOTTISWOODE, S.M. (2000), *Aftershocks and foreshocks of mine seismic events, Proc. Third Int. Workshop on the Application of Geophysics to Rock and Soil Eng.*, Melbourne, Australia, November 2000.

SPOTTISWOODE, S.M. and LINZER, L.M. (2004), *Improved seismic locations and location techniques*, SIM 020304 Final Project Report (Appendix 1), Mine Health and Safety Council, Johannesburg, South Africa.

SPOTTISWOODE, S.M., LINZER, L.M., and MAJIET, S. (2008), *Energy and stiffness of mine models and seismicity*. In *SHIRMS 2008* (Y. Potvin, J. Carter, A. Dyskin, R. Jeffrey, eds), Australian Centre for Geomechanics, Perth, Australia, vol. 1, pp. 693–707.

TURCOTTE, D.L., NEWMAN, W.I., and GABRIELOV, A. (2000), *A statistical physics approach to earthquakes*. In *Geocomplexity and the Physics of Earthquakes* (J.B. Rundel, D.L. Turcotte and W. Klein, eds), Geophysical Monograph *120*, American Geophysical Union, Washington, DC, U.S.A., pp. 83–96.

(Received September 26, 2008, revised April 9, 2009, accepted April 21, 2009, Published online December 22, 2009)

Pure Appl. Geophys. 167 (2010), 119–129

DOI 10.1007/s00024-009-0007-2

Pure and Applied Geophysics

Passive Seismic Monitoring of Mine-scale Geothermal Activity: A Trial at Lihir Open Pit Mine

X. Luo,[1] A. Creighton,[2] and J. Gough[3]

Abstract—The Lihir open pit mine in Papua New Guinea is located inside an old volcano where geothermal activity is strongly present. Outbursts of hot water and steam into the mining areas were a major safety concern. Passive seismic monitoring was carried out at the mine to investigate whether the geothermal activities could be detected and located using microseismic techniques in a mining environment. In this trial, sixteen triaxial geophones which can withstand temperature up to 200°C were used and installed in four deep boreholes inside the pit. The microseismic events were discriminated using the STA/LTA triggering criterion. During 6 weeks of monitoring, more than 17,000 events were recorded. Approximately 12% of the events showed harmonic vibration characteristics similar to those observed in other geothermal and volcanic areas, suggesting that the geothermal activity inside the pit was captured by the microseismic monitoring system. More than 75% of the events present both P and S waves and they were interpreted to be associated with rock fracturing due to stress release near the bottom of the pit. Many geothermal-type events were located in areas where shear events occurred, implying that the detected geothermal events were not far from the mining area below the pit and they may also be associated with mining. The borehole installation of the geophones significantly reduced the interference of mining noise and achieved good observation of the seismic events. However, equipment installation requires great attention as the geophones may be destroyed due to unexpected rising temperature within the boreholes.

Key words: Passive seismic, geothermal activity, minescale, harmonic events.

1. Introduction

Lihir Island is about 700 km northeast of the mainland of Papua New Guinea and is a volcanic sea mount of about 320 km^2 that rises from the sea floor to approximately 600 m above sea level. The gold deposits on Lihir are located in the heart of the Luise Caldera and the Lihir open pit is the first mine in the world that is inside the hot area of the collapsed crater of a dormant volcano. The presence of hot springs and fumaroles in the Luise Caldera indicates persistent geothermal activity in the mining region. Studies (White *et al.*, 2003) have indicated high temperatures (120–250°C) and a high geothermal pressure underneath the mining areas.

Gold production at Lihir commenced in 1997 and by 2002 more than 200 Mt of ore and waste rocks had been removed from the mining area. The bottom of the pit had reached more than 100 m below the sea level. The removal of overburden has significantly released the confining stresses to the underlying geothermal sources and this has resulted in a number of steam outbursts occurring within the pit. The pit is planned to ultimately reach more than 200 m below sea level. With the mining advancing into deeper levels and hotter areas, sudden gas and steam outbursts into the pit may occur if no cooling and depressurisation of the geothermal source associated with the gold mineralisation is conducted.

The occurrence of seismic events associated with geothermal/fluid activity has been widely observed at many geothermal and volcanic locations around the world (Aki *et al.*, 1977; Chouet, 1988; Feng and Lees, 1998). Previous studies have suggested that seismic monitoring is one of the most reliable and diagnostic techniques in mapping geothermal activities and gas migration and predicting impending outbursts or volcanic eruptions (McNutt, 1996; Bryan and Sherburn, 1999; Chouet, 2003). The geothermal activities and the overburden stress release in the Lihir open pit were expected to

[1] CSIRO Exploration and Mining, Brisbane, Australia. E-mail: xun.luo@csiro.au
[2] Rio Tinto Technical Services, Brisbane, Australia.
[3] Insite Geology Pty Ltd, Samford Valley, Australia.

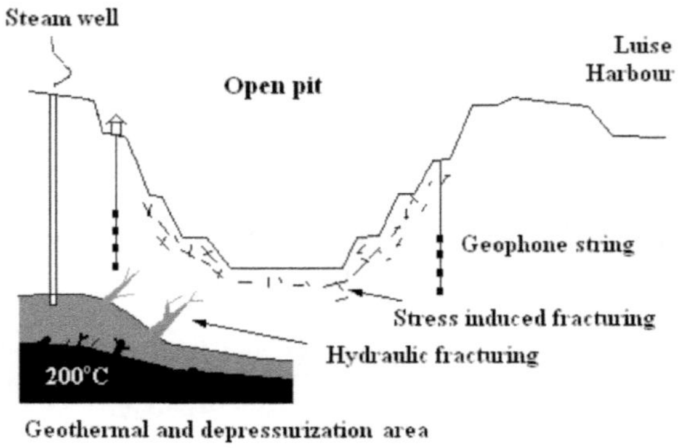

Figure 1
Sketch of passive seismic monitoring of seismic events associated with geothermal and mining activities

generate seismic events (Fig. 1), which may be observed using passive seismic methods for the prediction of hot water and steam outbursts. The mine intended to use microseismic monitoring as a tool for geothermal risk assessment for mining production.

No microseismic monitoring work was carried out previously for detecting geothermal activities at a small scale and in a mining environment. In 2003, an experiment was conducted at Lihir to investigate the applicability of microseismic techniques for seismic events associated with geothermal activity and rock geomechanic fracturing (Luo, 2004). It was expected that the results of the trial would help the mine to make a permanent microseismic monitoring system for mine geothermal risk management.

2. *Field Monitoring*

2.1. *Monitoring Plan and Geophone Installation*

At the time of the trial, the size of the pit was more than 1,000 m in diameter (Fig. 2). Ground surface and borehole temperature measurements showed that the west wall is considerably hotter than the east wall. Two microseismic arrays were then designed and deployed at the two walls in order to measure any difference in seismic characteristics.

At the west wall, boreholes A and B were drilled vertically from the pit bench to a depth of 210 m. The temperatures at the bases of boreholes A and B were measured as 120 and 130°C, respectively, before geophone installation. Another two boreholes, C and D were situated at the east wall and their temperatures were 99 and 72°C, respectively. All of the temperatures were well below the geophone working temperature, rated at 200°C.

The purpose of the borehole installation was to reduce the interference of mining noise on the seismic signals and enable small events associated with geothermal activity and rock fracturing to be detected. The borehole installation was found to be necessary after a brief microseismic experiment in the pit using seismic sensors at the ground surface; during the experiment, good evidence of microseismic events was not captured, probably because of strong ambient noise near the ground surface and seismic attenuation near the ground surface of the pit.

In each of the boreholes, four triaxial geophones spaced 15 m apart were lowered down and cemented in place to ensure good coupling between the geophones and ground (Fig. 3). The use of triaxial sensors allowed a complete seismic waveform to be recorded, which is very helpful for a detailed analysis of seismic characteristics. In addition, on the three geophone components, more accurate P- and S-wave arrival times and seismic ray directions can be determined for seismic event location. After the geophone installation, calibration shots were fired into the arrays to determine the geophone orientations and a seismic velocity model for event location.

2.2. *Microseismic System*

The microseismic data acquisition system used in this trial was built by the Commonwealth Scientific and Industrial Research Organisation (CSIRO). The

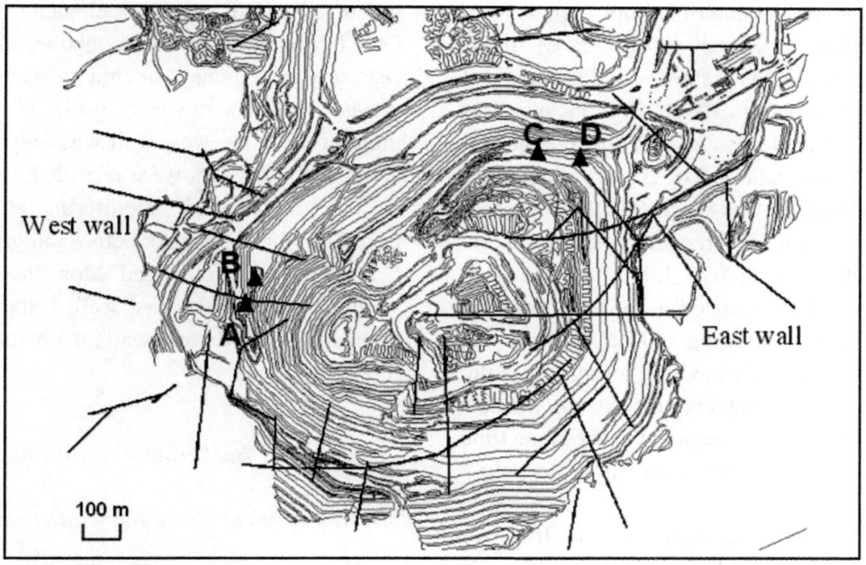

Figure 2
Plan view of Lihir open pit and geophone borehole locations (*triangles*). Boreholes A and B are located at the west wall and C and D are located at the east wall. The *red line* denotes shear structure or fault (*dark straight lines*)

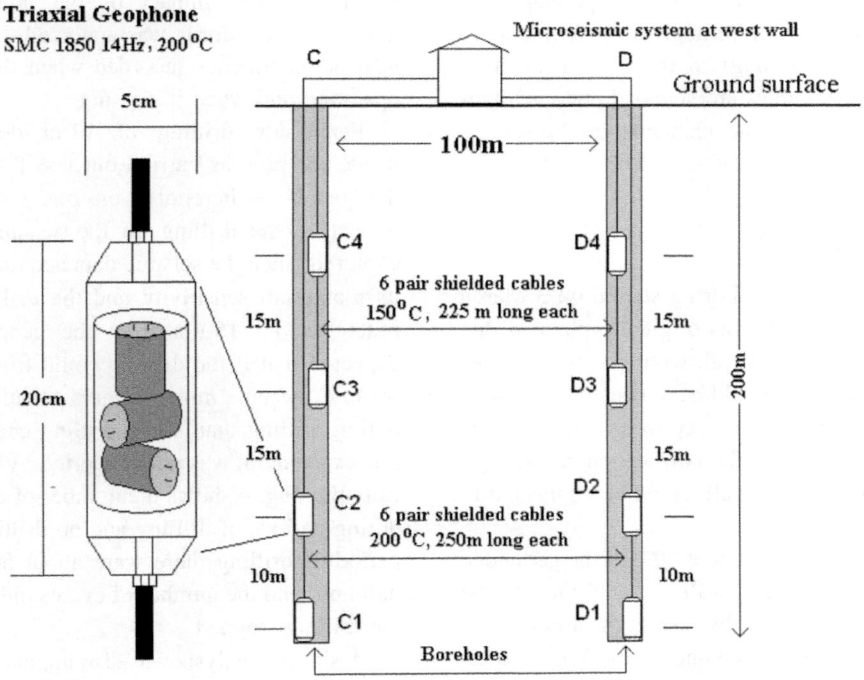

Figure 3
Triaxial geophone construction and geophone strings in the boreholes at the east wall. The geophone were cemented in the ground after the installation

system allows fast processing of seismic signals from as many as 30 triaxial sensors (90 channels) simultaneously. The system has a signal-conditioning unit consisting of preamplifiers (0–60 dB), an anti-alias filter and a 12-bit analog to digital conversion unit containing additional gain of 0 or 20 dB. In this trial, the signal amplification was set to 60 dB for each of the channels, the sampling frequency was 1,000 Hz and recording data length was 1,000 ms. The digitized signals were discriminated using the short term averaging/long term averaging (STA/LTA) method (ALLEN, 1978) with a predefined triggering threshold. Once a geophone signal triggered the system, the data of this geophone and those recorded at the same time frame by the other geophones were all saved onto a hard disk.

The triaxial geophones used in this trial are constructed using the GeoSpace SMC-1850-14 Hz single component sensors which can sustain temperatures up to 200°C. Inside each of the geophones a thermocouple sensor was installed to obtain the temperature changes to the geophone after the installation. The signals from the geophones were continuously transmitted through high temperature cables (up to 700 m long) to the data acquisition system. The signal cables are shielded and each pair of wires for the geophone element was twisted to reduce external electrical interference.

2.3. *Field Monitoring*

The microseismic monitoring started on 3 March 2003 at the east wall. It was originally planned that the monitoring at the east wall would be for 3 weeks after which the system would be relocated to the west wall to monitor for another three weeks. At any time during the monitoring only two geophone strings, either at the east or west wall, could be connected to the system.

However, on 6 March, about 80% of the geophone components at the west wall were found to be defunctional, apparently due to high temperature water ingress into the geophone cases. The temperatures in boreholes A and B unexpectedly increased to more than 180°C. A decision was made to disconnect the geophone strings at the east wall and relocate the microseismic system to the west wall before all of the geophone elements were damaged. On 11 March, all of the geophones at the west wall ceased working and the microseismic system was moved back to the east wall. The microseismic monitoring at the west wall was only for 4 days and about 4,000 triggers were recorded.

The microseismic monitoring at the east wall lasted for about 5 weeks before the geophones inside C and D were all damaged. More than 13,000 events were recorded at the east wall. Data collected from both the east and west walls were sufficient for this feasibility study.

3. *Background Noise Analysis*

3.1. *Background Noise Associated with Production*

Mining activity in the pit was carried out for nearly 24 h each day. It was a concern that the monitoring system might be falsely triggered by the noise sources originating from blast-hole drilling, heavy truck traffic, shoveling and dewatering pump operations. The impact of the noise sources on microseismic records was analyzed using a comparison of seismicities recorded when the mine was in operation and when it was not.

Percussion drilling of 12 m deep blast-holes inside the pit was carried out less than 300 m from the geophone boreholes on many occasions. The impact of the drilling on the seismic records was explored when the seismic data acquisition was set to its maximum sensitivity and the drilling was intermittent about 150 m from the geophones. It was expected that if the drilling could trigger the microseismic system, more triggers should be observed during drilling than when drilling ceased. However, this expectation was not supported by the results. For example, Fig. 4 shows the number of events recorded during periods of drilling and no drilling. During the period of drilling there were about 50 events every half hour and the number of events did not drop when the drilling stopped.

A similar analysis was also applied to the moving of fully loaded trucks, shovel operations and dewatering pumping near the geophone boreholes. Again, we did not find a direct correlation of these activities with the instrument triggering. These results

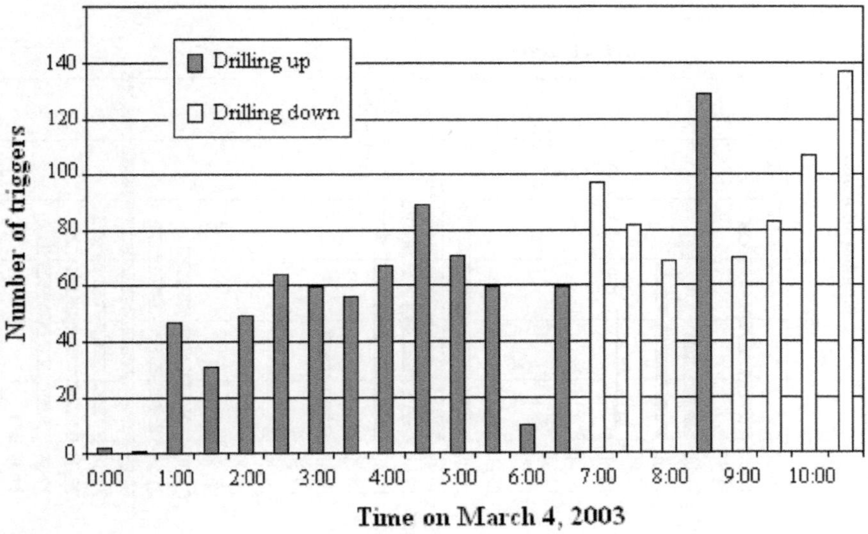

Figure 4

Number of seismic events recorded on 4 March when the production drilling was 150 m to the geophones. The *blue bar* shows the triggers recorded during drilling and the other is for the triggers recorded when drilling was stopped

suggested that the mining activities in the pit had an insignificant impact on seismic system triggering.

3.2. *Impact of Blasts and Earthquakes*

Production blasts were fired nearly every day in the pit. The blasts were arranged in a group (150–250) of 12-m shallow holes and fired sequentially within a short period of time. The total charge for every group of blasts varies from 5,000 to 25,000 kg. This results in 11,000–70,000 m^3 of overburden rocks being blasted. Rock fractures should be generated after the blasts due to the sudden change of the confining stress. However, no obvious increase in seismicity was recorded soon after the blasts. Figure 5 shows the number of events recorded on 11 March before and after the production blasts which were fired approximately 150 m from the geophone boreholes C and D. The lack of microseismicity immediately after the blasts was below our expectation. It is perhaps that the background noise level after the blasts may be high and any possible after-events were too small to trigger the presence of a large background noise level, based on the predefined STA/LTA threshold.

Lihir Island is located near a subduction zone. Regional and local earthquakes are frequently felt in the pit. Figure 5 also shows the seismicity after the strong Rabaul earthquake (ML 6.8) that occurred on the same day about 200 km from the pit. There is no evidence of increasing seismicity immediately after the earthquake.

4. *Seismic Event Characterization*

The microseismic events recorded at Lihir mine can be mainly divided into the following groups according to their characteristics.

4.1. *Shear Events*

This type of event presents strong P- and S-wave arrivals (Fig. 6). They have high frequency characteristics (up to 200 Hz). The strong shear-wave component suggests that the event source is associated with brittle shear fracturing. Most of the shear events triggered the top geophones first, suggesting that they are located near and above the top geophones and close to the bottom of the pit. This type of event accounts for about 70% of the data set.

The shear type events have been widely observed in volcano and geothermal areas (Chouet, 1996). They were interpreted to be caused by rock shear

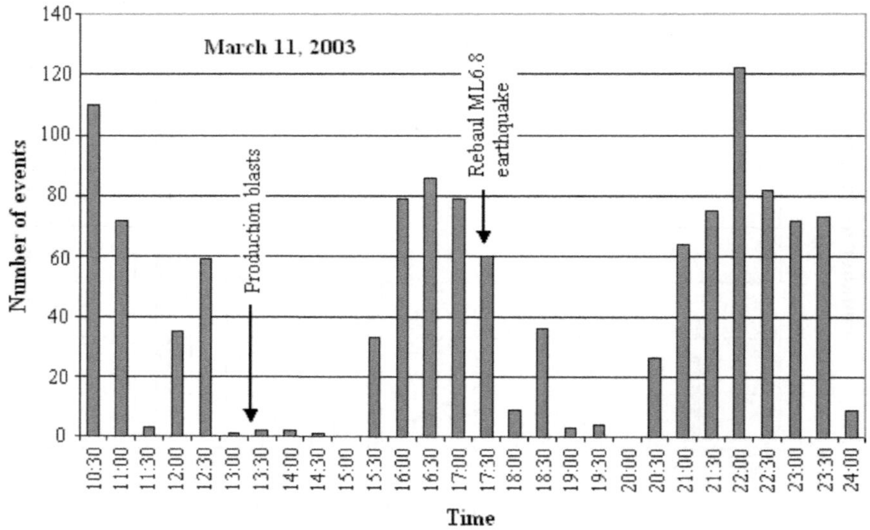

Figure 5
The number of seismic events in 30 min periods recorded from 10:30 a.m. to 14:00 p.m. on 11 March, 2003 at the east wall. The time of production blasts and the arrival time of the ML 6.8 Rabaul earthquake at the pit are indicated in the chat

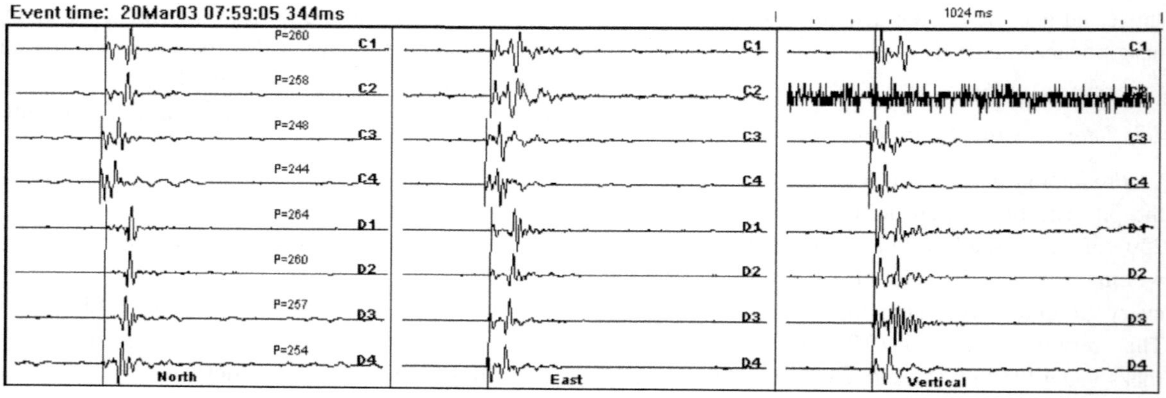

Figure 6
Waveforms of the north, east and vertical components of a shear event recorded by the geophones in boreholes at the east wall. C4 and D4 are the top geophones in boreholes C and D, respectively

fractures or slip on faults. Most of the shear-type events observed at Lihir are considered to be associated with rock fracturing caused by stress re-adjustment near the edge and bottom of the pit, similar to the events detected near the bottom edge of highwall pit (Luo *et al.*, 2001).

4.2. Harmonic Events

Other events observed at both the east and west walls are characterized by a dominant oscillation frequency and unclear S-wave phases. They are classified as harmonic events. The harmonic events observed at Lihir exhibit characteristics similar to the harmonic events widely observed in volcanic and geothermal areas. They were interpreted to be in relation to magmatic and hydrothermal activities (Chouet, 1996; Julian, 1994). This suggests the existence of gas and geothermal activities in the Lihir pit.

The majority of harmonic events recorded at Lihir are harmonic tremors which present a group of

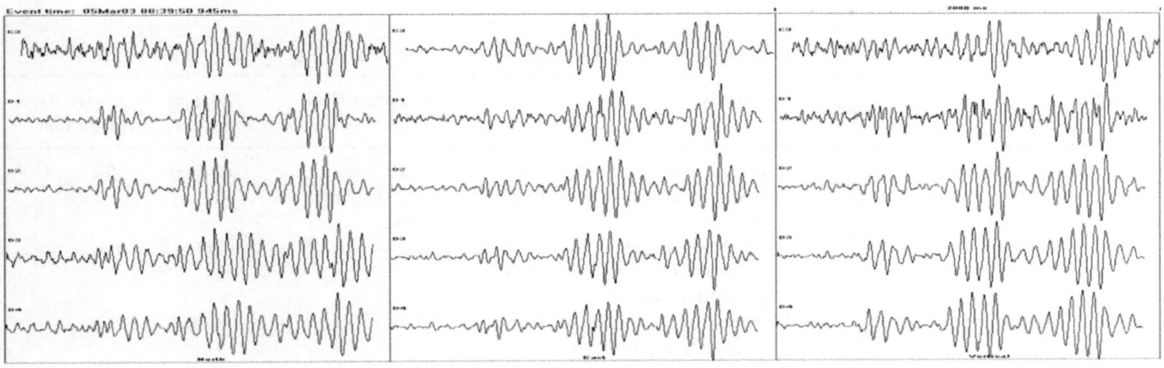

Figure 7
Harmonic tremors recorded at the east wall on 5 March 2003

harmonic events occurring successively. The tremors contain frequencies varying from 40 to 60 Hz with a duration exceeding 600 ms. The P-wave arrivals are very emergent and no S phases can be identified. Figure 7 shows typical harmonic tremor events recorded at Lihir. They account for 10% of the data set. Harmonic tremors have been widely recorded in volcanic areas (McNutt, 1994; Chouet, 2003). Many investigations have shown that the harmonic tremor has a certain correlation to the movement of magma, liquid and gas inside conduits and cracks (Tolstoy *et al.*, 2002) and hence can be used to predict impending volcanic eruptions.

There are several other types of events characterized. They only account for a small percentage in the data set. Table 1 lists the categories of all the seismic events recorded in this trial.

5. *Event location*

Understanding the locations of the seismic events is important for identifying potential hazard areas within the pit. In this trial, as at any time during the monitoring only geophones in two boreholes, either at the west (A and B) or east (C and D) wall could be connected to the system. The location of an event in plan view must then be determined using both seismic arrival times and ray directions. However, the majority of the seismic events could not be located accurately because they were either recorded by less than two geophone stations or have a very poor signal-to-noise ratio, so reliable arrival-time picking could not be made. In order to obtain information about the distribution of the recorded events we developed two strategies for determining event locations: accurate event location and event location estimation.

5.1. Accurate Event Location

Accurate event locations were determined only for seismic events that were recorded by at least four geophone stations in two boreholes and for which P-wave arrival times could be reliably obtained. Of the geophones, at least one had good waveforms on the horizontal components from which the seismic ray azimuth can be determined. There are more than 3,000 events in this category. In this study only 300 selected events from 13 March to 10 April recorded at the east wall, were located.

5.1.1 Velocity Model

The velocity model near the east wall was determined based on the P-wave arrival times of a calibration shot recorded at the geophone locations. The calibration shot was placed directly above borehole C. This shot allowed the average velocity between every two geophones to be approximately determined using

$$V_j = \frac{z_{j-1} - z_j}{t_{j-1} - t_j}, \tag{1}$$

where z_j is the elevation and t_j is the seismic arrival time of geophone j, respectively. The resolution of the velocity structure is dependent on the interval

Table 1

Classification of seismic events recorded at Lihir mine

Event type		Description	Frequency (Hz)	Duration (ms)	Approx proportion
Shear	Single	Strong S wave, impulsive P- and S-wave arrivals, high frequency, short duration	30–70, some can be as high as 200	200–400	70%
	Multiple	Strong S waves, impulsive P- and S-wave arrivals, high frequency, occurred at same location	30–80	>600	5%
Harmonic	Decaying type	Regular decay in amplitude, longer duration, low frequency	30–50	300–600	<1%
	Equal amplitude	Equal amplitude, short duration, low frequency	40–60	200–400	<1%
	Wave packet	Smooth increasing and dying amplitude, emergent P- wave arrivals, no S wave, low frequency	30–40	300–600	1%
	Tremor	Successive events, no clear P- and S-wave arrivals, long duration	40–60	>600	10%
Hybrid	Shear and low frequency	High frequency shear event followed by a low frequency event	30–70	>500	1%
	Shear and harmonic	Shear event followed by a high frequency harmonic event	40–60	>500	<1%
	Shear and tremor	Shear event followed by a harmonic tremor	40–60	>700	<1%
Low frequency		Very low frequency, no S-wave	15–30	>600	1%
Small earthquake		Clear P- and strong S-wave	50–100	>1,000	<0.1%
Blast		Occurred at specific time, long duration, waveform saturation	15–50	>1,000	About 50 blasts
Unknown		Noisy signals			10%

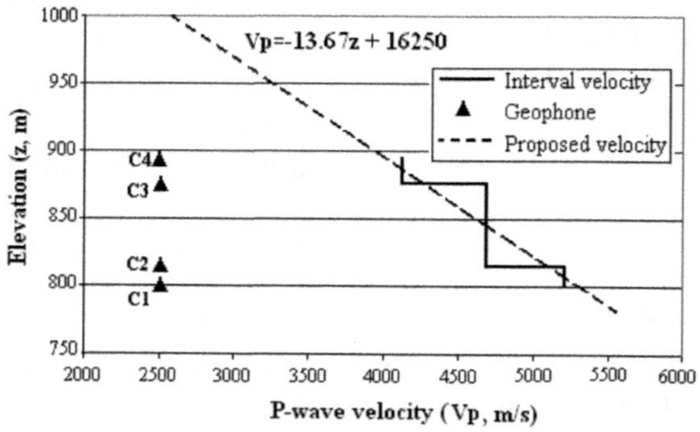

Figure 8
Seismic velocity model obtained using the P-wave arrival times of a calibration shot recorded in borehole C. The *vertical* gradient velocity model (*dashed line*) smoothed from the interval velocities (*solid line*) was used for event location

distance between the geophones. The interval velocities obtained clearly show that seismic velocity increases with depth (Fig. 8). As the drilling logs do not show a layered structure in the pit, the P-wave velocity structure is approximated using a vertical gradient velocity model for event location

$$V_p(z) = \alpha z + b = -13.67z + 16,250\,(\mathrm{m/s}) \quad (2)$$

where z denotes the elevation and $700\ \mathrm{m} < z < 980\ \mathrm{m}$ (ground surface elevation).

5.1.2 Location Methods

The area for the event location was defined as east: 9,660–9,900 m, north: 3,800–4,040 m and elevation: 750–950 m. In this velocity model, the seismic ray travels in a circular path and the P-wave travel time, t_{ij}, from a source (x_i, y_i, z_i) to a geophone station (x_j, y_j, z_j), can be calculated analytically. A grid search method was applied on the gradient velocity model, which calculates the arrival-time residuals for each of the assumed location point. The grid point that has the minimum residual value was taken as the event location:

$$\min \sum \left| t_{ij}^{cal}(V_p) - t_{ij}^{abs} \right|. \quad (3)$$

5.1.3 Event locations

Using geophones in two boreholes and the vertical ingredient velocity model for event location caused a problem of double solutions, which have the identical minimum time residual. This multiple solution problem was solved using seismic ray azimuths. For this velocity model a ray path in plan view is a straight line between the source and geophone station. A ray path was determined using the polarities of the P waves registered at the east and north geophone components at a geophone site. Only the solution that aligns in the direction of ray paths was considered to be the true location. Figure 9 shows the locations of the 300 events using the location methods described previously. It is estimated that the location errors of the events should be no greater than 30 m.

As seen in Fig. 9, most of the located events concentrate along the inner edge of the pit and less than 500 m from the seismic network. Many of the events to the south of the boreholes at the east wall appear to be located near the geological structures.

The events shown in Fig. 10 are shear, harmonic and unknown type events recorded on 24 March. There are no distinct areas for the shear and harmonic types of events. Their locations are mixed-up in general. The deepest located event is about 80 m from the bottom of the pit. Therefore, the located events near the east wall all are close to the bottom level of the pit.

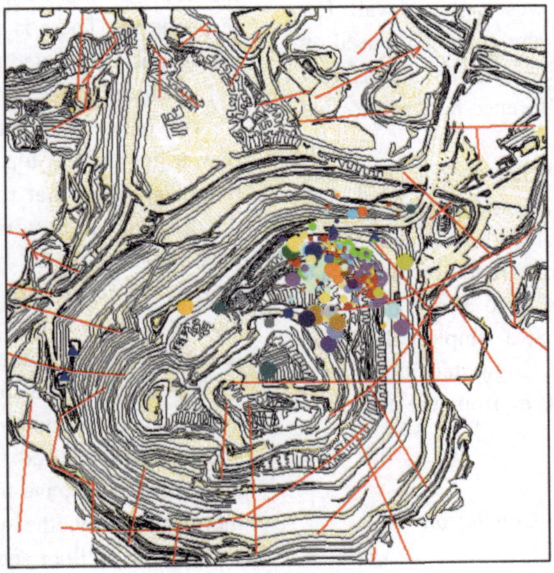

Figure 9
Plan view of the locations of the 300 seismic events (*coloured circles*) recorded at the east wall. The events are coloured on a daily basis from 13 March to 11 April 2003. The size of the *circles* represents the relative intensity of the events. The *red lines* are identified as geological structures

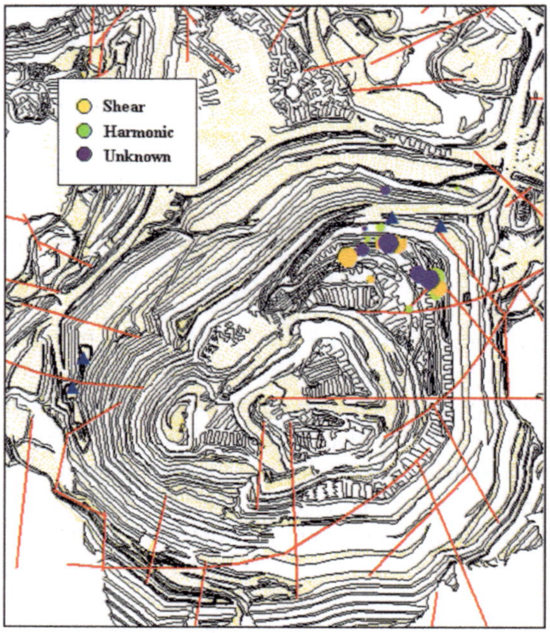

Figure 10
Plan view of seismic locations on 24 March 2003 with shear, harmonic and unknown events marked in different *colours*

5.2. Estimation of Event Location

The majority of events triggered geophones with incomplete triaxial components, due to the damage of geophone components at the west and east wall. For these events, their locations were estimated based on the sequence of seismic wave arrivals at different geophone stations and the difference between P- and S-wave arrival times.

The majority of the events were weak and triggered the top geophone stations first. This implies that these events occurred at elevations where the top geophones are located. The difference of the P- and S-arrival times varied from 2 to 6 ms for some events. Assuming $V_p/V_s = 1.8$, a simple calculation on the hypocenter distance suggested that these events were about 500–3,000 m from the triggered geophones.

6. Discussions and Conclusions

The open pit development and geothermal activity at Lihir create a unique environment for the generation of different types of microseismic events. The microseismic experiment at the mine detected more than 17,000 seismic events of different characteristics. The majority of the recorded events are shear and harmonic type events. They account for approximately 75 and 12% of the data set (Table 1), respectively. There is no evidence that mining noise significantly impacted the seismic acquisition carried out in this trial.

The majority of the events (shear and harmonic) first triggered the top geophones in the boreholes, thus suggesting that the events occurred at shallow depths near the bottom of the pit. The shear events may most likely be induced by brittle failures of underlying rocks and the harmonic events may be closely associated with hot water/gas flow, due to the reduction of the confinement stresses.

The results have shown that the geothermal-type events occurred within a zone where many shear events also occurred. Although the geomechanical events may not have a direct relationship with geothermal activity, the extensive shear fractures may weaken the pit floor and walls and create channels for the fluids to flow upwards towards the ground surface, under the condition of decreasing confinement stresses. An understanding of the shear event distribution may facilitate identification of potential hazard areas.

Studies in volcanology have shown that harmonic events may be associated with an opening and resonating crack associated with the movement of magma towards the surface (Chouet, 1996) or with the existence of pressure transients within the fluid-gas mixture causing resonance phenomena within the magma itself (Seidl *et al.*, 1981). Active seismicity (both shear and harmonic) always was found to occur prior to volcanic eruptions. The microseismicity associated with the geothermal activity observed in the pit is thus crucial information for a warning of impending steam and gas outbursts. As no significant steam and gas outbursts were observed in the pit during the trial period, these harmonic events may have reflected the normal level of geothermal activity inside the pit.

Lessons have been learned from geophone damage in this trial. The temperature unexpectedly increased in the boreholes after the geophones were cemented and destroyed the waterproof seals of the geophones. In order to significantly reduce the interference of mining noise and achieve good observation of the seismic events, it is strongly suggested to have borehole installations of the geophones for geothermal monitoring at minescale. However, greater care is required to ensure that the geophone strings can withstand the very high temperatures, high pressure and wet conditions.

Acknowledgments

This project was fully supported by Lihir Gold Limited. Thanks are extended to staff members of Lihir Mine who assisted with the borehole drilling, field installation, calibration shot firing and generator maintenance. Two reviewers of the paper are thanked for their valuable comments which enhanced this manuscript.

References

Allen R. (1978), *Automatic earthquake recognition and timing form single traces*, Bull. Seismol. Soc. Am. *68*, 1521–1532.

Aki, K., Fehler, M., and Das, S. (1977), *Source mechanism of volcanic tremor: Fluid driven crack models and their application to the 1963 Kilauea eruption*, J. Volcanol. Geotherm. Res. *2*, 259–287.

Bryan, C.J. and Sherburn, S. (1999), *Seismicity associated with the 1995-1996 eruption of Ruapehu volcano, New Zealand: Narrative and insights into physical processes*, J. Volcanol. Geotherm. Res. *90*, 1–18.

Chouet, B. (1988), *Excitation of a fluid-driven crack: Radiation properties and implication for the source of long-period events and harmonic tremor*, J. Geophys. Res. *93*, 4375–4400.

Chouet, B. (1996), *Long-period volcano seismicity: Its source and its use in eruption forecasting*, Nature *380*, 309–316.

Chouet, B. (2003), *Volcano seismology*, Pure Appl. Geophys. *160*, 739–788.

Feng, Q. and Lees, J. M. (1998), *Microseismicity, stress, and fracture in the Coso geothermal field, California*, Tectonophysics *289*, I 1–3, 221–238.

Julian, B.R. (1994), *Volcanic tremor: Nonlinear excitation by fluid flow*, J. Geophys. Res. *99*, 11859–11877.

Luo, X., Ross, J., Hatherly, P., Shen, B., and Duncan Fama, M. (2001), *Microseismic monitoring of highwall mining stability at Moura Mine, Australia*, Exploration Geophys. *32*, 075–079.

Luo, X. (2004), *The application of microseismics to assist risk management of mining in geothermal areas at Lihir Gold Mine, PNG—Field trial*, CSIRO Project Report 1125R.

McNutt, S.R. (1994), *Volcanic tremor from around the world: 1992 update*, Acta Volcanologica *5*, 197–200.

McNutt, S.R. (1996), *Seismic monitoring and eruption forecasting of volcanoes: A review of the state-of-the-art and case histories*. In: *Monitoring and Mitigation of Volcano Hazards* (Eds. R. Scarpa and R.I. Tilling). pp. 100–146.

Seidl, D., Schick, R., and Riuscetti, M. (1981), *Volcanic tremors at Etna: A model for hydraulic origin*, Bull. Volcanol. *44*, 43–56.

Tolstoy, M., Vernon, F.L., Orcutt, J.A., and Wyatt, F.K. (2002), *Breathing of the seafloor: Tidal correlations of seismicity at Axial volcano*, Geology *30*(6), 503–506.

White, S.P., Creighton, A.L, Bixley, P.F., and Kissling, W.M. (2003), *Modelling the dewatering and depressurisation of the Lihir open pit gold mine*, *In* Proc. TOUGH Symposium 2003, California, USA, 8 pages.

(Received July 24, 2008, revised May 5, 2009, accepted May 21, 2009, Published online December 9, 2009)

Pure Appl. Geophys. 167 (2010), 131–152

DOI 10.1007/s00024-009-0006-3

Pure and Applied Geophysics

Seismic Tomography by Monte Carlo Sampling

Wojciech Dębski[1]

Abstract—The paper discusses the performance and robustness of the Bayesian (probabilistic) approach to seismic tomography enhanced by the numerical Monte Carlo sampling technique. The approach is compared with two other popular techniques, namely the damped least-squares (LSQR) method and the general optimization approach. The theoretical considerations are illustrated by an analysis of seismic data from the Rudna (Poland) copper mine. Contrary to the LSQR and optimization techniques the Bayesian approach allows for construction of not only the "best-fitting" model of the sought velocity distribution but also other estimators, for example the average model which is often expected to be a more robust estimator than the maximum likelihood solution. We demonstrate that using the Markov Chain Monte Carlo sampling technique within the Bayesian approach opens up the possibility of analyzing tomography imaging uncertainties with minimal additional computational effort compared to the robust optimization approach. On the basis of the considered example it is concluded that the Monte Carlo based Bayesian approach offers new possibilities of robust and reliable tomography imaging.

Key words: Mining-induced seismicity, seismic tomography, Bayesian inversion, Markov Chain, Monte Carlo.

1. Introduction

Seismic velocity tomography is an inversion technique aimed at imaging the spatial distribution of velocity heterogeneities in global, regional, local or even laboratory-scale problems. It relies on high frequency approximation to the wave equation according to which seismic energy may be assumed to propagate along an infinitesimal "tube" between the source and the receiver, called the ray path. Hence, the wave travel time between two points provides information on the average seismic velocity along the ray path. If the travel-time data are available for a number of ray paths probing different parts of the studied area, it becomes possible to obtain a spatial map of local heterogeneities of the velocity distribution. The obtained velocity distribution (tomogram) is usually a starting point for further seismological, mineralogical, tectonic, or similar analysis. To make this interpretation as quantitative as possible, a quantitative estimation of the tomography uncertainties and their spatial distribution is required. Currently this can be done in two ways. Assuming a linear or linearized forward problem and Gaussian-type data, modeling and a priori errors, the imaging errors can be approximated by a covariance matrix calculated under the minimum least-squares assumption (Nolet *et al.,* 1999; Yao *et al.,* 1999). This is a very simple and efficient technique but its applicability is strongly limited by the underlying assumption of the linear/linearized forward problem. Also the assumption on the Gaussian character of the data or modeling errors is quite restrictive. The second approach relies on the Bayesian (also called probabilistic) approach which is capable of dealing with highly nonlinear forward problems and all types of errors. The limitation of the approach comes only from the available computational resources as it requires quite exhaustive numerical calculations. This is becoming an increasingly unimportant consideration, however. The advantages of using the Bayesian inversion technique for solving tomography problems are so great that it should be used whenever possible.

This paper examines new possibilities offered by applying Bayesian inversion using Monte Carlo sampling to passive seismic tomography in underground mine scenarios. First, the basic elements of the classical, travel-time based velocity tomography

[1] Institute of Geophysics, Polish Academy of Sciences, ul. Ksiecia Janusza 64, 01-452 Warsaw, Poland.
E-mail: debski@igf.edu.pl

are sketched. Next, two classical inversion algorithms, namely the algebraic and optimization techniques are briefly reviewed. Then, the Bayesian approach-including some elements of the Monte Carlo sampling technique—is discussed in detail. Finally, all three methods are applied to two-dimensional passive tomography imaging at the Rudna (Poland) copper mine.

The basic tomographic forward problem relates the observable travel-time data $\mathbf{d}^{\rm obs}$ with the slowness distribution $\mathbf{s}(\mathbf{r})$ and reads (Aki and Richards, 1985)

$$\mathbf{d}^{\rm th}(\mathbf{s}) = \int\limits_{{\rm ray}[\mathbf{s}]} \mathbf{s}(\mathbf{r})dl, \qquad (1)$$

where $\mathbf{s}(\mathbf{r})$ denotes slowness (inverse of velocity), $\mathbf{d}^{\rm th}$ is the predicted travel time, and the integral is taken along a seismic ray path. The ray path depends on the slowness distribution, and so Eq. 1 is highly nonlinear. However, this effect may often be neglected in the case of local seismic tomography if the slowness distribution does not have large gradients (Cardarelli and Cerrto, 2002). Such a simplification makes the tomography problem easier to solve at the cost of a slight loss of tomogram sharpness (Maxwell and Young, 1993). We follow this assumption throughout the analysis presented here.

Having a set of travel-time data $\mathbf{d} = (t_1, t_2, \ldots t_N)$ recorded for different source-receiver pairs and the discretized slowness distribution (for example, assuming that the imaged area may be described as a set of homogeneous cells), the linear tomographic forward problem can be written as a set of linear equations

$$d_i^{\rm th} = \sum_j^M G^{ij} s_j \quad i = 1 \ldots N, \qquad (2)$$

where $\mathbf{s} = (s_1, s_2, \ldots s_M)$ is the vector of the discretized slowness field and $\mathbf{G}$ is a matrix whose elements G^{ij} are equal fractions of the ith ray path in the jth cell.

2. *Inverse Algorithms*

Inversion of the slowness parameters $\mathbf{s}$ can be carried out in various ways, among which the direct algebraic solution and the optimization approaches are the most common (Iyer and Hirahara, 1993). Another relatively new approach, referred to as the Bayesian (or probabilistic) approach, is based on sampling of the a posteriori probability density and is a method based on the pioneering work of Tarantola and Vallete (1982) (see also Sambridge and Mosegaard (2002) for a review of the history of applying the Monte Carlo technique in geophysics). The Bayesian approach has been gaining more and more popularity recently (Curtis and Lomax, 2001; Bosch *et al.*, 2000; Mosegaard and Tarantola, 2002; Dębski, 2004). The reasons for this include the method's advantages (flexibility, ability to deal with highly nonlinear forward problems, and easy estimation of inversion errors at almost no additional cost) and the simultaneously rapid development of the computational resources which are required for large-scale tomographic tasks.

2.1. Algebraic Solution

If the forward modeling formula given by Eq. 1 is linear or can be linearized, the relation between the data $\mathbf{d}$ and model parameters $\mathbf{s}$ (slowness) takes the form of a set of linear equations

$$\mathbf{d} = \mathbf{G} \cdot \mathbf{s}. \qquad (3)$$

The problem of inverting for $\mathbf{s}$ from $\mathbf{d}^{\rm obs}$ can be viewed as the task of solving the set of linear equations using as $\mathbf{d}$ the results of measurements $\mathbf{d}^{\rm obs}$. This task can be accomplished within a simple algebraic approach (Menke, 1989; Tarantola, 1987; Parker, 1994) by a matrix manipulation, as follows.

In the first step, the so-called normal equation is formed by multiplying both sides of Eq. 3 by the transposed $\mathbf{G}^T$ operator (matrix)

$$\mathbf{G}^T \mathbf{d}^{\rm obs} = (\mathbf{G}^T \cdot \mathbf{G})\mathbf{s}. \qquad (4)$$

The matrix $\mathbf{G}^T\mathbf{G}$ is a square matrix which usually cannot be inverted yet because of possible singularity. To fix this problem, the $\mathbf{G}^T$ $\mathbf{G}$ matrix is regularized which, in the simplest case, is achieved by adding a small diagonal term:

$$\mathbf{G}^T \cdot \mathbf{G} \longrightarrow \mathbf{G}^T \cdot \mathbf{G} + \gamma \mathbf{I}, \qquad (5)$$

where $\mathbf{I}$ is a diagonal matrix. Finally, the analytical formula for the model $\mathbf{s}^{\rm lsqr}$ regarded as the solution can be cast into the form with the explicitly introduced initial (a priori) model $\mathbf{s}^{\mathrm{apr}}$ (MENKE, 1989; TARANTOLA, 1987)

$$\mathbf{s}^{\mathrm{lsqr}} = \mathbf{s}^{\mathrm{apr}} + (\mathbf{G}^T \cdot \mathbf{G} + \gamma \mathbf{I})^{-1} \mathbf{G}^T (\mathbf{d}^{\mathrm{obs}} - \mathbf{G}\mathbf{s}^{\mathrm{apr}}) + O(\gamma^2). \tag{6}$$

This form is particularly convenient when $\mathbf{s}^{\mathrm{lsqr}}$ is estimated iteratively (TARANTOLA, 1987) and is de facto the starting point of various linear inversion techniques (PARKER, 1994; LIMES and TREITEL, 1983).

The above formula for $\mathbf{s}^{\mathrm{lsqr}}$ in fact coincides with the solution obtained by the optimization approach (method 2), when the least-squares difference between observed and modeled travel times is minimized subject to an additional "smoothness" condition. That is why this method is often called the LSQR (Damped Least- Squares) solution (PARKER, 1994; MENKE, 1989).

The algebraic approach described above can be classified as a back-projection technique (DEANS, 1983) as it performs a direct projection of the data space $\mathcal{D}$ onto the model space $\mathcal{M}$ (PARKER, 1994). The apparent advantage of this technique is its ability to cope with extremely large-scale problems as various numerical algorithms can find $\mathbf{s}^{\mathrm{lsqr}}$ quickly and efficiently. However, the method requires the forward problem to be linear or approximately linear, which limits its application to weakly heterogeneous media. Moreover, the method, being equivalent to the l_2 norm-based optimization, is non-robust and requires very high quality data to yield acceptable results (TARANTOLA, 1987; MENKE, 1989; DĘBSKI, 2004).

2.2. Optimization Approach

Although the above approach is attractive due to its simplicity and ability to deal with large-scale problems, it suffers from a lack of generality as it is restricted to explicit linear problems. It also suffers from a lack of mathematical rigorousness because of ad hoc regularization and unclear physical meaning of the regularization procedure. Both drawbacks are eliminated within the so-called optimization technique (method 2).

The idea of the optimization method is to search for a model for which predicted data are as close as possible to the observed data with respect to a given norm. However, the optimization problem posed in this way is often non-unique, since the minimized differences between observed data and predictions for the current model can be multi-modal. Thus, there is often a need for *regularization* which could reduce the solutions non-uniqueness (MENKE, 1989). This is usually achieved by requiring the optimum model to be "smooth" or, more generally, to be somehow similar to an initial a priori model. Taking this into account, the discussed approach turns the inverse task into an optimization (minimization) problem for the so-called misfit function

$$S(\mathbf{s}) = ||\mathbf{d}^{\mathrm{obs}} - \mathbf{d}^{\mathrm{th}}(\mathbf{s})||_{\mathcal{D}} + ||\mathbf{s} - \mathbf{s}^{\mathrm{apr}}||_{\mathcal{M}}, \tag{7}$$

where $||\cdot||_{\mathcal{D}}$ and $||\cdot||_{\mathbf{M}}$ denote the norms in the data (**D**) and the model ($\mathcal{M}$) spaces. They measure the "distances" between observed and predicted data ($||\cdot||_{\mathcal{D}}$) and between current and a priori models ($||\cdot||_{\mathcal{M}}$) The choice of the norms and parameters through which they are defined determines the relative influence of both terms on the final solution, similar to the choice of the damping parameter γ in Eq. 6. Some of the most popular norms used in the context of geophysical data inversion are listed in Table 1.

The main difference between the various norms stems from how unexpected outliers in data or predictions are treated. The quadratic norm (l_2 norm) is not a robust norm in the sense that it cannot properly handle data sets that include even just a few large outliers. On the other hand, the Cauchy norm (l_c) is the most robust and can be used to invert data strongly contaminated by large errors. The performance of the l_1 norm is a case in between the l_2 and l_c norms (KIJKO, 1994). It can efficiently and accurately handle data sets with a moderate number of outliers (KIJKO, 1994). It is important to note that the more robust a given norm is, the less constrained is the final solution. In extreme situations this effect can lead to

Table 1

Examples of norms used to measure the "distance" in a data/model space. C^{ij} and C_i are the constants defining the norms

l_1	$\|\|\mathbf{d}\|\|_{l_1} = \sum_i \left\|\frac{d^i}{C^i}\right\|$	Absolute value norm
l_2	$\|\|\mathbf{d}\|\|_{l_2} = \frac{1}{2}\sum_{ij} d^i C^{ij} d^j$	Quadratic norm
l_c	$\|\|\mathbf{d}\|\|_{l_c} = \sum_i \log\left(1 + \left(\frac{d^i}{C^i}\right)^2\right)$	Cauchy norm

Reprinted from the journal

instabilities of numerical (optimization) algorithms used to find the "best" model (DĘBSKI, 2004). Hence, the choice of the "best performing" norm depends on data quality and modeling accuracy.

For a given data set the quality of the solution found by the optimization technique depends on the quality of the optimizer used, the norms selected and often how well the a priori model approximates the true solution. If the observational data set contains outliers, the l_2 norm will often fail to give any reasonable results (SCALES, 1996; DĘBSKI, 1997a, b; TARANTOLA, 1987). Also the choice of the initial (a priori) model $\mathbf{s}^{apr}$ requires some attention and is quite crucial if any local optimization approach (for example, a gradient-based technique) is employed to search for the best model. If $\mathbf{s}^{apr}$ is taken too far from the "true" model there is a great risk that the employed optimizer will converge to a model which corresponds to a local minimum and not the global minimum. This risk is significantly reduced when global optimization techniques such as Simulated Annealing, Genetic Algorithm, or Nearest Neighborhood algorithm are used (SEN and STOFFA, 1995; MICHALEWICZ, 1996; SAMBRIDGE, 1999).

Finally, let us note that this approach is not limited to the linear forward problem like the LSQR method. The optimization approach can cope with any forward problem. Furthermore, the optimization approach reduces to an analytical approach when the forward problem is linear and the norms $||\cdot||_{\mathcal{D}}$ and $||\cdot||_{\mathcal{M}}$ are both l_2 (TARANTOLA, 1987; MENKE, 1989; LIMES and TREITE, 1983).

2.3. *Bayesian Inversion*

Due to its generality, the optimization method reviewed above is often used in geophysical tomography problems (ZHAO, 2001; IYER and HIRAHARA, 1993; NOLET, 1987). The main drawback of this approach is a lack of reliable estimation of the solutions quality. This is because the optimization technique provides the solution of the inverse problem in the form of a single (optimum) model and cannot evaluate the "size" of the part of the model space around the optimum model consisting of models leading to similar residua as the optimum model. Exploration of the model space for mapping such a region is necessary for systematic error evaluation (TARANTOLA, 2005; DĘBSKI, 2004; CURTIS and LOMAX, 2001; SCALES, 1996; DUIJNDAM, 1988).

Actually, for the LSQR method there exists an error estimator called the LSQR posterior covariance matrix (MENKE, 1989; NOLET *et al.*, 1999; YAO *et al.*, 1999). However, this error estimate is restricted by the assumed linear forward problem, Gaussian-type distribution of the data and forward modeling errors or approximations, to name a few factors, which often result in misleading values in real cases. The above drawbacks of the optimization approach are conveniently overcome by the application of Bayesian inverse theory to the tomography problem (BOSCH, 1999; DĘBSKI, 2004).

The solution of the tomographic inverse problem using the Bayesian inverse theory consists in building the a posteriori probability distribution $\sigma(\mathbf{s})$ over the model space which describes the probability of a given model ($\mathbf{s}$) being the true one (TARANTOLA, 1987; DĘBSKI, 2004).

It has been shown by TARANTOLA and VALLETE, (1982), TARANTOLA (1987), JACKSON and MATSU'URA (1985) that, in general, $\sigma(\mathbf{s})$ is the product of a distribution $f(\mathbf{s})$ describing a priori information by a *likelihood function* $L(\mathbf{s})$ which measures to what extent theoretical predictions fit the observed data:

$$\sigma(\mathbf{s}) = \text{const.} f(\mathbf{s}) L(\mathbf{s}), \tag{8}$$

where the constant represents a normalization of the probability density and the *likelihood function* $L(\mathbf{s})$ reads (see, for instance, TARANTOLA, (2005); DĘBSKI (2004))

$$L(\mathbf{s}) = \exp\left(-\|\mathbf{d} - \mathbf{d}^{th}(\mathbf{s})\|\right). \tag{9}$$

Knowing the $\sigma(\mathbf{s})$ distribution allows not only to find the most likely model $\mathbf{s}^{ml}$ for which $\sigma(\mathbf{s}^{ml})$ = max but also other characteristics like, for example, the average model and the variance which provides a convenient measure of imaging accuracy. These two basic characteristics of the a posteriori distribution $\sigma(\mathbf{s})$ are very important.

The most important operational difference between the probabilistic approach and those discussed previously consists in the different form of the solutions. While the algebraic and optimization techniques provide a single, in a sense optimum estimation of the sought parameters, the probabilistic solution is a probability distribution over the model space which quantifies the "chance" that a given model is the true one. In this way it provides a natural framework for comparing different models and allows a very general estimation of inversion uncertainties.

At this point I would like to offer two comments on a popular use of the Bayesian inversion technique.

Very often, instead of sampling of the a posteriori *PDF* only the global maximum of the $\sigma(\mathbf{s})$ is sought (e.g., DĘBSKI and YOUNG, 2002) and claimed to be the solution of the Bayesian inversion. It is obviously a correct procedure, but it reduces the full Bayesian inversion to the optimization method, disregarding the potential ability of the Bayesian method to deal with error analysis.

A similar situation occurs when the a posteriori *PDF* is a Gaussian distribution due to a specific assumption about the a priori *PDF* and the form of the likelihood function (linear or linearized forward problem and Gaussian error statistics). In such a case, the a posteriori *PDF* $\sigma(\mathbf{s})$ is fully described by two parameters, namely the likelihood model $\mathbf{s}^{ml}$ and the covariance matrix $\mathbf{C}^p$ equivalent to the algebraic solution given by Eq. 6. (MENKE, 1989; TARANTOLA, 1987; PARKER, 1994). Solving the inverse problem is then often understood as finding their numerical values by directly solving Eq. 6 (e.g. TAYLOR *et al.*, 2003) and building the covariance matrix (MENKE, 1989), which is obviously correct but is more of an algebraic method than a general Bayesian technique.

3. *Solving Tomography Problems—Bayesian Point of View*

The important question at this point is which of the inverse methods should be chosen to solve a particular tomography problem. I argue that whenever available computational resources allow, the probabilistic approach should be used. The arguments are as follows.

First of all, it should be recognized that when solving any tomography (inverse) problem we face various types of uncertainties (TARANTOLA, 1987; SCALES, 1996).

Any piece of information we use to solve an inverse task always represents some degree of uncertainty. For example, any travel-time measurement we perform has a finite accuracy either due to the applied method or due to the finite resolution of the hardware (usually both). Thus, it is quite natural to describe the output of a measurement by means of the appropriate probability density function (*PDF*) describing the statistics of the experimental errors.

Besides the observational uncertainties discussed above there are also theoretical errors caused by an approximate (or simplified) calculation of theoretical predictions necessary for a comparison with observational data. Numerical calculations are also a source of additional modeling errors. As a consequence we face a situation where the theoretical predictions are also disturbed by some errors which can be modeled conveniently by a probability distribution.

The solution to the inverse problem also includes some a priori assumptions and information. By a priori I mean all the information that is not connected with the current measurement but results from previously acquired knowledge, experience, previous measurements, and so forth. This a priori knowledge is never perfect. It is usually quite vague. If for some reason the a priori uncertainties were small, we would not need to make any inversion; we would simply know the solution. The a priori information can also be described in terms of the probability distribution interpreted according to the Bayesian point of view as a measure of "confidence" about the sought parameters (JEFFREYS, 1983; GELMAN *et al.*, 1997; SCALES and SNIEDER, 1997).

The Bayesian (probabilistic) inverse theory allows for consistent treatment of all the above-mentioned uncertainties. This is one of the most important advantages of this technique which allows incorporation of an exhaustive error analysis into solving the inverse problem.

One very important fact concerning the probabilistic technique is that it allows calculation of more statistical estimators of the true model $\mathbf{s}^{\mathrm{true}}$ than just the simplest, best-fitting maximum likelihood model $\mathbf{s}^{\mathrm{ml}}$, as discussed below.

Finally, a typical tomography problem is characterized by a highly nonlinear forward modeling relation (ray-path bending) which leads to a complex form of the misfit function often having a number of secondary minima. In such a situation solving the tomography problem by seeking the model which best fits the data within the optimization technique requires a very careful choice of the numerical optimization algorithms: Local, gradient-based optimization algorithms should generally be avoided. In the case of the Bayesian approach the complex, non-Gaussian form of the a posteriori *PDF* function is also a problem, especially when local maxima of the a posteriori *PDF* are significantly separated in the model space, however there exist many quite efficient Monte Carlo samplers that can sample quite complicated "non-bell-shaped" functions (ROBERT and CASELLA, 1999).

3.1. Bayesian Solutions

More than just a technical problem emerges at this point. Specifically, the problem is how to inspect the a posteriori *PDF* distribution to extract the required information when in most practical cases the a posteriori *PDF* is a complicated, multi-parameter function. Basically, there are three different strategies to explore the a posteriori probability density function: evaluating the point estimators, calculating the marginal*PDF* distributions, or sampling the a posteriori *PDF*.

3.1.1 Point Estimators

The first approach relies on searching for the point estimators among which the most popular are the lowest-order moments of the a posteriori *PDF* (JEFFREYS, 1983):

1. The maximum likelihood model

$$\mathbf{s}^{\mathrm{ml}} : \sigma(\mathbf{s}^{\mathrm{ml}}) = \max, \tag{10}$$

2. The average model

$$\mathbf{s}^{\mathrm{avr}} = \int_{\mathcal{M}} \mathbf{s}\sigma(\mathbf{s})d\mathbf{s}, \tag{11}$$

3. The covariance matrix

$$\mathbf{C}^{p}_{ij} = \int_{\mathcal{M}} (\mathbf{s}_i - \mathbf{s}^{\mathrm{avr}}_i)(\mathbf{s}_j - \mathbf{s}^{\mathrm{avr}}_j)\sigma(\mathbf{s})d\mathbf{s}. \tag{12}$$

If a more comprehensive description of $\sigma(\mathbf{s})$ is required, higher-order moments can also be calculated (JEFFREYS, 1983).

Solving the inverse problem by searching for the $\mathbf{s}^{\mathrm{ml}}$ model in fact reduces the probabilistic approach to the optimization technique (TARANTOLA, 1987; DĘBSKI, 2004). Since this approach does not require full knowledge of $\sigma(\mathbf{s})$, obtaining the $\mathbf{s}^{\mathrm{ml}}$ solution is usually easier than full sampling of $\sigma(\mathbf{s})$. Two other point estimators, namely the average model $\mathbf{s}^{\mathrm{avr}}$ and the covariance matrix $\mathbf{C}^p$, provide significantly more information than $\mathbf{s}^{\mathrm{ml}}$ but their evaluation requires full knowledge of $\sigma(\mathbf{s})$ in order to calculate the appropriate integrals. This is trivial when $\sigma(\mathbf{s})$ is a Gaussian distribution (this implies that $\mathbf{G}(\mathbf{s}) = \mathbf{G}{\cdot}\mathbf{s}$). In such a case $\mathbf{s}^{\mathrm{avr}} = \mathbf{s}^{\mathrm{ml}}$ is given by Eq. 6 and the covariance a posteriori matrix reads (TARANTOLA, 1987; MENKE, 1989; PARKER, 1994):

$$\mathbf{C}^p = (\mathbf{G}^T\mathbf{C}_d^{-1}\mathbf{G} + \mathbf{C}_m^{-1}))^{-1} \tag{13}$$

where $\mathbf{C}_d$ describes the variance of the sum of the observational and modeling errors and $\mathbf{C}_m$ stands for the variance of the a priori uncertainties.

The important point is to recognize the significant difference between the first two point-like estimators, namely $\mathbf{s}^{\mathrm{ml}}$ and $\mathbf{s}^{\mathrm{avr}}$.

The interpretation of $\mathbf{s}^{ml}$ is obvious—this is the model which maximizes the a posteriori *PDF* function which, roughly speaking, means that $\mathbf{s}^{\mathrm{ml}}$ is the model which best explains the data and a priori expectations. The importance of the average model $\mathbf{s}^{\mathrm{avr}}$ stems from the fact that it provides not only information on the best-fitting model but also includes information about other plausible models from the neighborhood of $\mathbf{s}^{\mathrm{ml}}$. If sub-optimum models defined as those for which $\sigma(\mathbf{s}) \sim \sigma(\mathbf{s}^{\mathrm{ml}})$ are similar to $\mathbf{s}^{\mathrm{ml}}$, then $\mathbf{s}^{\mathrm{avr}} \sim \mathbf{s}^{\mathrm{ml}}$. Otherwise both estimators can be quite different. Thus, a simple comparison of $\mathbf{s}^{\mathrm{avr}}$ and $\mathbf{s}^{\mathrm{ml}}$ provides a qualitative evaluation of the reliability of the inversion procedure: The more $\mathbf{s}^{\mathrm{avr}}$ differs from $\mathbf{s}^{\mathrm{ml}}$, the more complex and non-Gaussian is the form of the a posteriori *PDF*. This immediately implies that in such a case more care must be taken when interpreting the inversion results, especially inversion uncertainties. In fact, using confidence levels instead of the covariance matrix is highly recommended in such situations (JEFFREYS, 1983). In addition, the average model $\mathbf{s}^{avr}$ is usually a much considerably reliable estimator of the true values $\mathbf{s}^{true}$ than a single, even best fitting model like $\mathbf{s}^{ml}$. The reason for this is that the integration (averaging) procedure acts in the manner weighted stacking (CLAERBOUT, 1985) which filters out the random part of the inversion errors which contributes to $\mathbf{s}^{ml}$. In other words, averaging smoothes out a posteriori noise by removing individual variations among all sub-optimum models and leaves only the features common for all reasonable models if such a common part exists at all. This reasoning assumes that the a posteriori distribution is somehow reasonably concentrated around the optimum model. If this is not the case and the a posteriori *PDF* is, for example, multi-modal, the average model has no meaning and eventually a more exhaustive analysis is required (JEFFREYS, 1983; BRANDT, 1999).

The diagonal elements of the a posteriori covariance matrix $\mathbf{C}^p$ are convenient estimators of the inversion uncertainties for each component of $\mathbf{s}$ while the non-diagonal elements measure the degree of correlation between pairs of parameters (MENKE, 1989; JEFFREYS, 1983). In fact, $\mathbf{C}^p$ given by Eq. 12 is a generalization of the LSQR covariance matrix to the case of an arbitrary statistics $\sigma(\mathbf{s})$ including possibly nonlinear forward problems. As in the case of the average model, the posterior covariance matrix is meaningful only if the $\sigma(\mathbf{s})$ distribution is unimodal. In cases of multi-modality, the existence of non-resolved directions in the model space, or other "pathologies", a more exhaustive error analysis is necessary by a full inspection of the a posteriori distribution (TARANTOLA, 1987; WIEJACZ and DĘBSKI, 2001).

3.1.2 Marginal PDF Distributions

While the point estimators discussed previously provide a very convenient means of synthetic representation of the inversion results, the a posteriori marginal distributions give a deeper insight into the form of $\sigma(\mathbf{s})$ and their inspection is extremely important for the correct interpretation of the inversion results.

The one-dimensional (1-D) marginal a posteriori *PDF* distribution is defined by integrating out all but one parameter from $\sigma(\mathbf{s})$

$$\sigma_i(s_i) = \int \sigma(\mathbf{s}) \prod_{j \neq i} \mathrm{d}s_j. \quad (14)$$

Multi-dimensional marginals are defined similarly. It is important to realize that marginal *PDF* distributions contain the same information on s_i as $\sigma(\mathbf{s})$ except for the correlation with other parameters (JEFFREYS, 1983). However, inspections of the full a posteriori *PDF* $\sigma(\mathbf{s})$ and marginal *PDF* $sigma_i(s_i)$ are not equivalent (DĘBSKI, 1997a). To see the difference let us consider two maximum likelihood solutions derived from the $\sigma(\mathbf{s})$ and $\sigma_i(s_i)$ a posteriori distributions. In the first case, when seeking the $\mathbf{s}^{ml}$ model we consider the set of N_m parameters which simultaneously maximizes $\sigma(\mathbf{s})$. In the second case we solve the 1-D optimization problem seeking the optimum value of only one parameter s_i, regardless of the values of the other components of $\mathbf{s}$. In other words, in the former case we obtain information about all the parameters simultaneously while the inspection of the marginal distributions provides information only about selected parameters, ignoring any relations with the remaining ones.

In the most frequently encountered circumstances, point estimators calculated from the marginal distributions and from $\sigma(\mathbf{s})$ differ slightly (DĘBSKI *et al.*, 1997; WIEJACZ and DĘBSKI, 2001). A difference significantly larger than the corresponding elements of the a posteriori covariance matrix indicates a very strong correlation among parameters which, besides the case of intrinsically correlated parameters, may indicate that some parameters or their combinations are not resolved by the data (WIEJACZ and DĘBSKI, 2001).

Inspection of marginal *PDF*'s is always recommended to verify if $\sigma(\mathbf{s})$ is a multi-modal distribution or not. This is especially important if $\mathbf{s}^{avr}$ is used as the numerical estimator of $\mathbf{s}^{true}$. Moreover, the inspection of the marginal *PDF* provides conclusive verification of whether estimating the inversion errors by a posteriori covariance is justified or not.

3.1.3 Sampling a posteriori PDF

Whenever we look for a solution of the inverse problem other than the maximum likelihood model, we face the problem of sampling of $\sigma(\mathbf{s})$, usually in the context of calculating multidimensional integrals as in Eqs. 11 and 12 or 14. More generally, we may need to calculate the integrals represented in a general form as

$$\langle F \rangle = \int_{\mathcal{M}} F(\mathbf{s})\sigma(\mathbf{s})d\mathbf{s}, \tag{15}$$

where $F(\mathbf{s})$ stands for a generic function.

If the number of parameters is very small (smaller than, say, 10), the integral in Eq. 15 can be calculated by sampling $\sigma(\mathbf{s})$ over a predefined regular grid, such as, for example, in the case of the seismic source location problem (WIEJACZ and DĘBSKI, 2001; LOMAX *et al.*, 2000; SAMBRIDGE and KENNETT, 2001). Otherwise, the stochastic (Monte Carlo) sampling technique has to be used (MOSEGAARD and TARANTOLA, 1995; BOSCH *et al.*, 2000; DĘBSKI, 2004).

The mathematical aspects of Monte Carlo (MC) sampling of the a posteriori *PDF* and some algorithms are discussed later and here I would like to point out only a few advantages of this approach in solving inverse problems.

Generally speaking, the MC technique allows generation an ensemble of models which can be regarded as a set of samples drawn from the $\sigma(\mathbf{s})$ distribution so the average integrals $\langle F \rangle$ can be approximated by (ROBERT and CASELLA, 1999)

$$\langle F \rangle \approx \frac{1}{N}\sum_{\mathbf{s}^\alpha} F(\mathbf{s}^\alpha) \tag{16}$$

where the sum is taken over an ensemble of all (N) generated models $\mathbf{s}^\alpha$.

One of the biggest advantages of the possibility of calculating the statistical averages with respect to the a posteriori *PDF*, as in Eq. 16, is that inversion errors can be easily estimated at no additional cost. If they can be approximated by variance of the a posteriori *PDF* then the choice

$$F = (\mathbf{s} - \mathbf{s}^{\mathrm{ml}})^2 \quad \text{or, better} \quad F = (\mathbf{s} - \mathbf{s}^{\mathrm{avr}})^2 \tag{17}$$

leads to an extremely simple, easy to calculate estimator of the a posteriori errors

$$\epsilon_m^2 = \frac{1}{N}\sum_{\mathbf{s}^\alpha}(\mathbf{s}^\alpha - \mathbf{s}^{\mathrm{ml}})^2 \tag{18}$$

or

$$\epsilon_a^2 = \frac{1}{N}\sum_{\mathbf{s}^\alpha}(\mathbf{s}^\alpha - \mathbf{s}^{\mathrm{avr}})^2. \tag{19}$$

Moreover, if the variance is not a satisfactory estimator of the inversion errors due to the complexity of the a posteriori *PDF*, higher-order moments of $\sigma(\mathbf{s})$ or the marginal distributions such as, for example,

$$\sigma_i(s) = \frac{N_i}{N} \tag{20}$$

where N_i is the number of generated models whose ith component has value s can easily be calculated. This means that even an exhaustive error analysis becomes almost trivial, no matter how many parameters are estimated. Meanwhile, classical error analysis based on calculating the covariance matrix is neither general, due to approximation of $\sigma(\mathbf{s})$ by the multivariate Gaussian distribution, nor easy to obtain, especially in the case of tomographic high dimensional problems (YAO *et al.*, 1999, 2001; NOLET *et al.*, 2001; ZHANG and MCMECHAN, 1996; DEAL and NOLET, 1996).

Having generated a set of model samples drawn from $\sigma(\mathbf{s})$, one can inspect them one by one or, better, by making a "movie", as suggested by MOSEGAARD and TARANTOLA, (1995), explore their common properties. Such an inspection can provide deep insight into plausible solutions and allows for the discovery of some hidden correlations, particularly macroscopic features, some trends and other features which can even be difficult to quantify. This is rather an important point because searching for correlations is quite a difficult task, especially for numerical algorithms. In fact, any feature which appears in a large fraction of generated models leads to non-vanishing average values of some functionals F such as, for example, multi-parameter correlation moments $F_{i,j,\ldots n} = s_i\, s_j \ldots s_n$. However, finding the F which is the best indicator of a particular "feature" is usually extremely difficult (VAN KAMPEN, 1992).

3.2. *Markov Chain Monte Carlo Sampling*

The choice of a scheme of sampling $\sigma(\mathbf{s})$ is by no means a trivial task, especially when the number of estimated parameters is large (MOSEGAARD and TARANTOLA, 2002; BOSCH *et al.,* 2000). This is due to the fact that with an increasing number of parameters the model space becomes extremely large but only a very small part contributes to the a posteriori distribution (CURTIS and LOMAX, 2001). In such a case any geometrical sampler, meaning a deterministic algorithm which explicitly depends on the size of the model space, such as the neighborhood algorithm for example (SAMBRIDGE, 1999), becomes prohibitively inefficient. A similar situation occurs if blind (random) stochastic sampling is used. This difficulty can, however, be overcome if a posteriori *PDF* is sampled by the Markov Chain Monte Carlo (MCMC) technique (MOSEGAARD and TARANTOLA, 1995; TIERNEY, 1994; ROBERT and CASELLA, 1999).

The MCMC technique is a class of numerical algorithms which originally aimed to simulate so-called short-memory stochastic processes (Gillespie, 1992; Robert and Casella, 1999). By a short-memory stochastic process I mean here a dynamical process (an evolution of a given physical system in time) which fulfills the following two conditions.

1. The future state of the system depends only on its current state and not on the entire evolution history-previous states are "forgotten". This feature is commonly called a short-memory of the process.
2. The next state of the system cannot be predicted deterministically but only a probability of transition to other states can be provided (stochastic process).

As can be deduced from the above conditions, the state of the short-memory stochastic process is never uniquely determined at any time. Instead, it assumes the form of a probability distribution evolving in time over the space of all the possible states of the system (GILLESPIE, 1992). If some conditions are fulfilled (ROBERT and CASELLA, 1999) this probability is stationary-it does not change in time. As a consequence, the subsequent states of such a stationary process are distributed according to the same probability distribution. Actually, they can be regarded as correlated samples drawn from the same distribution. Short-memory processes with a discrete evolution time are called Markov chains and have been used in analyses of various scientific and engineering problems (e.g. GAMERMAN, 1997; GILKS *et al.,* 1995; TIERNEY, 1994; CASELLA, 1999; DĘBSKI, 2008).

A mathematical procedure for building a Markov chain with the prerequisite stationary distribution $\sigma(\mathbf{s})$ was put forward by HASTINGS (1970) who generalized the algorithm proposed by METROPOLIS *et al.* (1953). The simplest form of such an algorithm (the original Metropolis algorithm) is shown in Fig. 1 (TIERNEY, 1994; MOSEGAARD and TARANTOLA, 2002; ROBERT and CASELLA, 1999).

The important feature of the Metropolis algorithm (also the more general Metropolis-Hastings algorithms) is that it does not depend explicitly on the size of the sampled space. Such a dependence appears only when solving the forward problem necessary for calculating $\sigma(\mathbf{s}^{\beta})$. This independence gives the Metropolis-Hastings sampling algorithms very high efficiency when applied to large parameter problems, such as seismic tomography. More details on the application of the Metropolis or Metropolis-Hastings algorithms, setting the parameters, monitoring convergence, and so forth, can be found in the literature of the subject (ROBERT and CASELLA, 1999; GAMERMAN, 1997; GILKS *et al.,* 1995; DĘBSKI, 2004).

4. *Rudna Copper Mine Case*

The three algorithms described above, namely LSQR, the optimization technique based on genetic algorithm optimization and MC sampling, were used to image the velocity distribution in a part of the Rudna copper mine in southwestern Poland. The same data set and the same initial (a priori) velocity model was used in each case to make a comparison between the different methods as straightforward as possible. The remaining inversion parameters, namely the damping factor γ in the LSQR method, the weights C_d and C_v in the optimization technique, and the form of the a priori and likelihood functions, were chosen to assure the methods' maximum compatibility. Their choice is briefly discussed below.

- Initialize $\mathbf{s}^0$
- Repeat
  - generate uniform random number $u \sim U(0,1)$
  - generate new test sample $\mathbf{s}^\beta = \mathbf{s}^\alpha + \delta\mathbf{s}$ where $\delta\mathbf{s}$ is a small perturbation to the current state
  - if $u < P(\mathbf{s}^\beta, \mathbf{s}^\alpha) = \min\left[1, \frac{\sigma(\mathbf{s}^\beta)}{\sigma(\mathbf{s}^\alpha)}\right]$
    $\mathbf{s}^{\alpha+1} = \mathbf{s}^\beta$
    otherwise
    $\mathbf{s}^{\alpha+1} = \mathbf{s}^\alpha$
- Continue until sufficient number of samples $\{\mathbf{s}^\alpha\}$ is generated

Figure 1
The Metropolis algorithm for sampling a posteriori *PDF* $\sigma(\mathbf{s})$

4.1. Algorithms

The algebraic approach to the tomography task, referred to as the LSQR method, requires setting of the damping factor γ. So that the optimization and Monte Carlo approaches are consistent with each other, I define the damping factor as the ratio

$$\gamma = C_d/C_v, \tag{21}$$

where the parameters C_d and C_v can be interpreted as the estimator of data uncertainties (C_d) and the expected variation of velocity around the a priori model (C_v).

In the case of the optimization technique I assumed the l_2 norms in both the data ($\|\cdot\|_D$) and model ($\|\cdot\|_M$) spaces with diagonal covariance matrices. Specifically, I assumed that

$$\begin{aligned} C_d^{ij} &= C_d^{-2}\delta^{ij} \quad \text{for the data space} \\ C_v^{ij} &= C_v^{-2}\delta^{ij} \quad \text{for the model space} \end{aligned} \tag{22}$$

where δ^{ij} is the Kronecker delta function ($\delta^{ij} = 1$ if $i = j$, otherwise 0).

To solve the optimization problem, the Genetic algorithm-based optimizer was used with parameter settings assuring maximum robustness of the calculation, but at the cost of a rather slow convergence rate (for details see Dębski (2002) and references therein). The velocity tomograms obtained by this method are marked from here forward as the GA solutions.

Finally, the Monte Carlo (Metropolis) sampler was run assuming Gaussian probability densities for both the a priori distribution $f(\mathbf{s})$ and the likelihood function $L(\mathbf{s})$ with the same covariance matrices as in the GA case defined in Eq. 22. The Metropolis sampler was used to generate 100,000 models according to the a posteriori distribution. The update step was always selected to keep the acceptance ratio at around 50% (see flowchart in Fig. 1). The generated models (samples) were used to approximate the a posteriori distribution and to calculate the maximum likelihood model (MC-ML), the average velocity model (MC-AV), and the a posteriori covariance matrix.

Within this selection the velocity images provided by each method should be identical, since from a mathematical point of view the solutions provided by all three methods are equivalent with the analytical solution given in Eq. 6. In reality, they will differ and the difference can be used to evaluate the performance of each of the inversion methods.

4.1.1 Data and Velocity Parameterization

The three described algorithms were applied to a set of events occurring in 1998 at the Rudna copper mine, Poland. This mine, located in southwestern Poland, is monitored by a digital seismic network composed of 32 vertical Willmore MK-II and MK-III sensors located underground at depths from 550 to 1,150 m. The data is digitized with a frequency band ranging from 0.5 to 150 Hz, and the recording dynamics is about 70 dB. The sampling frequency is $F_s = 500$ Hz and determines the absolute accuracy of travel-time estimate ($C_d = 2$ ms). The uncertainty in event location is better than 100 m, typically around 50 m.

For the current study I selected a set of 36 events which were recorded by at least 4 of the 8 nearest stations located approximately at the same depth as the estimated hypocenters. This resulted in a subset of 177 travel times used in the P-wave velocity inversion. The distribution of selected events, employed stations and considered ray paths is shown in Fig. 2. The events were located assuming a constant background velocity, $V_a = 5{,}800$ m/s. This value was also assumed as the initial a priori velocity distribution for the inversion.

The studied area of around 5 $\times$ 5 km was discretized into 250 m by 250 m square cells, for a total of 400 blocks. However, the ray paths did not sample most of these cells, as shown in Fig. 2. In fact, only 109 cells marked in Fig. 2 in gray colors were touched by at least one ray path. Only such cells were effectively considered during the inversion, and the a priori velocity $V_a = 5{,}800$ m/s was assigned to the remaining, untouched cells.

4.1.2 Damping Parameters

Inversion of travel-time data requires two parameters to be specified, namely C_d and C_v, or at least their ratio $\gamma = C_d/C_v$ in the case of the LSQR method. Usually, the optimum γ coefficient can be determined by analyzing the trade-off between the variance (complexity) of the final model and the reduction of the data variance with respect to the initial (a priori) model (MENKE, 1989; PARKER, 1994). In this study the analysis encountered a problem: because of the very short distances between the event epicenters and the recording stations, even very complex models led to a reduction of the residua by less than 10%, as shown in Fig. 3. In such cases the standard trade-off analysis is meaningless, therefore I introduced physical constraints to estimate C_d and C_v. Firstly, I assumed on the basis of mining practice that velocity changes generally do not exceed 20% of background velocity at a 95% confidence level. This a priori assumption corresponds approximately to the choice $C_v = 400$ m/s. The value of C_d which expresses the joint accuracy of the travel-time measurements and travel-time modeling (DĘBSKI, 2004) is limited by the sampling

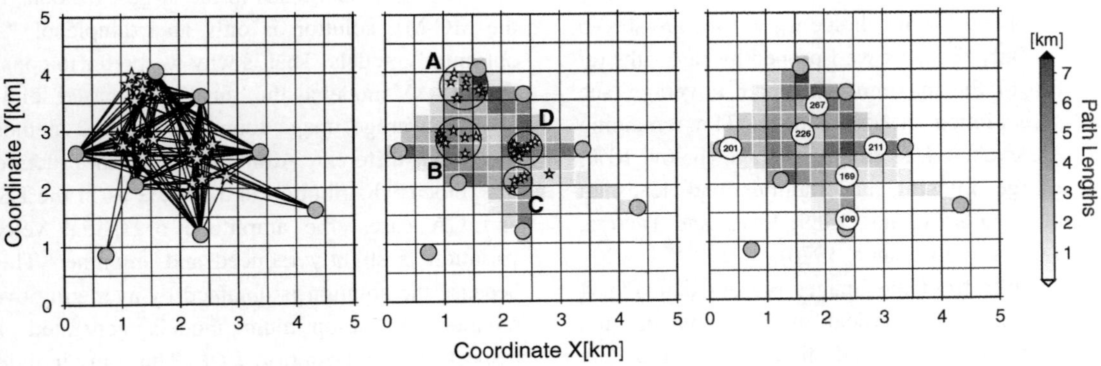

Figure 2
Ray paths (*left*), ray path coverage—sum of fractions of lengths of all rays passing through cells used to parameterize the studied region (center) and cells selected for a posteriori *PDF* analysis (*right*). Seismometer positions are marked by *circles* and the locations of seismic sources used in the current study are depicted by *stars*. The events used in the current study formed four spatial clusters marked A, B, C and D, respectively

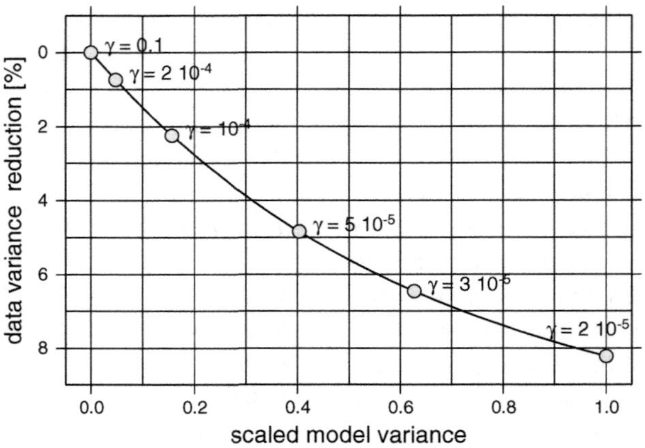

Figure 3
Trade-off curve: the correlation between reduction of data and model variance for different values of damping factor γ

period (picking accuracy) which is 2 ms as well as by the location uncertainties (travel-time modeling uncertainties). Location errors of about 100 m lead to modeled travel-time inaccuracies of about 17 ms.

For the analysis presented here I used C_v = 400 m/s and two limiting values of data variances: C_d = 20 ms and C_d = 4 ms, which roughly corresponds to the uncertainties due to location errors and those due to the travel time onset picking accuracy.

4.2. *Tomography Images*

The main results of the tomography imaging are shown in Figs. 4 and 5.

Figure 4 shows the images obtained assuming data uncertainties C_d = 20 ms, while the velocity images obtained for the choice C_d = 4 ms are shown in Fig. 5. Only the cells well probed by ray paths (at least 10% of the maximum ray path coverage, see Fig. 2) are plotted in both figures. The remaining cells were excluded, as small coverage (below 10%) creates large imaging uncertainties and leads to unreliable results (Menke, 1989; Limes and Treitel, 1983; Backus and Gilbert, 1970).

The results from the images presented in Fig. 4 show that the two classical methods, namely the LSQR approach (left-hand figure) and the GA optimization approach (center figure) led to practically identical images. This suggests that the a posteriori imaging errors have a Gaussian distribution, and thus the use of the LSQR formula (Eq. 6) with the damping factor $\gamma = 5 \times 10^{-5}$ is justified and consistent with the result provided by the GA optimization.

The maximum likelihood solution (MC-ML) which maximizes the a posteriori *PDF* function has a structure similar to the GA and the LSQR solutions, but differs in terms of small-scale variations. The reason for this might be that the Metropolis sampler was unable to find the global optimum solution found by the genetic algorithm. The MC-ML solution represents the optimum solution among all the MC generated models, but only approaches the true global optimum found by the GA approach. Indeed, a comparison of the RMS residua for the LSQR GA and the MC-ML models listed in Table 2 shows that the MC-ML model leads to the largest residua. Thus the MC-ML solution is only an example of "sub-optimum" models. That is why we prefer to consider the MC-AV model as the "final" MC-based solution.

The average image found by the MC method is somewhat different. Although the main structure of the velocity distribution is the same as in the LSQR and GA cases, the amplitude of spatial velocity variation is slightly reduced and smoother. This is because the solution is obtained by averaging over a number of sub-optimum models generated with respect to the a posteriori *PDF*. Thus, any individual variations among the models which are almost equally probable as the optimum solution (for example, the MC-ML model) average out and only the basic structure common for all the models

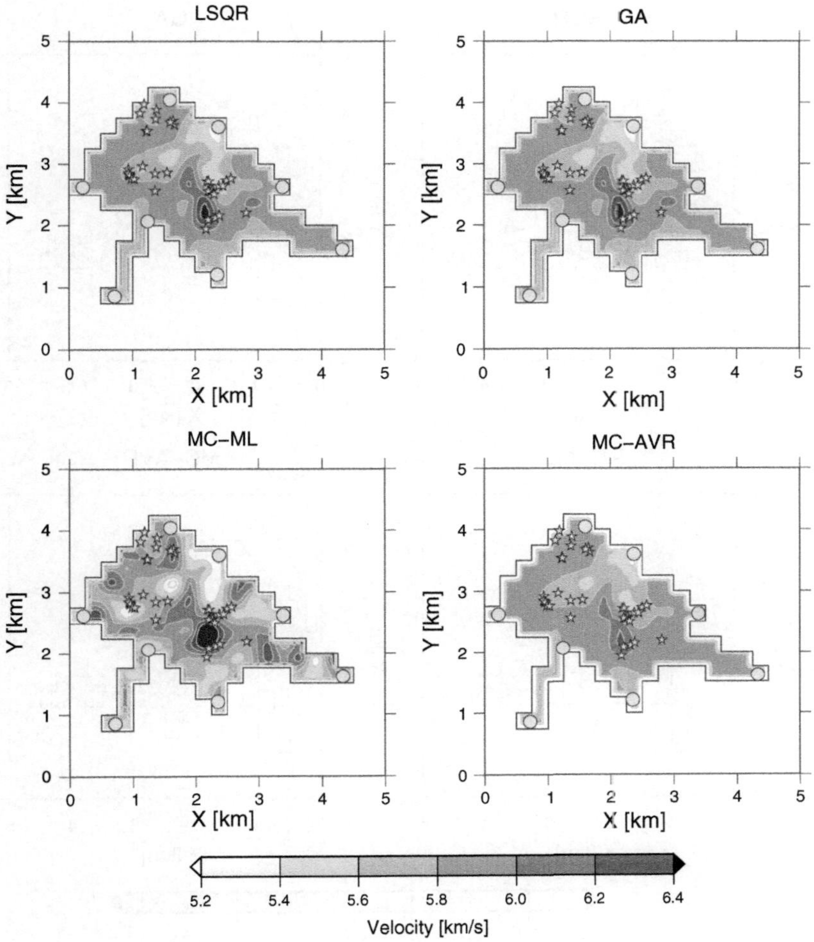

Figure 4
Velocity images obtained by the LSQR method (*upper-left*), genetic algorithm (GA) optimization approach (*upper-right*) and the maximum likelihood (MC-ML) and average (MC-AV) models obtained by the Monte Carlo sampling technique (*lower row*). All models were calculated assuming data uncertainties C_d = 20 ms, which corresponds to $\gamma = 5 \times 10^{-5}$. Note that the LSQR and GA tomograms are practically identical, while the MC tomogram exhibits the same pattern as the LSQR and GA distributions but with a reduced amplitude of velocity variations

remains. If the sampled models are strongly dispersed, the average procedure will naturally lead to a smoother solution. The degree of smoothing in the MC-AV solution with respect to the MC-ML solution is a relative measure of the accuracy and reliability of the tomogram. The smoother the average solution is with respect to the MC-ML, the less reliable is the MC-ML tomogram.

Figure 5 shows the velocity images obtained when data uncertainties were taken near the limit of travel-time picking accuracy, namely $C_d = 4$ ms. This choice corresponds to a much lower damping factor $\gamma = 10^{-5}$, as shown in Fig. 3, consequently a better fit to the data at the cost of a "rougher" model is expected. The upper-left panel in Fig. 5 shows that the velocity model obtained by the LSQR approach is significantly different from the previous result calculated for $\gamma = 5 \times 10^{-5}$ ms. It exhibits very large velocity variations, in some places exceeding the imposed physical limit of a 20% range. Thus this model cannot be regarded as physically reasonable.

The image provided by the GA procedure (upper-right panel) still resembles the smooth model from Fig. 4, but is characterized by smaller-scale variations

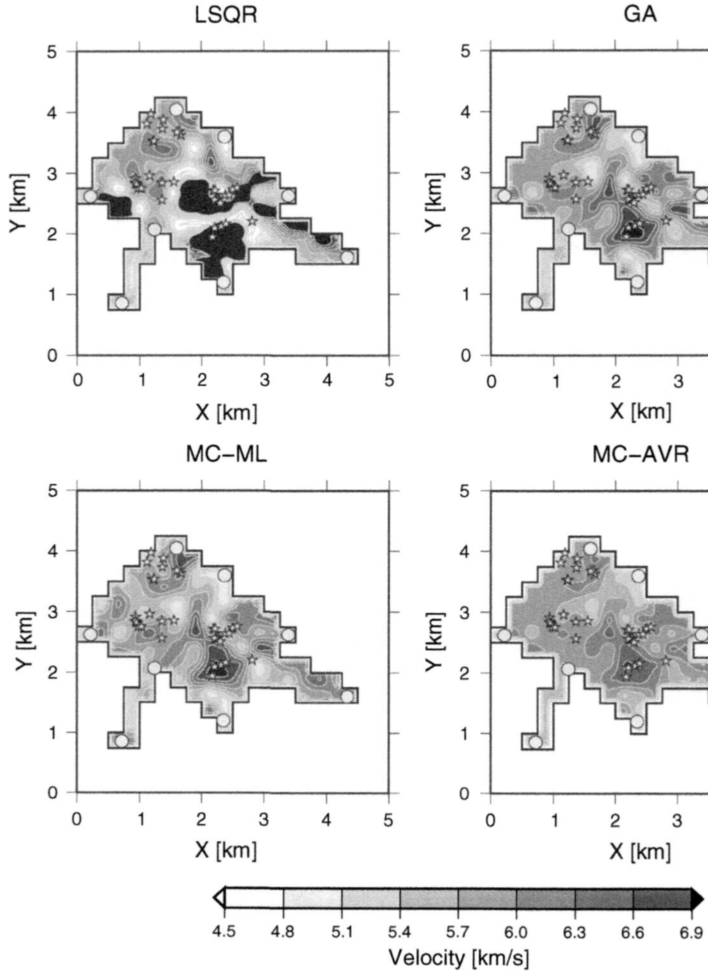

Figure 5
Velocity images obtained by the LSQR method (*upper-left*), genetic algorithm (GA) optimization approach (*upper-right*) and the maximum likelihood (MC-ML) and average (MC-AV) models obtained by the Monte Carlo sampling technique (*lower row*). All models were calculated assuming data uncertainties $C_d = 4$ ms, which corresponds to $\gamma = 10^{-5}$. The LSQR image is apparently destroyed by numerical artifacts, nonetheless due to "insufficient" damping. The GA image has also been seriously contaminated by numerical artifacts, but due to the robustness of the GA optimizer the basic structure of the velocity distribution visible in Fig. 4 is preserved. The MC-based tomogram has only been slightly influenced by the change of damping and shows only slightly larger velocity variations when compared to the previous solution (see Fig. 4)

likely due to noise. Thus, the optimization method still produces a reasonable velocity distribution although with exaggerated velocity contrasts. The fact that the GA and LSQR solutions are so different in this case is a consequence of the GA optimization method's much greater robustness. The MC-ML solution (lower-left panel) also behaves similarly to the GA method, where decreasing the damping leads to an increased roughness of the solution.

The situation is different for the MC-AV solution (lower-right panel). It is the smoothest model among the considered four solutions and, like the other models obtained for $C_d = 4$ ms, it shows a broadening of the high velocity zone located around the region $x = 2$ km, $y = 2$ km. In contrast to the GA and MC-ML models, changing the damping factor had very little influence on the "roughness" of the model. In fact, comparing the MC-AV solutions for the two

Table 2

Root mean square (RMS) values of residua calculated for the models obtained by LSQR, GA, and MC (the maximum likelihood (MC-ML) and average (MC-AV) models) for two considered data uncertainties

C_d (ms)	LSQR (ms)	GA (ms)	MC-ML (ms)	MC-AV (ms)
4	26.15	28.32	28.42	28.46
20	29.96	29.96	30.02	30.09

Note that for both C_d all models lead to very similar RMS values with slightly larger values obtained for the MC-ML solutions. The residua obtained for "under-damped" inversion (C_d = 4 ms) are smaller than for C_d = 20 ms as expected

damping factors, the main differences in the solutions are the aforementioned broadening of the high velocity zone as well as an increase in the velocity amplitude heterogeneities. Since choosing a proper regularization (damping) factor is always a difficult task and often subjective in any tomographic imaging problem, it is obvious why the weak sensitivity of the MC-AV solution to the damping factor is an important issue.

To explore this in more detail, let us consider the following. Let us assume that we can divide the inverted velocity image into a "noisy part" ($\mathbf{v}_n$) due to all the uncertainties, such as data noise, forward modeling inaccuracy, optimization errors, etc., and the remaining "true" distribution ($\mathbf{v}_b$) we would expect in the ideal case. What type of behavior would we expect for these two parts when γ or C_d change? A change in the damping factor will influence $\mathbf{v}_b$ and $\mathbf{v}_n$ in different ways. While $\mathbf{v}_b$ is expected to interpolate more or less smoothly between the a priori ($\gamma \to \infty$) model and the model which exactly fits the observational data ($\gamma = 0$), $\mathbf{v}_n$ will behave in an unpredictable way due to the randomness of all the uncertainties. This type of expected behavior is visible in the considered cases. In all but the LSQR solutions, the region of high velocity broadened when the damping factor was decreased. However, for the GA and MC-ML solutions, there is an additional increase in roughness . This increase in roughness or "noise" is much weaker in the case of the MC-AV solution. This suggests that the average model is a considerably more robust estimator of the true (noise-free part $\mathbf{v}_b$) velocity distribution than any of the maximum likelihood models. When averaged, this noise suppression effect is, in fact, very similar to noise reduction in the case of seismogram stacking (Claerbout, 1985) and follows from the incoherent nature of the noise (Jeffreys, 1983; Claerbout, 1985). Assuredly, any optimization algorithm which searches for a single, best solution, cannot perform noise reduction.

4.3. *Imaging accuracy*

The very first step in evaluating tomography results relies on inspecting the residuals for the "best" fit models. Figs. 6 and 7 display the residuals for the models obtained for C_d = 20 and 4 ms, respectively. Table 2 displays the root mean squares of the residuals calculated for all the models.

Note that all the residua histograms (for both C_d = 20 ms and C_d = 4 ms) are very similar, with a variance around 30 ms. The residuals exceed 50 ms only for a few data points. Thus the residuals suggest that the approximation of the sum of the travel time measurement and ray tracing errors by a Gaussian distribution (Gaussian Likelihood function) with variance C_d = 20 ms is justified. Furthermore Figs. 6 and 7 indicate that the choice C_d = 4 ms as the variance for the sum of observational and model errors is incompatible with the calculated residuals.

In the case of the well-damped inversion (C_d = 20 ms), all the methods led to similar residuals and hence, similar velocity images. It is interesting to note, however, that in the under-damped case (C_d = 4 ms) the differences between the velocity models are very substantial, yet the residuals are still similar for all four models. This demonstrates that evaluating the accuracy of a tomographic solution by inspecting the a posteriori residua alone is not sufficient. As in the studied case, there exist completely different models leading to, similar travel-time predictions.

A more comprehensive insight into imaging accuracy is obtained by inspecting the a posteriori covariance matrix which can be calculated easily as part of the MC technique. Figure 8 shows the spatial distribution of the velocity imaging errors estimated from the diagonal elements of the covariance matrix calculated from the a posteriori *PDF* for the MC solutions. Note that the a priori value C_v = 400 m/s is

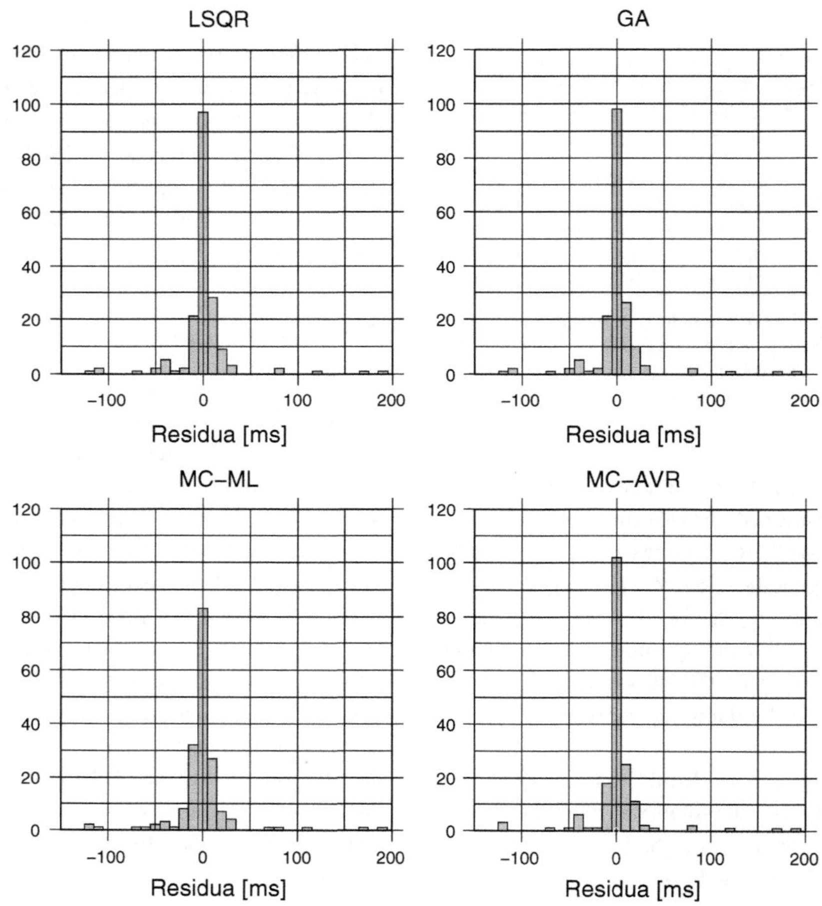

Figure 6
Travel time residua for the LSQR (*upper left*), GA (*upper right*), MC-ML (*lower left*), and MC-AV (*lower right*) velocity models obtained using $C_d = 20$ ms setting. In this case the GA and LSQR algorithms performed equally well. The average MC solution provides slightly more scattered residua. The MC-ML model leads to the largest residua

attached to cells probed by no rays. The obtained distribution shows that within the applied velocity parameterization (cell size) the imaging errors are almost homogeneously distributed over the imaged area and are approximately $\delta_v \sim 160$ m/s for the image obtained, assuming $C_d = 20$ ms and $\delta_v \sim 50$ m/s for $C_d = 4$ ms.

Comparing these results with the images in Figs. 4 and 5 indicates that great care must be taken when interpreting covariances as imaging errors. While such an interpretation seems to be justified in the case of the well-damped solution (Fig. 4), it is meaningless for the GA and LSQR in the case of the under-damped solutions, in which the final velocity images are strongly influenced by additional numerical artifacts far exceeding δ_v. The interpretation of the covariance matrix as a "measure" of inversion errors still seems to be justified (at least for this case) when the average model is taken as an estimator of the sought velocity distribution.

Estimating the imaging errors by the covariance matrix is justified if the a posteriori *PDF* has a "bell-shaped" and approximately symmetric, unimodal distribution. To verify whether this assumption is fulfilled in this study, I have selected a few cells (see Fig. 2) for which 1-D a posteriori marginal *PDF*'s were calculated according to Eq. 20. The results are shown in Figs. 9 and 10 for well-damped and under-damped cases, respectively. In both cases the marginal distributions resemble the Gaussian *PDF*,

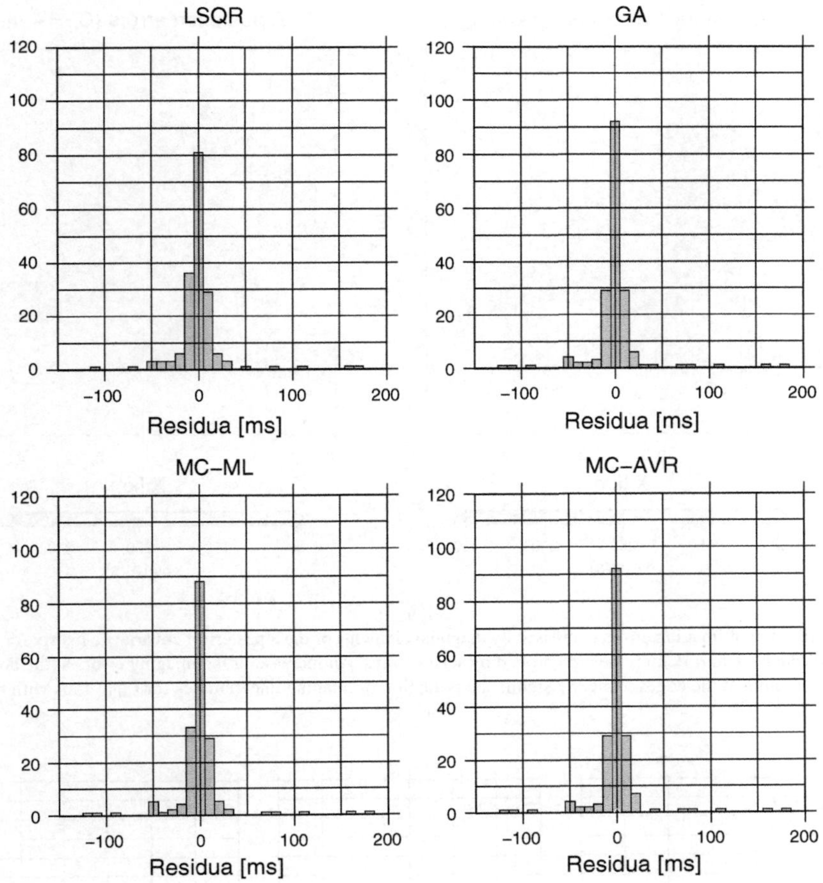

Figure 7
Travel time residua for the LSQR (*upper left*), GA (*upper right*), MC-ML (*lower left*), and MC-AV (*lower right*) velocity models obtained using C_d = 4 ms setting. Although all histograms are very similar, the corresponding models are quite different (see Fig. 5)

which means that the assumptions regarding the data and modeling error having a Gaussian distribution as well as the specific choice of the Gaussian a priori *PDF* were justified and consistent.

Finally, a question that remains to be addressed is which of the tomographic inversion methods is the preferred approach to estimate the true velocity distribution. Figures 11 and 12 compare the differences between the LSQR, GA, MC-ML models and the MC-AV model obtained for the two damping factors. For the case of a properly chosen damping factor all the models are compatible within one standard deviation range. For the LSQR solution, decreasing the damping factor leads to an unphysical solution. The remaining GA, MC-ML and MC-AV models are still compatible. However, for the reasons discussed above, the GA solution is expected to be a more reliable MLL estimator of the true velocity distribution than the MC-ML solution.

5. *Discussion*

Two types of conclusions can be drawn on the basis of the analyzed case study. The first concerns the performance of different inversion algorithms when applied to small-scale seismic tomography tasks. The second concerns a very preliminary quantitative interpretation of the obtained velocity images.

The MC sampling technique seems to provide the most robust estimation of the velocity distribution

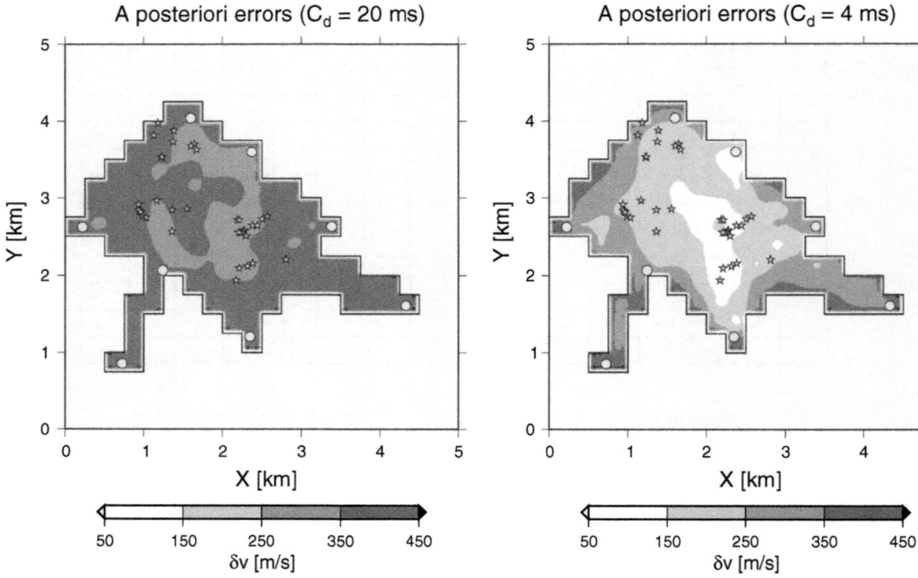

Figure 8
Plot of the spatial distribution of imaging errors expressed by diagonal elements of the a posteriori covariance matrix for $C_d = 20$ ms (*left*) and $C_d = 4$ ms (*right*) The *black contour* defines the area probed by seismic rays. An increase of the imaging errors at the boundary of the imaged area is due to poorer sampling at the edges. Note the significant reduction of imaging uncertainties for $C_d = 4$ ms with respect to the $C_d = 20$ ms case

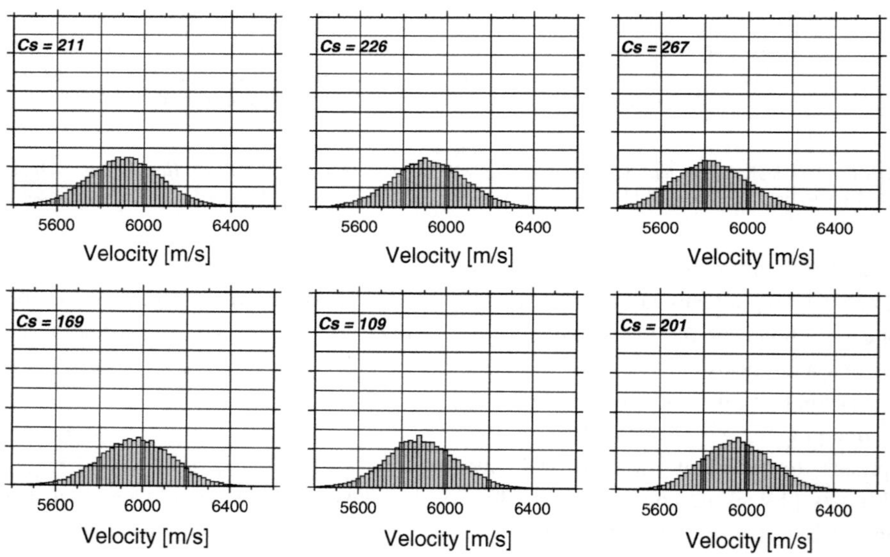

Figure 9
Examples of the a posteriori 1D *PDF* functions for a few selected cells (see Fig. 2) for the $C_d = 20$ ms case. Note that the velocity heterogeneities shown in Fig. 4 are lower than the imaging errors

compared to any of the other inversion techniques based on the maximum likelihood estimator. In fact, the average image found by the MC technique was much smoother and, more importantly, provided a more realistic solution than the other two methods.

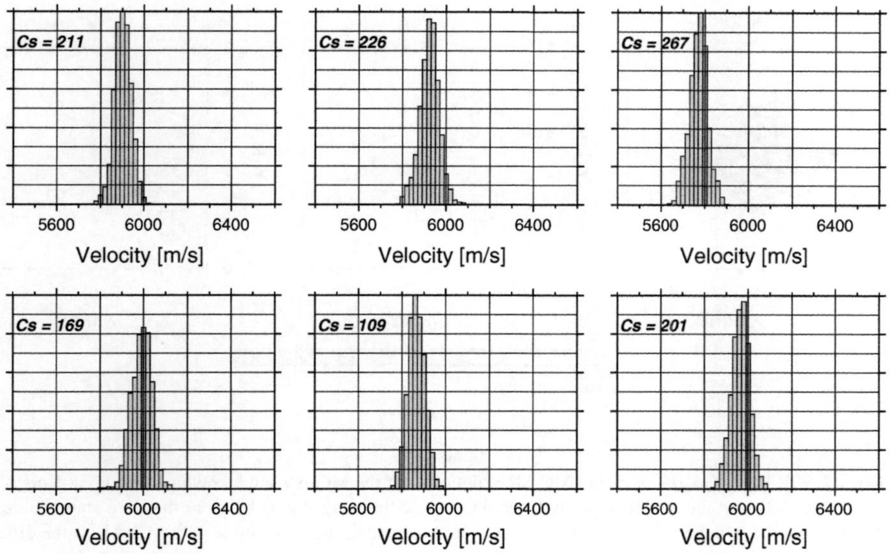

Figure 10
Examples of the a posteriori 1-D marginal *PDF* functions for a few selected cells (see Fig. 2) for the $C_d =$ 4 ms case. In this case the velocity resolution is sufficient to identify regions with velocity contrasts larger than the imaging errors

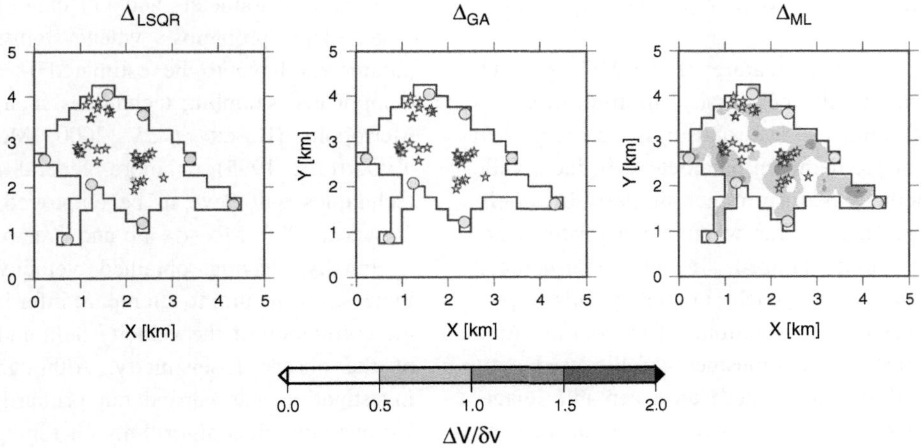

Figure 11
The difference between LSQR (*left*), GA (*middle*) and MC-ML solutions and the MC-AV model obtained for C_d = 20 ms scaled down by the a posteriori imaging errors. Note that the differences between the LSQR, GA and MC-AV models are smaller than half the standard deviation (imaging error) in the whole imaged area. This means that the LSQR, GA, and MC-AV models are compatible with each other at a 68% confidence level. Only in the case of the MC-ML solutions the difference occasionally increases to one standard deviation

This indicates that sub-optimum models which lead to similar residua such as the GA or LSQR solutions and provide the main contribution to the MC-AV model are quite different. However, when summed up, the individual variations among them are smoothed out and the basic structure common to all the models remains. This is an important point and suggests that the MC technique is the most favorable and robust inversion approach.

In fact, this noise suppression effect is similar to that encountered in seismic stacking procedures (CLAERBOUT, 1985). This possibility of enhancing the

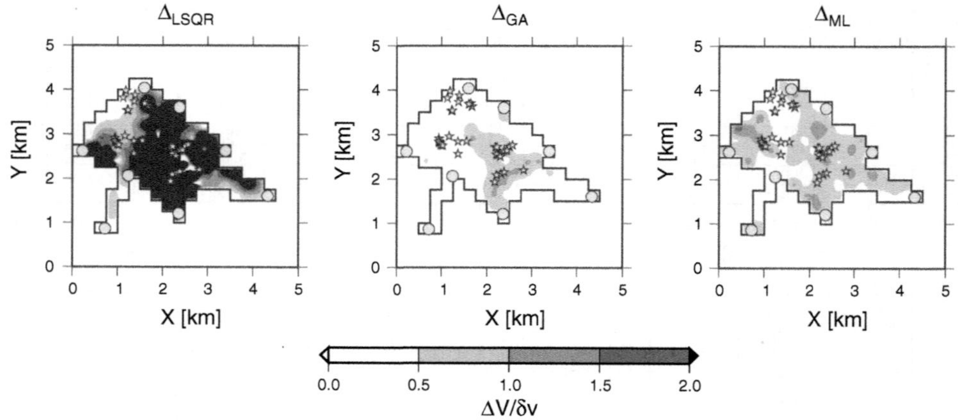

Figure 12
The difference between LSQR (*left*), GA (*middle*) and MC-ML solutions and the MC-AV model obtained for $C_d = 4$ ms scaled down by the a posteriori imaging errors. Note that the LSQR solution differs in most of the imaged area by more than two standard deviations (imaging errors) from the MC-AV solution. On the other hand the GA and MC-ML models are compatible with MC-AV as the differences reach one standard deviation (value of the imaging errors) only in part of the imaged area. This means that the three solutions, namely GA, MC-ML and—MC-AV are statistically equivalent at a 68% confidence level

imaging accuracy by the proper choice of the inversion approach makes the MC technique more versatile than any other tomography technique known today.

Another important feature of the MC approach when compared with the classical (optimization and algebraic) techniques is its robustness with respect to the choice of the damping parameters. In fact, while all three methods perform rather similarly in a well-damped case, they differ when the algorithms are significantly under-damped. In an under-damped case, the algebraic approach completely fails to produce a reasonable solution. The optimization technique, due to the robustness of the used optimizer, is still able to produce an acceptable image. However, the solution is strongly influenced by numerical noise. On the other hand, in spite of apparently insufficient damping, the MC technique provides an image which differs only slightly from the result obtained for an optimum choice of γ.

Following these arguments, one can state that the Bayesian probability inversion approach with Monte Carlo sampling for the tomographic imaging problem is a very promising technique which is able to provide robust and generally more reliable images than currently used techniques. Its application is only limited by the size of the problem in hand. In the studied case, the simple Metropolis sampling algorithm was efficient enough to sample the a posteriori *PDF* due to the very small scale of the problem (109 model parameters and 177 data). In the case of larger-scale problems, when significantly more parameters have to be estimated (sampled), more complicated sampling techniques such as multi-step Metropolis (Bosch *et al.*, 2000; Mosegaard and Tarantola, 1995) or more general random walk techniques will have to be employed (Robert and Casella, 1999; Mosegaard and Tarantola, 2002).

Finally, having obtained velocity distribution images, it is natural to attempt to infer information on the correlation of the velocity field and the structure of the observed seismicity. Although the current investigation was carried out primarily to evaluate various numerical algorithms for tomographic inversion, the obtained MC results seem to be robust enough to draw preliminary conclusions. Firstly, note that the events used in the current studies formed four different spatial clusters, as shown in Fig. 2. Excessively large uncertainties in clusters A and C prevent any reasonable correlation of seismicity in these two clusters with velocity distribution. In clusters B and D the MC velocity image has a high enough resolution to distinguish between fine velocity heterogeneities with sufficient precision. The obtained tomograms indicate that clusters B and D are located in regions of high spatial velocity gradients. The geomechanical interpretation of this fact is not quite clear. One possible explanation of this fact is the hypothesis on the existence of a spatially complex zone consisting of "hard" and "weak" parts. Physically, the "hard" part may consist of intact rock masses which have the possibility of accommodating larger stresses and thus showing higher velocities. The weaker part could be partially crushed rock masses. In such a condition we can expect a transfer of large stresses from the hard part to the weak one which cannot withstand them and will crush. Such a process would result in localized (clustered) seismicity.

Acknowledgments

I would like to thank Prof. S. J. Gibowicz for years of very fruitful collaboration, many interesting discussions and all the assistance he has given me. C. Trifu, the editor of this special issue, and the anonymous reviewers are acknowledged for their efforts in significantly enhancing the paper. This work was partially supported by the Polish Committee for Scientific Research under Grant No. 2P04D 033 30.

References

AKI, K. and RICHARDS, P.G., *Quantitative Seismology* (Freeman and Co, San Francisco 1985).

BACKUS, G. and GILBERT, F. (1970), *Umiqueness in the inversion of inaccurate gross earth data*, Phil. Trans. Roy. Soc. Lond. *266*:123–192.

BOSCH, M. (1999), *Lithologic tomography: From plural geophysical data to lithology estimation*, J. Geophys. Res. *104*, 749–766.

BOSCH, M., BARNES, C., and MOSEGAARD, K. *Multi-step samplers for improving efficiency in probabilistic geophysical inference.*In: *Methods and Application of Inversion* (eds. P. C. Hansen, B. H. Jacobsen, K. Mosegaard) (Springer, Berlin 2000), vol. 92 of *Lecture Notes in Earth Sciences*, pp. 50–68.

BRANDT, S., *Data Analysis. Statistical and Computational Methods for Scientists and Engineers.* (Springer-Verlag 1999), third ed.

CARDARELLI, E. and CERRTO, A. (2002), *Ray tracing in elliptical anisotropic media using the linear traveltime interpolation (LTI) method applied to traveltime seismic tomography*, Geophys. Prosp. *50*, 55–72.

CASELLA, R., *Monte Carlo Statistical Methods*, Springer Texts in Statistics. Springer-Verlag, New York 1999).

CLAERBOUT, J.F., *Imaging the Earth's Interior*. (Blackwell Sci. Publ., Boston 1985).

CURTIS, A. and LOMAX, A. (2001), *Prior information, sampling distributions and the curse of dimensionality*, Geophysics *66*, 372–378.

DĘBSKI, W. (1997a), *The probabilistic formulation of the inverse theory with application to the selected seismological problems*, Publs. Inst. Geophys. Pol. Acad. Sc. *B19*, 1–173.

DĘBSKI, W. *Study of the image reconstruction accuracy of active amplitude tomography*, in: *Rockburst and Seismicity in Mines* (eds. S. J. Gibowicz and S. Lasocki) (Balkema, Postbus 1675, Rotterdam, Nederland 1997b), pp. 141–144.

DĘBSKI, W. (2002), *Seismic tomography software package*, Publs. Inst. Geophys. Pol. Acad. Sci. *B-30*, 1–105.

DĘBSKI, W. (2004), *Application of Monte Carlo techniques for solving selected seismological inverse problems*, Publs. Inst. Geophys. Pol. Acad. Sci *B-34*, 1–207.

DĘBSKI, W. (2008), *Estimating the earthquake source time function by Markov Chain Monte Carlo sampling*, Pure Appl. Geophys. 165:1263–1287. doi:10.1007/s00024-008-0357-1.

DĘBSKI, W. and YOUNG, R.P. (2002), *Tomographic imaging of thermally induced fractures in granite using Bayesian inversion*, Pure Appl. Geophys. *159*, 277–307.

DĘBSKI, W., GUTERCH, B., LEWANDOWSKA, H., and LABAK, P. (1997), *Earthquake sequences in the Krynica region, Western Carpathians, 1992 – 1993*, Acta Geophys. Pol. *XLV*, 255–290.

DEAL, M. NOLET, G. (1996), *Comment on 'Estimation of resolution and covariance for large matrix inversions' by Zhang J. and McMechan G*, Geophys. J Int *127*, 245–250.

DEANS, S. R., *The Radon Transform and Some of Its Applications* (John Wiley and Sons, New York 1983).

DUIJNDAM, A. (1988), *Bayesian estimation in seismic inversion. part II: Uncertainty analysis*, Geophys. Prosp. *36*, 899–918.

GAMERMAN, D. *Markov Chain Monte Carlo: Stochastic Simulation for Bayesian Inference.* (Chapman and Hall 1997).

GELMAN, A., CARLIN, J.B., STERN, H.S., RUBIN, D. B., *Bayesian Data Analysis* (Chapmann & Hall 1997).

GILKS, W., RICHARDSON, S., SPIEGELHALTER, D., *Markov Chain Monte Carlo in Practice* (Chapman& Hall/CRC Press 1995).

GILLESPIE, D. T. *Markov Processes - An Introduction for Physical Scientists.* (Academic Press, Inc., San Diego 1992).

HASTINGS, W. K. (1970), *Monte Carlo sampling methods using Markov chains and their applications*, Biometrica *57*, 97–109.

IYER, H. HIRAHARA, K., *Seismic Tomography, Theory and Practice* (Chapman and Hall, London 1993).

JACKSON, D. D. and MATSU'URA, M. (1985), *A Bayesian approach to nonlinear inversion*, J. Geophys. Res. *90*, 581–591.

JEFFREYS, H. *Theory of Probability* (Clarendon Press, Oxford 1983).

KIJKO, A. (1994), *Seismological outliers: L1 or adaptive Lp norm application*, Bull. Seismol. Soc. Am. *84*, 473–477.

LIMES, L. R. and TREITEL, S. (1983), *Tutorial, a review of least-squares inversion and its application to geophysical problems*, Geophys. Prospect. *32*, 159–186.

LOMAX, A., VIRIEUX, J., VOLANT, P., BERGE, C. *Probabilistic earthquake location in 3D and layered models: Introduction of a Metropolis-Gibbs method and comparison with linear location.* In: *Advances in Seismic Event Location* (eds C. Thurber, E. Kissling, and N. Rabinovitz) (Kluwer, Amsterdam 2000).

MAXWELL, S. C. YOUNG, R. P. (1993), *A comparison between controlled source and passive source seismic velocity images*, Bull. Seismol. Soc. Am. *83*, 1813–1834.

MENKE, W. *Geophysical Data Analysis: Discrete Inverse Theory*, International Geophysics Series (Academic Press, San Diego 1989).

METROPOLIS, N., ROSENBLUTH, A., ROSENBLUTH, M., TELLER, A., and TELLER, E .(1953), *Equation of state calculations by fast computing machines*, J. Chem. Phys. *21*, 1087–1092.

MICHALEWICZ, Z. *Genetic Algorithms + Data Structures = Evolution Programs*. (Springer-Verlag, Berlin 1996).

MOSEGAARD, K. and TARANTOLA, A. (1995), *Monte Carlo sampling of solutions to inverse problems*, J. Geophys. Res. *100*, 12431–1247.

MOSEGAARD, K., and TARANTOLA, A. *International Handbook of Earthquake & Engineering Seismology*, (Academic Press 2002), chap. Probabilistic Approach to Inverse Problems, pp. 237–265.

NOLET, G. *Seismic Tomography* (D. Reidel Publishing Company, Dordrecht 1987).

NOLET, G., MONTELLI, R., and VIRIEUX, J. (1999), *Explicit, approximate expressions for the resolution and a posteriori covariance of massive tomographic systems*, Geophys. J. Int. *138*, 36–44.

NOLET, G., MONTELLI, R., and VIRIEUX, J. (2001), *Replay to comment by Z. S. Yao and R. G. Roberts and A. Tryggvason on 'Explicit, approximate expressions for the resolution and a posteriori covariance of massive tomographic system'*, Geophys. J. Int. *145*, 315.

PARKER, R. L. *Geophysical Inverse Theory* (Princeton University Press, New Jersey 1994).

ROBERT, C. P. and CASELLA, G. *Monte Carlo Statistical Methods* (Springer Verlag 1999).

SAMBRIDGE, M. (1999) *Geophysical inversion with a neighbourhood algorithm - I. Searching a parameter space*, Geophys. J. Int. *138*, 479–494.

SAMBRIDGE, M. and MOSEGAARD, K. (2002), *Monte Carlo methods in geophysical inverse problems*, Rev. Geophys. *40*, 3.1–3.29.

SAMBRIDGE, M. S. and KENNETT, B. L. N. (2001), *Seismic event location: Nonlinear inversion using a Neighbourhood Algorithm*, Pure Appl. Geophys. *158*, 241–257.

SCALES, J. A. *Uncertainties in seismic inverse calculations*. In: *Inverse Methods Interdisciplinary Elements of Methodology, Computation, Application* (eds B. H. Jacobsen, K. Moosegard, and P. Sibani) (Springer-Verlag, Berlin 1996), vol. 63 of Lecture Notes in Earth Sciences, pp. 79–97.

SCALES, J. A. and SNIEDER, R. (1997), *To Bayes or not to Bayes?* Geophysics *63*, 1045–1046.

SEN, M. and STOFFA, P. L., *Global Optimization methods in Geophysical Inversion*, vol. 4 of *Advances in Exploration Geophysics* (Elsevier, Amsterdam 1995).

TARANTOLA, A., *Inverse Problem Theory: Methods for Data Fitting and Model Parameter Estimation*, (Elsevier, Amsterdam 1987).

TARANTOLA, A., *Inverse Problem Theory and Methods for Model Parameter Estimation* (SIAM, Philadelphia 2005).

TARANTOLA, A. and VALLETE, B. (1982), *Inverse Problems = Quest for Information*, J. Geophys. *50*, 159–170.

TAYLOR, S. R., YANG, X., and PHILIPS, S. (2003), *Bayesian l_g attenuation tomography applied to Eastern Asia*, Bull. Seismol. Soc. Am. *93*, 795–803.

TIERNEY, L. (1994), *Markov chains for exploring posterior distributions*, Ann. of Stat. *22*, 1701–1762.

VAN KAMPEN, N., *Stochastic Processes in Physics and Chemistry* (Elsevier 1992).

WIEJACZ, P. and DĘBSKI, W. (2001), *New observation of Gulf of Gdansk Seismic Events*, Phys. Earth Planet Int. *123*, 233–245.

YAO, Z. S., ROBERTS, R. G., and TRYGGVASON, A. (1999) *Calculating resolution and covariance matrices for seismic tomography with the LSQR method*, Geophys. J. Int. *138*, 886–894.

YAO Z. S., ROBERTS, R. G., TRYGGVASON, A. (2001) *Comment on 'Explicit, approximate expressions for the resolution and* a posteriori *covariance of massive tomographic system' by G. Nolet, R. Montelli and J. Virieux*, Geophys. J. Int. *145*, 307–314.

ZHANG, J. MCMECHAN, G. (1996), *Replay to comment by M. Deal and G. Nolet on 'Estimation of resolution and covariance for large matrix inversions'*, Geophys. J. Int. *127*, 251–252.

ZHAO, D. (2001), *New advances of seismic tomography and its applications to subduction zones and earthquake fault zones: A review*, The Island Arc *10*, 68–84.

(Received August 29, 2008, revised April 14, 2009, accepted April 30, 2009, Published online November 25, 2009)

Pure Appl. Geophys. 167 (2010), 153–167

DOI 10.1007/s00024-009-0011-6

Pure and Applied Geophysics

Characterizing and Locating Very Weak ($-2.2 \geq M_L \geq -3.4$) Induced Seismicity in Unstable Sandstone Cliffs by Nanoseismic Monitoring

HILLEL GILLES WUST-BLOCH[1]

Abstract—Unloaded natural rock masses are known to generate seismic signals (GREEN *et al.*, 2006; HAINZL *et al.*, 2006; HUSEN *et al.*, 2007; KRAFT *et al.*, 2006). Following a 1,000 m^3 mass failure into the Mediterranean Sea, centimeter-wide tensile cracks were observed to have developed on top of an unstable segment of the coastal cliff. Nanoseismic monitoring techniques (WUST-BLOCH and JOSWIG, 2006; JOSWIG, 2008), which function as a seismic microscope for extremely weak seismic events, were applied to verify whether brittle failure is still generated within this unconsolidated sandstone mass and to determine whether it can be detected. Sixteen days after the initial mass failure, three small-aperture sparse arrays (Seismic Navigation Systems-SNS) were deployed on top of this 40-m high shoreline cliff. This paper analyzes dozens of spiky nanoseismic ($-2.2 \geq M_L \geq -3.4$) signals recorded over one night in continuous mode (at 200 Hz) at very short slant distances (3–67 m). Waveform characterization by sonogram analysis (JOSWIG, 2008) shows that these spiky signals are all short in duration (>0.5 s). Most of their signal energy is concentrated in the 10–75 Hz frequency range and the waveforms display high signal similarity. The detection threshold of the data set reaches M_L −3.4 at 15 m and M_L −2.7 at 67 m. The spatial distribution of source signals shows 3-D clustering within 10 m from the cliff edge. The time distribution of M_L magnitude does not display any decay pattern of M_L over time. This corroborates an unusual event decay over time (modified Omori's law), whereby an initial quiet period is followed by regained activity, which then fades again. The polarization of maximal waveform amplitude was used to estimate spatial stress distribution. The orientation of ellipses displaying maximal signal energy is consistent with that of tensile cracks observed in the field and agrees with rock mechanics predictions. The M_L– surface rupture length relationship displayed by our data fits a constant-slope extrapolation of empirical data collected by WELLS and COPPERSMITH (1994) for normal fault features at much larger scale. Signal characterization and location as well as the absence of direct anthropogenic noise sources near the monitoring site, all indicate that these nanoseismic signals are generated by brittle failure within the top section of the cliff. The atypical event decay over time that was observed suggests that the cliff material is undergoing post-collapse bulk strain accommodation. This feasibility study demonstrates the potential of nanoseismic monitoring in rapidly detecting, locating and analyzing brittle failure generated within unconsolidated material before total collapse occurs.

Key words: Nanoseismic monitoring, induced seismicity, microseismicity, brittle failure, unstable cliffs, event decay.

This paper is dedicated to my father, whose unconditional support and intellectual curiosity I dearly miss.

[1] Department of Geophysics and Planetary Sciences, Tel Aviv University, 69978 Tel Aviv, Israel.
E-mail: hillelw@post.tau.ac.il

1. *Introduction*

The transfer of tidal energy into the Israeli Mediterranean seashore results in sustained coastal erosion. The coastline has been observed to recede at an overall rate of 50 cm per year (ALMAGOR, 2005; PERATH and ALMAGOR, 1996; ZVIELY and KLEIN, 2004). This process is more dramatic where the coast consists of weakly consolidated sandstone (*kurkar*) cliffs (KATZ *et al.*, 2007; WISEMAN *et al.*, 1981) as it exposes the vulnerability of numerous public structures, private housing and archeological sites (KATZ *et al.*, 2007). At the Apollonia archaeological site, significant sections of a Roman fortress built on a 40-m high *kurkar* cliff can be observed to hang in the air or have already been washed out by the waves. On June 29, 2006, a 1,000-m^3 mass of *kurkar* failed into the sea (Fig. 1), along an inhabited coast section north of Apollonia (pers. comm. Dr. S. Marco 2006). This failure left a large scarp with a sizable (200 m^3) overhanging block at the top of the 40 m high cliff (Fig. 2), on which an asphalted patrol road and a fence were partly destroyed and left hanging in the air (Fig. 1). Centimeter-wide tension cracks could be observed in the soil above this section of the cliff (Fig. 3). Given the hazard that this situation presents for the public, the wildlife and the remaining

Figure 1
View (in the north direction) of the cliff 16 days after the initial collapse. The collapsed material can be seen in the talus below the overhanging cliff section. The three SNS were deployed on top of the cliff, beyond the destroyed fence

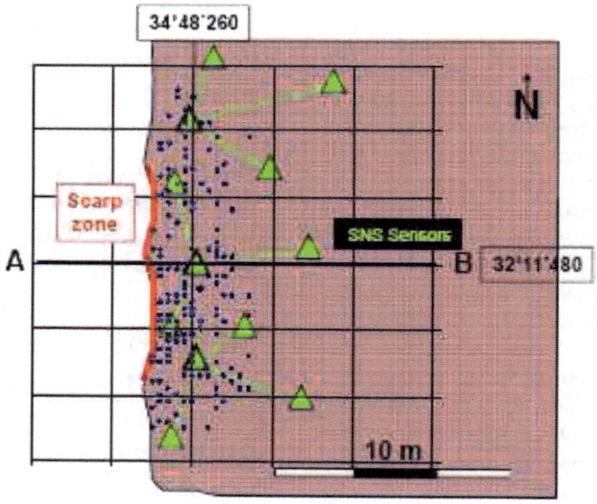

Figure 2
Map view of the experimental setup: Location of the three SNS monitoring systems (*normal thickness triangles* are 1D sensors; *thick triangles* are 3-D sensors), scarp zone and nanoseismic events (*black dots*). Note location of profile A-B (Fig. 4)

infrastructures, it was imperative to rapidly assess the stability of the remaining cliff section.

Standard geodetic and optical monitoring techniques require extended monitoring, and rely on the detection of small surface movements, as they are unable to monitor the true source process. Induced seismicity has been detected for decades in the most varied environments but detection has been limited primarily to source processes generated within hard rocks. At local to close (10^2–10^3 m) distances, magnitude thresholds rarely reach above M_L 0.0. Such a condition is perfectly suited to the application of nanoseismic monitoring (Joswig, 2008), a type of seismic microscope that locates, in space and time,

Figure 3
N-S oriented tensile cracks visible in the soil above the cliff. Crack length of individual segments varies between 15 and 35 cm and their opening can reach 25 mm

very weak source signals. Nanoseismic monitoring, which has been employed in detecting subsurface impacts in active sinkholes (WUST-BLOCH and JOSWIG, 2006) and rain-induced failure in landslides (WALTER and JOSWIG, 2008) has overcome two obstacles: extremely weak source processes ($M_L \geq -4.0$) and unconsolidated host material. The need for quick, inexpensive and reliable monitoring of pre-collapse failure within the cliff makes nanoseismic monitoring a particularly attractive option.

This paper presents a complete analysis of dozens of nanoseismic ($-2.2 \geq M_L \geq -3.4$) signals generated by brittle failure within a weakly consolidated sandstone (*kurkar*) cliff.

2. *Data Acquisition*

The data were acquired by deploying three Seismic Navigation System (SNS) on top of the unstable cliff section (Fig. 2). SNS are portable sparse arrays consisting of three vertical-component seismometers (1 Hz) arranged as a tripartite array around a single three-component center sensor (1 Hz) (WUST-BLOCH and JOSWIG, 2006). The experiment layout, which was characterized by a 10-m array aperture and an imbricate geometry (Fig. 2), was selected to minimize slant distance from source processes potentially occurring nearest the scarp left in the cliff. This is a much denser sensor coverage than that being typically used for cavitation (WUST-BLOCH and JOSWIG, 2006) or landslide (WALTER and JOSWIG, 2008) monitoring. Three SNS were deployed simultaneously during one night (12.07.2006) on the extreme edge of the IMI compound (N32°11′480; E34°48′260). Sixteen hours of continuous monitoring were recorded at the sampling rate of 200 Hz. The instrumental deployment and our investigations were guided by a series of test nanoseismic monitoring surveys that had been carried out on the same coastal cliff section, 2,300 m further to the north at Kibbutz Ga'ash (WUST-BLOCH, 2008). These provided general background noise conditions of the site and helped identify waveform patterns generated by a series of common noise sources (waves, agricultural vehicles, pumping). As most of the energy generated by sea waves is located below 1 Hz, it does not affect our ability to detect high frequency noise spikes associated with material failure within the cliff. On-site signal characterization enabled the unequivocal identification and screening-out of the following noise sources: material collapse along the cliff (avalanche and rockfall); material impact on the beach; anthropogenic (pedestrians and cyclists) or animal activity; low flying aircraft. Nighttime peak-to-peak ground noise amplitudes of 300–350 nm/s for 0.3–10 Hz are higher than the stable ground noise level of 100 nm/s recorded at the sub-surface EIL station of the Israel Seismic Network (ISN). However, these values compare well with the ambient 550 nm/s recorded during nanoseismic monitoring along the Dead Sea (WUST-BLOCH and JOSWIG, 2006; WUST-BLOCH and JOSWIG, in review; WUST-BLOCH *et al.*, in review). The region is considered seismically inactive, as it is located more than 50 km away from the earthquake source regions of the Dead Sea Fault zone and of the Carmel Fault zone. However, given the detection capacity of our SNS, a very weak (M_L −0.2) local earthquake was detected 2,700 m off shore.

The monitoring site is located on the top of the 40-m high cliff at the edge of the closed IMI compound. The cliff material consists of weakly consolidated sandstone, displaying minor and random signs of fracturation. Although the sandstone displays irregular cross-bedding structures throughout, their thickness (ranging between single digits to several dozen centimeters) and dip varies greatly, while the overall dominant bedding is horizontal. The cliff-face appears dry and stands near vertical. If it was not for the violent sea-driven winter erosion undermining their base and causing a yearly retreat of the coastline (50–100 cm/year) (ZVIELY and KLEIN, 2004), these tall cliffs would only undergo a slow natural weathering process. Seismic velocities were estimated along the cliff by recording hammer impacts at regular slant distances, up to 150 m. *Vp* distribution as a function of depth is: 280 m/s (0–5 m); 350 m/s (5–15 m) and 450 m/s below 15 m (Fig. 4). Note that these ultralow *Vp* values are in good agreement with those of seismic field monitoring at very small-scale (GREEN *et al.*, 2006; WUST-BLOCH and JOSWIG, 2006), whereby a medium with 17% of its volume with air-filled voids can see its original *Vp* reduced by a factor 2–5 (*Vp* intact sandstone ~2,000 m/s).

3. Data Processing

Initial events screening was carried out through sonogram analysis: SonoDet (JOSWIG, 1995), a pattern recognition-based scheme providing an auto-adaptive broadband optimum filter discriminator for signal energy (JOSWIG, 1990; 1999). Discrimination between noise signals and those potentially generated within the cliff was guided by previous knowledge of local noise sources and by past nanoseismic monitoring campaigns (WALTER and JOSWIG, 2008; WUST-BLOCH, 2008; WUST-BLOCH and JOSWIG, 2006; WUST-BLOCH *et al.*, 2006). Waveform discrimination tests screened out dozens of signals and left a catalog of 175 good quality events. The event analysis program HypoLine (JOSWIG, 2008) was used to analyze, locate and determine the energy of these signals.

The source of each event was located in network mode (12 stations) with a three-layer velocity model (280; 350; 450 m/s) that was customized on-site (Fig. 4) and which is consistent with values used for nanoseismic monitoring within weak material at shallow depth (WALTER and JOSWIG, 2008; WUST-BLOCH, 2008; WUST-BLOCH and JOSWIG, 2006). *P*- and *S*-phase onsets on 12 traces as well as depth were displayed and updated in real time, using the interactive HypoLine software. Dozens of alternative solutions were tested within minutes, enabling the selection of the most plausible combination of parameters for events just marginally above ambient noise level (JOSWIG, 2008). This immediate parameter constraint technique is an extremely useful tool when analyzing events at such short slant distances as the slightest material heterogeneity may drastically filter waveforms in specific paths. The spatial control (visible in Fig. 2) is set to steps of 1 m, which is adequate for engineering purposes. Controlled impacts by hammer at the surface of the cliff were used to further benchmark lateral location accuracy and verify solution stability. The direct location of the sensors right above the source further stabilizes the depth determination. In addition, the straightforward structure of the cliff and the absence of significant material discontinuities make it permissible to disregard solutions that would have to include considering non-direct paths.

Signal energy was estimated by a procedure developed for nanoseismic monitoring (WUST-BLOCH and JOSWIG, 2006). First, as the standard distance correction relationship given by RICHTER (1958) does not extend for events with slant distances below 10 km, it has been explicitly extended three orders of magnitude, down to 10 m. Because of increasing material-waveform interactions and material inhomogeneities at such scale, the stability of the relationship is verified by in situ calibration. Second, M_L magnitudes are estimated for the overall waveform as *P* and *S* phases are concatenated at extremely

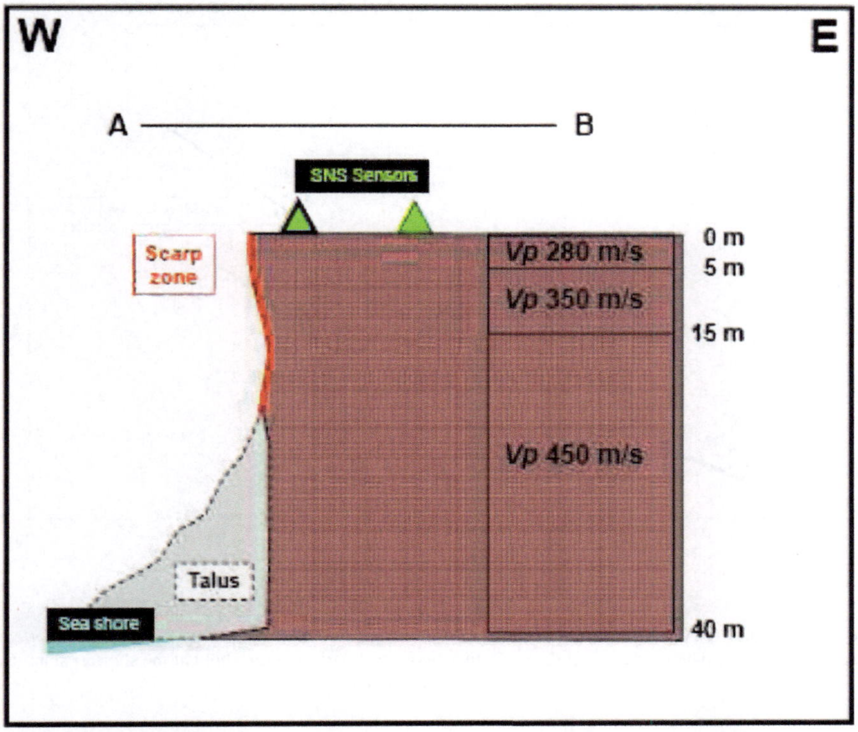

Figure 4
Cross section of the experimental setup (Location of profile A-B on Fig. 2): Location of one SNS monitoring system (*normal thickness triangles* are 1D sensors; *thick triangles* are 3-D sensors), scarp zone and talus material generated by the initial cliff failure. A profile (*right*) shows the distribution of *Vp* (m/s) as a function of depth

short slant distances (<100 m), thus departing from standard practice using exclusively P and P_n phases (IDC 2002). M_L magnitudes are computed on horizontal traces (N-S & E-W) for the maximum peak-to-peak amplitude of a full, broad-band (0.5–75 Hz) waveform cycle over the total signal duration.

The M_L–distance relationship is used to assess the event detection threshold and evaluate potential sensitivity bias on data interpretation (Fig. 5). It defines constant signal-to-noise ratio (SNR) conditions (amplitude corrected for distance) for a given reference area where it was calibrated (Wust-Bloch and Joswig, 2006). The detection threshold of the data collected on the cliff (shown by the green surface, bottom left) reaches M_L −3.4 at 15 m and M_L −2.7 at 67 m. The plain green curve marks the lower instrumental sensitivity limit of the SNS for cliff-generated events. The M_L–distance relationship confirms that the location of signals stronger than M_L −3.2 is unbiased by instrument sensitivity within a radius of 37 m. In practical terms, it means that if events of the same type were naturally generated along the cliff, beyond the scarp zone, one of the two SNS deployed at the periphery would also have detected them. In other words, the signals analyzed further are undoubtedly triggered by material instabilities, which occurred within this specific section of the cliff at the time of the monitoring. This relationship has been extended with constant-slope (green dashed curve) to larger source-to-sensor distances to allow for comparison with the detection threshold of four other nanoseismic monitoring campaigns:

1. natural seismicity [purple curve] (Häge and Joswig, 2004),
2. immediate aftershocks [red curve] (Wust-Bloch and Joswig, 2009),
3. sinkhole collapse activity [blue curve] (Wust-Bloch and Joswig, 2006),

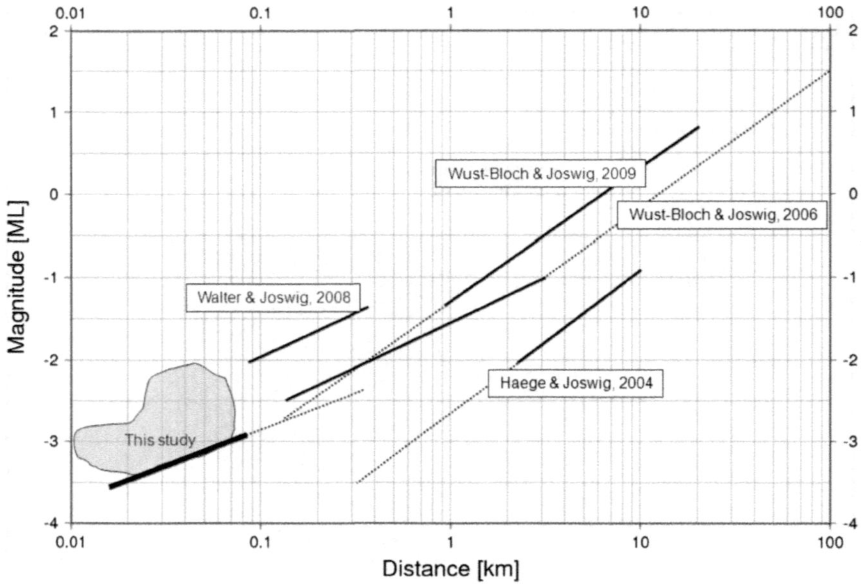

Figure 5
M_L–distance relationship at local (>20 km) slant distances for different nanoseismic monitoring experiments (*dotted curve sections* are constant slope extrapolations). Cliff data (surface in *pale gray*) display optimal detection threshold at the shortest slant distances and for very low M_L magnitudes

4. landslide activity [yellow curve] (Walter and Joswig, 2008).

Two data sets (Häge and Joswig, 2004; Wust-Bloch and Joswig, 2009) acquired at slant distances ranging 10^0–10^1 km display steeper slopes (~ 1.5) than the slopes near unity of those (Wust-Bloch and Joswig, 2006; Walter and Joswig, 2008) recorded at very short slant distances (10^{-1}–10^0 km). Distinct local detection thresholds are accountable for the offset between the curves. SNR conditions were significantly better in open fields within the Betic Cordillera (Häge and Joswig, 2004) than within the confines of a kibbutz at the Dead Sea (Wust-Bloch and Joswig, 2009). At shorter distances (>10^0 km), it appears that the coastal cliff (thick black curve) was quieter than Dead Sea sinkholes (Wust-Bloch and Joswig, 2006) and the active landslide site in the Austrian Alps (Walter and Joswig, 2008). In the M_L-distance space of Fig. 5, where empirical curves are indicative of local SNR conditions and their slopes reflect true attenuation conditions, cliff data presented in this study (thick black curve) display optimal detection threshold at the shortest slant distances.

4. *Signal Characterization*

The monitoring of failure processes taking place within an unstable cliff section (Fig. 4) offers a unique environment in which natural source processes can be observed in the field at true scale, under controlled conditions and with minimal logistical constraints (Wust-Bloch and Joswig, 2006). Nanoseismic monitoring of subsurface impacts (Wust-Bloch and Joswig, 2006; Wust-Bloch, 2008); landslide activity (Walter and Joswig, 2008), brittle failure in subsurface chalk cavities (Wust-Bloch, 2008) as well as the breaking of gypsum plates, all provide an abundant reference library of waveform patterns. This previous knowledge, together with field simulations, has a dual purpose. First, it links unequivocally source processes to recorded waveforms; second, it helps isolate the contributions of the source process from those of the transfer function to the final signal. In the coastal region investigated, two types of source processes might be expected to take place at such short source-to-sensor distance (<100 m):

- The initial brittle failure within the unstable material on top of the cliff.

- The subsequent falling of material along the cliff face and its impact on the beach below.

4.1. Material Failure

On the top of the cliff, rock mechanic predictions are in perfect agreement with field observations (Fig. 3). Taking into consideration the N-S trending scarp and the free boundary to the west of the cliff (Fig. 2), it is expected that failure within the source media, caused by E-W unloading, would be generated by tensile cracks propagating either in mode-I or in mode II (SCHOLZ, 2002) in the E-W direction (Fig. 3). The remaining challenge consists of verifying whether the energy that may be generated by these tensile cracks is released at an amplitude, time-span and frequency domain, which can be detected by nanoseismic monitoring. Note that previous short-term SNS deployment along stable sections of the same coastal cliff (2,300 m to the north), failed to detect noise spikes that could possibly be associated in any manner with brittle failure within the cliff material. To further verify that the signals analyzed here were indisputably generated by brittle failure in this unstable section of the cliff, SNS were subsequently deployed on a neighboring section of the same cliff, which is considered to be naturally stable. Only a couple of similar but barely detectable weak noise spikes were recorded over an identical monitoring period. At least two orders of magnitude in the frequency of noise spike generation seem to separate an unstable cliff section from one that degrades naturally.

In terms of waveforms, it is expected that material instabilities within the cliff top would generate brittle failure signals, such as those encountered when breaking commercial gypsum plates used for construction. Such tests were conducted under controlled conditions, whereby gypsum plates were clamped on one side and loaded on the free side. In order to achieve progressive loading conditions, a fine but constant flow of water was let into a holding tank placed on top of the plate until it failed. The waveform generated shortly before and during gypsum failure is a broadband noise spike, with an extremely impulsive onset and coda (Fig. 6a). Incidentally, the signals recorded on top of the cliff present striking similarities and they are characterized (Fig. 6b) by a similar very short signal duration (0.1–0.5 s). The waveforms display both a very impulsive onset and a short coda. They are characteristically broadband (5–75 Hz) with a frequent concentration of their energy towards higher frequencies (20–75 Hz). The high frequency of the signals recorded within the cliff confirms that the source signal is not generated in water-saturated media, as low-energy ($M_L \geq -2.0$) source signals triggered within a liquid phase have been observed to be low-pass filtered (WUST-BLOCH and JOSWIG, 2006). Given the unusually short source-to-sensor distances (3–67 m), P and S phase onsets cannot always be unequivocally picked on individual traces but ambiguities can be resolved by analyzing waveforms recorded by the 3-D sensors (one per SNS). Some rare signals, whose magnitudes are larger ($-2.2 \geq M_L \geq -2.7$), have been observed to be less spiky and last up to 1.5 s (Fig. 6d). When analyzed manually by HypoLine software, most signals display striking similarities. However, self-similarity tests show substandard results even when carried out on a handpicked selection of 60 events. First, it may outline a scaling issue, indicating that the range of source-to-sensor distances considered by the sample may still be too broad to be stable. An identical scaling challenge has to be met when using pattern recognition for event discrimination purposes at local distances (BARZILAY *et al.*, 2006). Second, at the meter scale, the role played by material anisotropies (1–100 mm) on the dispersion of the waveform may be more dominant than anticipated.

4.2. Material Impact

Due to their much higher energy content, their extended duration and random occurrence, waveforms generated by impacts are relatively easy to detect and identify (WUST-BLOCH and JOSWIG, 2006; WUST-BLOCH, 2008). To validate this work hypothesis, several types of rockfall and debris avalanches were simulated from the top of the cliff. Source analogs consisted either of individual blocks of varying mass (5–30 kg) or of scree material, which was detached and pushed off the edge of the cliff.

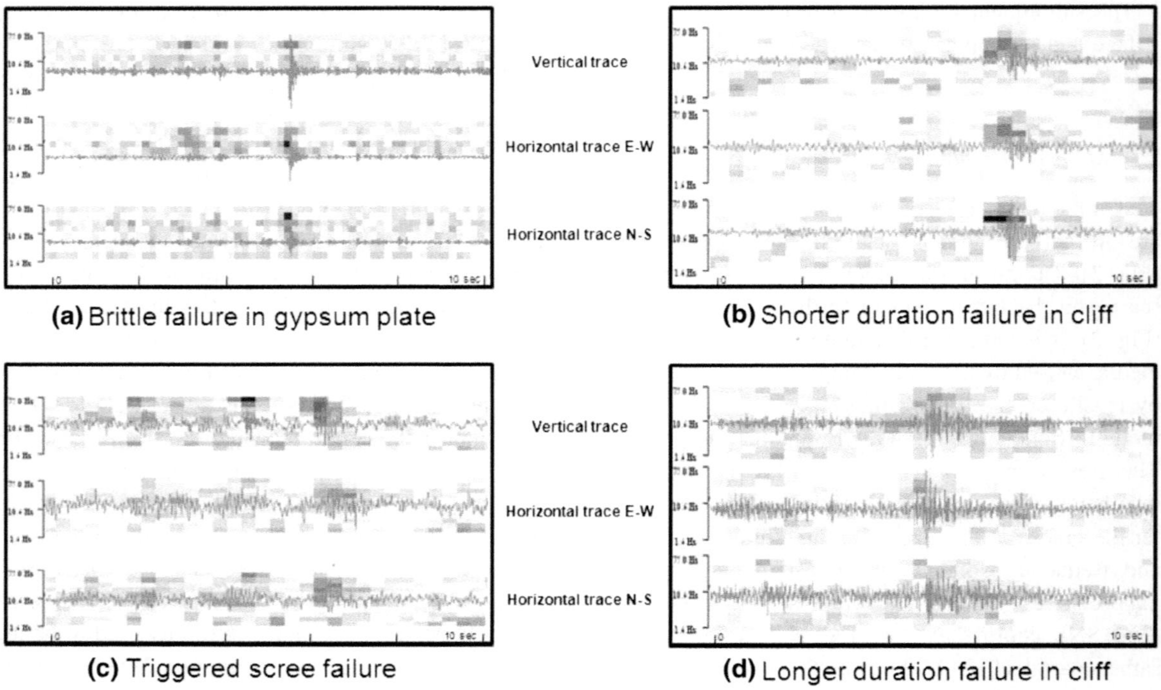

Figure 6
Signal characterization: Sonogram and superimposed waveform (*grey trace*) of three traces [vertical (*top*), horizontal E-W (*middle*) and horizontal N-S (*bottom*)] displayed in 10 s window: **a** Brittle failure of gypsum plates (1–75 Hz). **b** Material failure in cliffs: shorter duration signal (1–75 Hz). **c** Triggered avalanche/scree type of material failure (1–75 Hz). **d** Material failure in cliffs: longer duration signal (1–75 Hz)

Signals artificially generated are emergent, irregular and last over several seconds (Fig. 6c). Given the uniqueness of each process (mass, trajectory, number of impacts), the waveform produced by each such rockfall presents great disparity.

5. Results

The Gutenberg–Richter relationship of a data set can be used to assess catalog completeness. It was estimated for this series of 175 cliff-generated events over one order of magnitude ($-2.2 \geq M_L \geq -3.4$). Figure 7 shows the leveling out of the power-law size distribution around magnitude M_L −3.0 and a sharp decay below M_L −3.2; two observations which are in good agreement with detection threshold values (Fig. 5). A *b* value of 1.05 (dashed curve) obtained for an immediate series of weak aftershocks ($2.0 \geq M_L \geq -1.0$) by nanoseismic monitoring (Wust-Bloch and Joswig, 2009), which is typical of data sets including a higher percentage of smaller-magnitude events (Turcotte and Malamud, 2002), is given for comparative purpose. The corroboration between the event detection threshold analysis (Fig. 5) and the *b* value confirms that above M_L −3.0 the aftershock catalog is complete and the analysis suffers no sensitivity bias.

The spatial distribution of the signal sources shows that they cluster within a short distance (>10 m) from the cliff scarp (Fig. 8). No outliers can be seen beyond 17 m from the edge. The sources are distributed randomly throughout the volume, not displaying specific alignment or internal structure. In the cross section, many events are located within a volume of the cliff, which is partly overhanging and displays open cracks in the soil (Fig. 3). As shown on Fig. 5, the spatial distribution of sources is not biased by detection threshold: One of the two peripheral SNS (in the north and in the south) would have detected any other active sources. Figure 8, together with field observations, suggests that the volume outlined by the sources detected delimits an actively

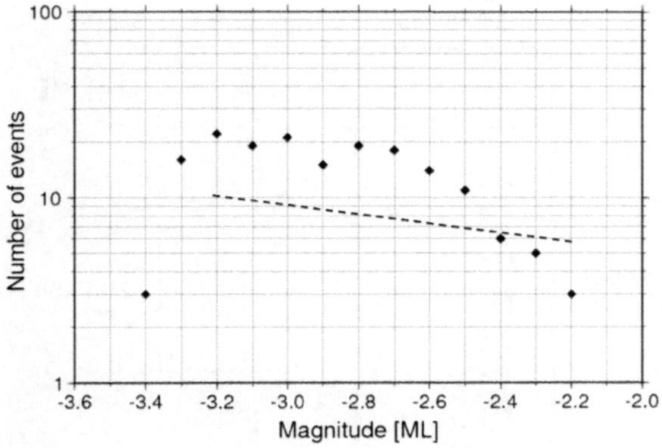

Figure 7
Gutenberg–Richter relationship for the cliff events showing catalog completeness down to M_L −3.0. *b* value (1.05) obtained by nanoseismic monitoring of immediate low magnitude aftershock series (Wust-Bloch and Joswig, 2009) is given for comparative purposes

unloading section of the cliff. When plotted in 3-D, the location of events magnitude is random throughout this active volume. Further, the spatial distribution of sources is equally random within the duration of the monitoring period (16 h). Both observations would suggest that, at that stage, the cliff was undergoing a phase of post-collapse bulk strain accommodation, rather than developing specific failure zones. Material failure appears to be confined to the top 20 m of the cliff, confirming predictions of a dynamic model in which strength and stress drop increase with depth, thereby preventing rupture to migrate to lower levels (Scholz 2002).

Figure 9 shows the time distribution of event M_L magnitudes. No regular event magnitude decay can be observed during a 12-h monitoring period. The time gap between individual events varies between 0.7 and 37 min. However, the event distribution over time is not totally random as one can see swarms of more sustained activity and periods of quiescence. The distribution of M_L magnitudes over time can also be utilized to verify the presence of second or third-order aftershock series (Wust-Bloch and Joswig, 2009). Thus, the initial intention is to check whether failure processes taking place at identical M_L magnitude cluster in time. A second aim is to investigate whether failure swarms exist at such low magnitude and small scale. Figure 9 displays a random time distribution of M_L magnitudes ranging between −2.2 and −3.4. Again, no second- or third-order series of decreasing magnitudes comparable to those of aftershocks can be seen. The time distribution of events magnitude, which is diffuse within the duration of the monitoring period, corroborates well its spatial distribution, which is equally random (Fig. 8), further supporting the hypothesis that, at that stage, the cliff was undergoing a phase of post-collapse bulk strain accommodation, rather than developing specific failure zones.

The distribution of event frequency over time (Fig. 10) is an additional way of quantifying the time decay of these cliff-generated events. Modified Omori's law (Utsu, 1961) is utilized here to evaluate whether similarities exist between aftershock decay and what happens after sudden material unloading in nature (cliff collapse). A representative time window of 400 min was selected for the analysis, together with a sliding dynamic sampling window (Dieterich, 1994). The calculated rate for a large sample of events displays a standard decay with a decay exponent, whose value is close to unity (thin black curve). It would be tempting to interpret the outliers to the left as either being a bias caused by data deficiency towards the detection threshold (Utsu, 2002), or indicating that both the data sample and the time window are not suitable for the analysis. However, this unusual event decay relationship is in good agreement with the time distribution of events displayed in Fig. 9, whereby an initial quiet period is followed by regained activity, which then fades again. Because the sampling window considered in Fig. 10 extends to only 12 h, the event frequency

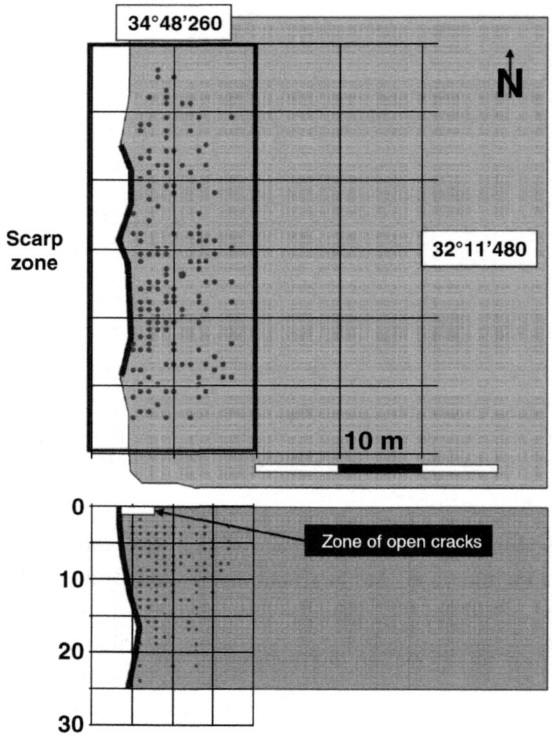

Figure 8
Spatial distribution of the signal sources generated within the cliff showing clustering within a short distance (>10 m) from the cliff scarp and below the zone of open cracks (*white box*). Map view (*top*) is the same as in Fig. 2 and the cross section (*bottom*) shows the truncated top section of Fig. 4 (0–25 m)

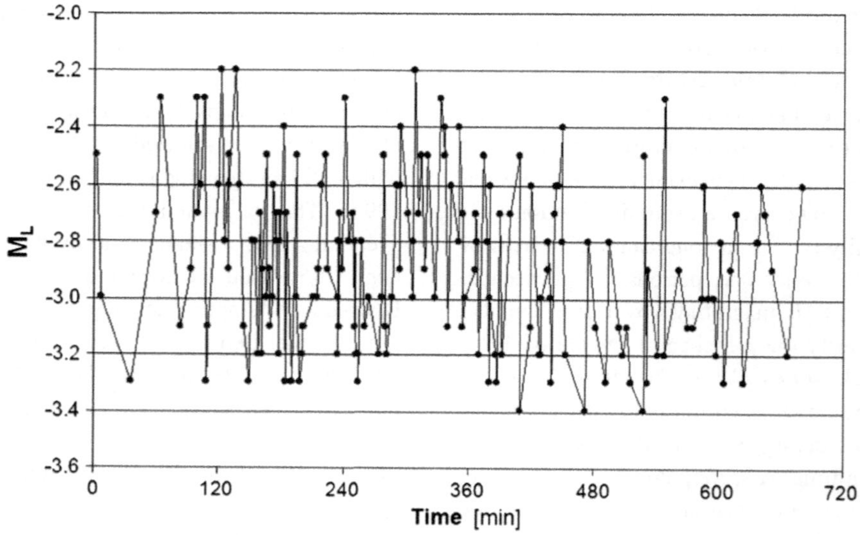

Figure 9
Time distribution of event magnitudes ($-2.2 \geq M_L \geq -3.4$) displaying an irregular decay with light clustering in time and a random distribution of M_L magnitudes over time. Time gaps between individual events vary between 0.7 and 37 min

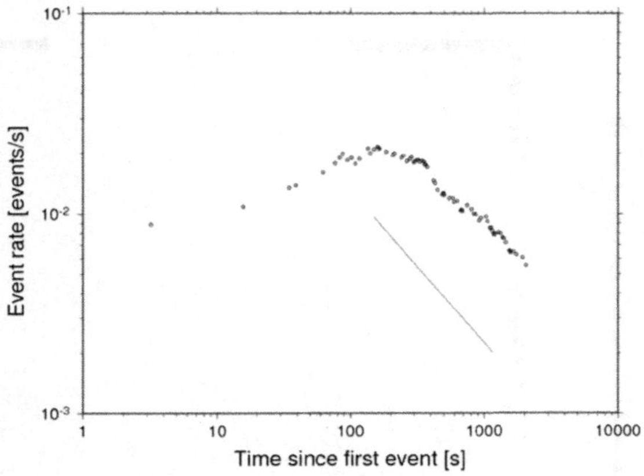

Figure 10
Modified Omori's law (Utsu 1961) displaying the event rate over time. A reference decay with an exponent whose value is close to unity is shown by the *black curve*

over time that we recorded can be interpreted as a transient of enhanced brittle failure activity (>10^{-2} events/s) within a long-term decay.

The polarization of maximal signal amplitude was utilized to estimate the stress field within the cliff top. Three traces (vertical, horizontal N-S and horizontal E-W) of the 3-D sensor nearest the source were selected. For each event, the maximal amplitudes of the entire waveform were estimated and coded relatively: 1 for minimal amplitude, 2 for intermediate amplitude and 3 for maximal amplitude. As the orientation of the cliff (N-S) coincided with that of the 3-D sensors, no coordinate transformation was required. Figure 11 shows a map view and a cross section of a section of the cliff closest to the scarp (see bold white box in Fig. 8). The ellipses display the spatial orientation of the maximal signal energy, whereby the ellipse axes are proportional to the signal amplitude. The map view (top) shows that 61% of the ellipses are orientated E-W, 30 are circles and 9% are orientated N-S. The majority of E-W oriented ellipses are located nearest the scarp zone and circles trend to concentrate at the periphery. The cross section (bottom) shows 55% of circles, 30% of E-W orientated ellipses and 15% of ellipses in the vertical plane. Most of the E-W orientated ellipses concentrate at shallow depth, nearest the scarp, a feature that corroborates well with field observations (Fig. 3).

The M_L–surface rupture length relationship is displayed in Fig. 12. The gray box (lower left) indicates the upper and lower boundaries for brittle failure (mode I–II) occurring at the cliff: M_L magnitudes range between −2.2 and −3.4; surface ruptures observed in the field (Fig. 3) consist of irregular crack segments of various lengths (1–35 cm) and width (1–20 mm). The thick black curve (top right) marks the limits of empirical data compiled for surface ruptures of normal faults (Wells and Coppersmith, 1994). The thin black curve shows a constant-slope extrapolation to lower magnitudes. Although seven orders of magnitude in M_L and five orders of magnitude in size separate them, both data sets plot roughly on a common slope, thereby suggesting that they may share mechanical behavior and material properties. However, the stability of such features over many orders of magnitude and the exact relationship between different event populations still needs to be verified systematically at all scales.

6. Discussion

The signal analysis (Figs. 5, 6, 7, 9, 10, 12) and locations (Figs. 8, 11) of these nanoseismic events, together with field evidence (Fig. 3), demonstrate that these signals are unequivocally generated within a specific section of the sandstone cliff (Figs. 2, 4) by stress release caused by sudden unloading at the cliff interface (cliff collapse). The diffuse spatial distribution of the events and the unusual event decay

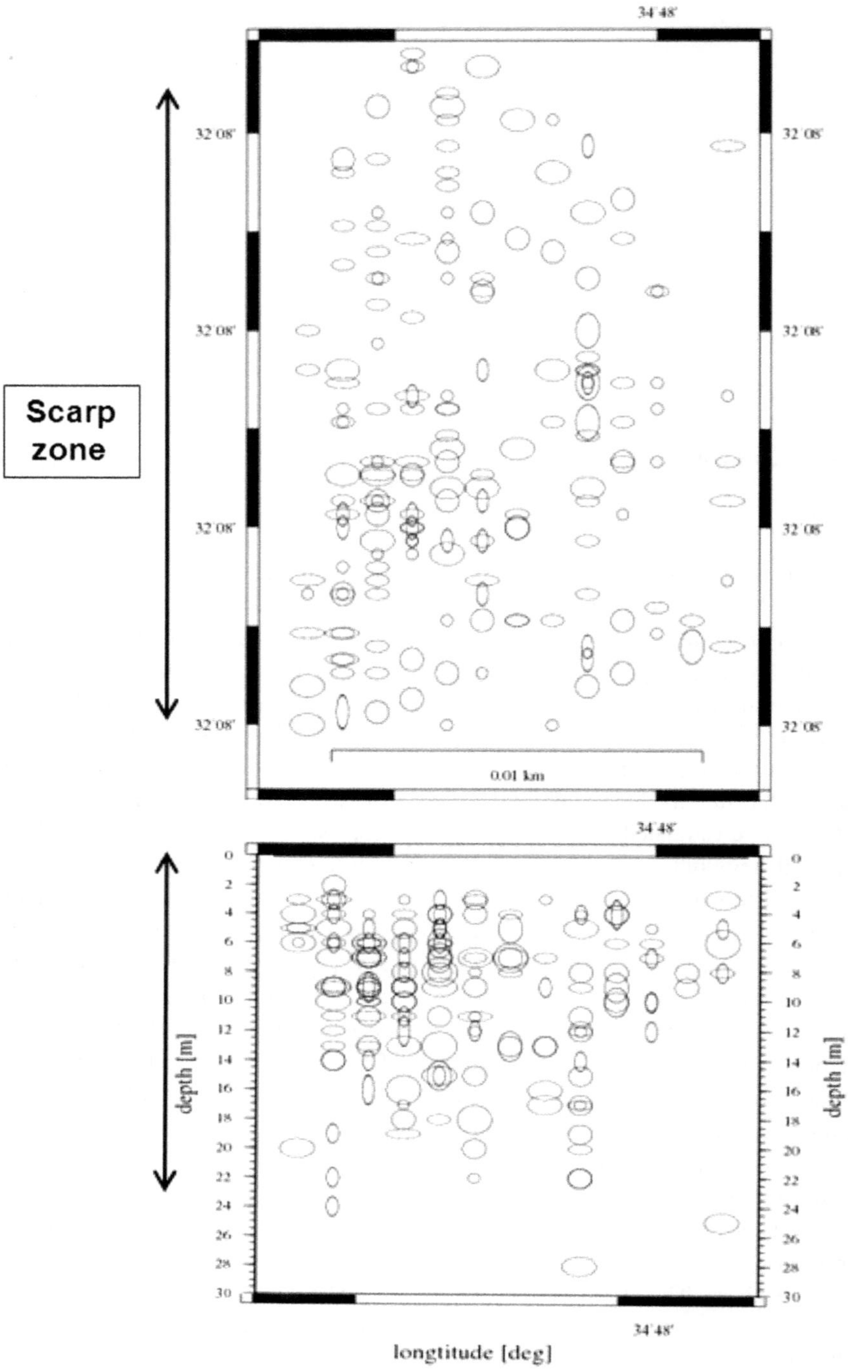

Figure 11
Spatial distribution of maximal signal amplitude shown in map view (*top*) and cross section (*bottom*). The outline of the map is shown by the *bold box* in Fig. 8. The location of the scarp zone is indicated by a *black arrow*

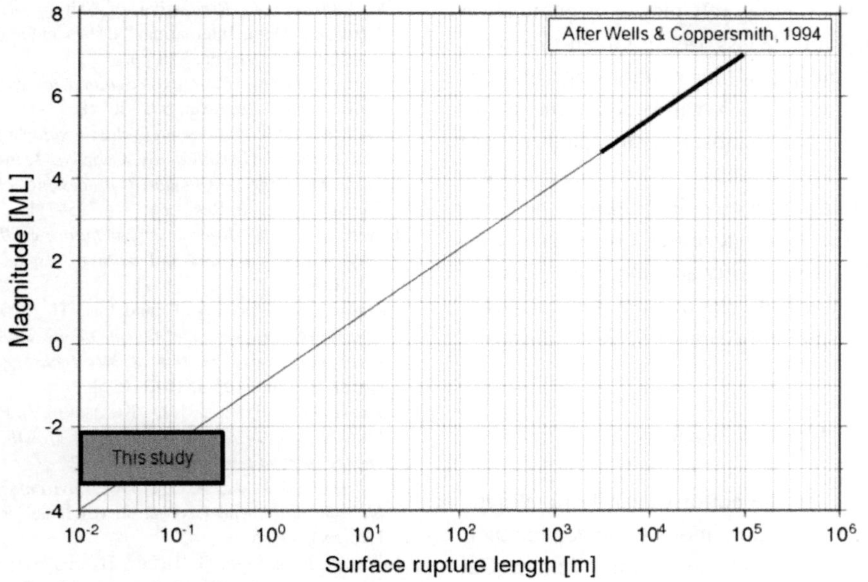

Figure 12
Magnitude [M_L]–surface rupture length [m] relationship for cliff-generated events (*gray box, lower left*). The *thick solid black line* (*top, right*) is the empirical relationship for large magnitude earthquakes (Wells and Coppersmith 1994) and the *thin black line* is its constant slope extrapolation to lower magnitudes

(Fig. 10) at the time of the monitoring (16 days after the collapse), suggests that the signals are generated by bulk stress accommodation within this cliff section rather than by localized failure along discrete zones. This is corroborated by the spatial polarity of maximum signal amplitude (Fig. 11) displaying a bulk tensile regime in the E-W direction.

This feasibility study demonstrates that nanoseismic monitoring can be utilized to detect and accurately locate brittle failure generated by unstable cliffs. Although the host material is a weakly consolidated layered sandstone, attenuation is not such that it prevents the detection of very weak signals (M_L −2.7) at source-to-sensor distance up to 60 m. Waveform analysis suggests that material scattering is standard and no evidence was found for erratic wave propagation and wave-guiding within the cliff material.

Extending the recording time of future nanoseismic monitoring within cliffs would provide the robust database required to identify pre-collapse failure processes without ambiguities. In the case of sustained seismicity, the spatial clustering of failure signals would help locate nucleation zones (LOCKNER *et al.*, 1991; SCHOLZ, 2002), thus estimating the potential characteristics of future failure zones. In the event of episodic seismicity, whose signals are randomly distributed in space and time, it would confirm that the cliff is undergoing bulk strain accommodation; a natural coastal cliff degradation process.

Signal energy concentrated in the high frequencies (Fig. 6b, d) confirms that the sandstone is not water-saturated (WUST-BLOCH and JOSWIG, 2006). This corroborates field observation that the cliff material is drained and that the source region is in a regime of negative pore-water pressure (FREDLUND and RAHARDJO, 1993). Because longer-duration waveforms (Fig. 6d) do not display a preferential polarization of their signal amplitudes and because they are located deeper within the cliff, they may be generated under higher confining pressure, by a different failure process.

Although extended monitoring would be required to obtain a reliable stress-field map or to identify failure localization zones in a reliable manner, this 16-h data set validates the results of previous feasibility studies (WALTER and JOSWIG, 2008; WUST-BLOCH and JOSWIG 2006). Nanoseismic monitoring appears to be a promising technique to detect and survey material failure at engineering scale (10^{-1}–10^{2} m) as it is not affected by the determining role played by attenuation on acoustic emission (WEISS 1997). Given that most analytical tools and prevailing relationships of standard seismology have been validated at the meter scale, it appears that nanoseismicity patterns are surprisingly similar to those of microseismicity. Although confining pressure is minimal and the effects of material heterogeneities are enhanced at such scale, SparseNet software modules (JOSWIG, 2008) were able to identify and locate ambient fracture processes despite unfavorable SNR conditions.

Acknowledgments

I would like to thank the editor Cezar Trifu, Wayne D. Pennington and an anonymous reviewer for their constructive comments and suggestions. Yehuda Gai of IMI-Israel granted permission to deploy the SNS. Joel Ettinger provided support, advice and friendship in the field. Yael Szerman pre-processed part of the data. Asaf Inbal designed Fig. 11. Figures were prepared using GMT (WESSEL and SMITH, 1998).

REFERENCES

ALMAGOR, G. (2005), *The Mediterranean coast of Israel, Geological Survey of Israel*, Report GSI/13/02 (in Hebrew), 53 pp.

BARZILAY, S., WUST-BLOCH, G.H., BEN-HORIN, Y., and BEN-AVRAHAM, Z. (2006), *Discrimination of local events in the northern Dead Sea fault region using pattern recognition schemes*, First European Conference on Earthquake Engineering and Seismology, Geneva, Switzerland, 447.

DIETERICH, J.H. (1994), *A constitutive law for rate of earthquake production and its application to earthquake clustering*, J. Gephys. Res. *99*/B2, 2601–2618.

FREDLUND, D.G. and RAHARDJO, H. Soil Mechanics for Saturated Soils (Wiley & Sons, New York 1993), 517 pp.

GREEN, A., MAURER, H., SPILLMANN, T., HEINCKE, B., and WILLENBERG, H. (2006), *High-resolution geophysical techniques for improving hazard assessments of unstable rock slopes*, Leading Edge *25*, 311–316.

HAINZL, S., KRAFT, T., WASSERMANN, J., and IGEL, H. (2006), *Evidence for rainfall-triggered earthquake activity*, Geophys. Res. Lett. *33*, doi:10.1029/2006GL027642.

HÄGE, M. and JOSWIG, M. (2004), *Mapping active faults in the Murcia region, Spain, by nanoseismic monitoring*, European Seismological Commission XXIX General Assembly, GFZ - Potzdam, Potzdam, Germany, 154.

HUSEN, S., BACHMANN, C., and GIARDINI, D. (2007), *Locally triggered seismicity in the central Swiss Alps following the large rainfall event of August 2005*, Geophys. J. Int. *171*, 1126–1134.

IDC (2002), IDC *Processing of Seismic, Hydroacoustic, and Infrasonic Data*, International Data Center (IDC) Documentation, CTBTO, Vienna, 326 pp.

JOSWIG, M. (1990), *Pattern recognition for earthquake detection*, Bull. Seismol. Soc. Am. *80*, 170–186.

JOSWIG, M. (1995), *Automated classification of local earthquake data in the BUG small array*, Geophys. J. Int. *120*, 262–286.

JOSWIG, M. (1999), *Automated processing of seismograms by SparseNet*, Seism. Res. Lett., *70*, 705–711.

JOSWIG, M. (2008), *Nanoseismic monitoring fills the gap between microseismic networks and passive seismic*, First Break *26*/6, 117–124.

KRAFT, T., WASSERMANN, J., and IGEL, H. (2006), *High-precision relocation and focal mechanism of the 2002 for rainfall-triggered earthquake swarms at Mt Hochstaufen, SW Germany*, Geophys. J. Int. *167*/3, 1513–1528.

LOCKNER, D.A., BYERLEE, J. D., KUKSENKO, V., PONOMAREV, A., and SIDORIN, A. (1991), *Quasi-static fault growth and shear fracture energy in granite*, Nature *350*, 39–42.

PERATH, I. and ALMAGOR, G. (1996), *Environmental risk along the Sharon escarpment*, Geolog. Survey Israel, Report GSI/5/96 (in Hebrew), 47 pp.

KATZ, O., PETRANKER, G., HECHT, H., and ALMOG, E. (2007), *Estimation of coastal cliff retreat rate and cliff location in 2050 and 2100*, Geolog. Survey Israel, Report GSI/21/07. 51 pp.

RICHTER, C.F. *Elementary Seismology* (Freeman, San Francisco CA 1958), 298 pp.

SCHOLZ, C.H. *The mechanics of earthquakes and faulting* (Cambridge University Press, Cambridge, UK 2002), 471 pp.

TURCOTTE, D.L. and MALAMUD, B. D., *Earthquakes as a complex system, In International Handbook of Earthquake and Engineering Seismology* (eds. Jennings, P., Kanamori, H., Lee, W. H. K. and Kisslinger, C.) (Academic Press, San Francisco, CA, 2002) pp. 209–227.

UTSU, T. (1961), *A statistical study on the occurrence of aftershocks*, Geophys. Mag. *30*, 521–605.

UTSU, T., *Statistical Features of Seismicity*. In *International Handbook of Earthquake and Engineering Seismology* (eds. Jennings, P., Kanamori, H., Lee, W. H. K. and Kisslinger, C.) (Academic Press, San Francisco, CA, 2002) pp. 719–746.

WALTER, M. and JOSWIG, M. (2008), *Seismic monitoring of fracture processes generated by a creeping landslide in the Vorarlberg Alps*, First Break *26*/6, 131–135.

WELLS, D.L. and COPPERSMITH, K.J. (1994), *New empirical relationships among magnitudes, rupture length, rupture width, rupture area, and surface displacement*, Bull. Seismol. Soc. Am. *84*/4, 974–1002.

WEISS, J. (1997), *The role of attenuation on acoustic emission amplitude distributions and b-values*, Bull. Seismol. Soc. Am. *87*, 1362–1367.

WISEMAN, C., HAYATI, G., and FRYDMAN, S. (1981), *Stability of heterogeneous sandy coastal cliff*, Proc. *Soil mechanics and Foundation Engineering*, 10th Conf., Stockholm, AA Balkema/Rotterdam *11*/59, 568–574.

WUST-BLOCH, G.H. (2008), *Detecting pre-collapse failure signals in unstable landslides, cliffs and cavities by nanoseismic monitoring*, Israel Geolog. Soc., Annual Meeting, Nazareth, Israel, 115.

WUST-BLOCH, G.H. and JOSWIG, M. (2006), *Pre-Collapse identification of sinkholes in unconsolidated media at Dead Sea area by "nanoseismic monitoring"* (Graphical Jackknife location of weak sources by few low-SNR records), Geophys. J. Int. *167*/3, 1220–12232.

WUST-BLOCH, G.H., HÄGE, M., WALTER, M., and JOSWIG, M. (2006), *Monitoring and characterizing natural or man-made nanoseismic ($M_L < 0.0$) sources for engineering purposes*, First European Conf. on Earthquake Engineering and Seismology, Geneva, Switzerland, 282.

WUST-BLOCH, G.H. and JOSWIG, M. (2009), *Location and characterization of short-term aftershock generated by the February 11th 2004, Northern Dead Sea (M_L 5.1) earthquake by nanoseismic monitoring*, Bull. Seismol. Soc. Am. (submitted).

ZVIELY, D., and KLEIN, M. (2004), *Coastal cliff retreat rates at Beit-Yannay, Israel in the 20th century*, Earth Surface Processes and Landforms *29*, 12–17.

(Received September 7, 2008, revised January 25, 2009, accepted February 3, 2009, Published online December 15, 2009)

Pure Appl. Geophys. 167 (2010), 169–182
© 2009 Birkhäuser Verlag, Basel/Switzerland
DOI 10.1007/s00024-009-0003-6

Pure and Applied Geophysics

Interpretation of Microseismicity Resulting from Gel and Water Fracturing of Tight Gas Reservoirs

C. Dinske,[1] S. A. Shapiro,[1] and J. T. Rutledge[2]

Abstract—We provide a comparative analysis of the spatio-temporal dynamics of hydraulic fracturing-induced microseismicity resulting from gel and water treatments. We show that the growth of a hydraulic fracture and its corresponding microseismic event cloud can be described by a model which combines geometry- and diffusion-controlled processes. It allows estimation of important parameters of fracture and reservoir from microseismic data, and contributes to a better understanding of related physical processes. We further develop an approach based on this model and apply it to data from hydraulic fracturing experiments in the Cotton Valley tight gas reservoir. The treatments were performed with different parameters such as the type of treatment fluid, the injection flow rate, the total volume of fluid and of proppant. In case of a gel-based fracturing, the spatio-temporal evolution of induced micro-seismicity shows signatures of fracture volume growth, fracturing fluid loss, as well as diffusion of the injection pressure. In contrast, in a water-based fracturing the volume creation growth and the diffusion controlled growth are not clearly separated from each other in the space-time diagram of the induced event cloud. Still, using the approach presented here, the interpretation of induced seismicity for the gel and the water treatments resulted in similar estimates of geometrical characteristics of the fractures and hydraulic properties of the reservoir. The observed difference in the permeability of the particular hydraulic fractures is probably caused by the different volume of pumped proppant.

Key words: Hydraulic fracturing, Microseismicity, Pressure diffusion, Permeability.

1. Introduction

Hydraulic fracturing is the process of initiation and propagation of a fracture by injecting fluid at high flow rates and pressures higher than the minimum horizontal stress S_{hmin} through a perforated domain of a wellbore. It is one of several techniques for creating fractures which are used to enhance the productivity of hydrocarbons, to stimulate geothermal energy reservoirs and to develop waste disposal sites (Fehler *et al.*, 1987; Block *et al.*, 1994; Keck and Withers 1994).

Since its introduction in the 1970s, the passive seismic monitoring has been used widely and successfully to estimate geometric parameters of hydraulic fractures (Albright and Pearson 1982; Fehler *et al.*, 1987; Warpinski *et al.*, 1995; Urbancic and Rutledge 2000; Fischer *et al.*, 2008). The fracture azimuth, its lateral extension from the wellbore and its vertical growth can be obtained using micro-earthquakes recorded from the fractured area during and after the injection. Relative source location techniques applied to the induced seismicity give high resolution images which allow us to resolve discrete fracture zones and to better understand source mechanisms (Phillips *et al.*, 1997; Rowe *et al.*, 2002).

The microseismic imaging provides not only a spatial map of the final geometry of the fracture, it also has significant potential to evaluate dynamic growth processes and to characterize physical processes related to fluid forced fracturing of reservoir rocks (Shapiro *et al.*, 2006). Shapiro *et al.*, (1997) introduced the SBRC (*Seismicity Based Reservoir Characterization*) method for analyzing spatio-temporal characteristics of microearthquakes induced by fluid stimulations. Hydraulic properties of the seismically active rock volume can be determined using this method. The assumption of the SBRC is that the triggering mechanism of stimulation-induced

[1] Freie Universität Berlin, Fachrichtung Geophysik, Malteserstr. 74-100, 12249 Berlin, Germany.
E-mail: carsten@geophysik.fu-berlin.de;
shapiro@geophysik.fu-berlin.de;
[2] Geophysics Group, Los Alamos National Laboratory, Mail Stop D443, Los Alamos, NM 87545, USA.
E-mail: jrutledge@lanl.gov

seismicity is a diffusional process of propagation of a pore pressure perturbation initiated at the injection source. The increased pore pressure reactivates pre-existing faults and fractures by triggering slip events and thereby releasing previously accumulated shear stress in the stimulated rock volume. However, in the case of hydraulic fracturing the SBRC approach requires modifications (SHAPIRO *et al.,* 2006). An analysis of the dynamical evolution of induced seismicity in the distance–time domain shows that triggering of microearthquakes can be related to different processes: Creation of new fracture volume, loss of fracturing fluid due to its infiltration into reservoir rocks, and diffusion of injection pressure into the surrounding rock and inside the fracture.

In this paper, we further develop the method of SHAPIRO *et al.,* (2006) and SHAPIRO and DINSKE (2007) for an interpretation of hydraulic fracturing-induced seismicity in order to quantify both geometric and hydraulic parameters of the fracture system. Simultaneously, the hydraulic properties of the virgin reservoir can be obtained as well. In contrast to SHAPIRO *et al.* (2006), here we determine the fluid loss not by estimating an apparent diffusivity from microseismic data but rather by considering a volume balance between the injected treatment fluid on the one hand and the fracture volume on the other hand. We apply this approach to two distinct situations: gel-based and water-based treatments. The microseismic data were recorded in hydraulic fracturing experiments in the Carthage Cotton Valley gas field (TX, USA). These well-documented fracturing experiments (e.g., WALKER 1997; ARCO EXPLORATION and PRODUCTION TECHNOLOGY 1997; PINNACLE TECHNOLOGIES 1999; RUTLEDGE and PHILLIPS 2003; RUTLEDGE *et al.,* 2004) allow testing of our findings. The influence of treatment parameters on the fracturing and on the induced seismicity can be studied by comparing the results of the two situations.

2. *Review of the Interpretational Model of Fracture Growth and Seismicity*

We will firstly review the approach by SHAPIRO *et al.* (2006) and SHAPIRO and DINSKE (2007). There, microseismic data from Cotton Valley experiments were also analyzed, but only the gel treatments (Stage A and Stage B). This approach was also applied by FISCHER *et al.* (2008) to several data sets obtained by gel-based fracturing in Canyon sandstone formation in Texas.

A hydraulic fracture will be initiated if the downhole pressure becomes larger than the minimal principal stress. The direction of propagation of a fracture is in accordance with the orientation of the local stress field. In a homogeneous medium, the fracture grows perpendicular to the axis of minimal principal stress (HUBBERT and WILLIS 1957). The hydraulic fracturing of a rock activates microearthquakes. Their triggering mechanism is similar to those by stimulation-induced seismicity. The injected fluid causes an increase of the pore pressure and hence a decrease in the effective stress. In addition, the propagating fracture affects the stress state in its immediate vicinity due to tensile opening. Microearthquakes are then triggered along pre-existing natural fractures that are favorably oriented for slip (PEARSON 1981; RUTLEDGE and PHILLIPS 2003; SHAPIRO *et al.,* 2006). Recent studies by SILENY *et al.* (2008) have shown that tensile events are also observed.

For investigation of fracturing processes and corresponding signatures of microseismicity we first consider a particular fracture geometry model. The microseismic data that will be discussed here are from hydraulic fracturing in horizontally layered sands and shales. The lateral extension of seismicity clouds is far longer than the vertical extent. For such conditions, PERKINS and KERN (1961) and NORDGREN (1972) derived a 2-D fracture geometry model (the PKN model, Fig. 1). In this model, fracture mechanics and fracture tip effects are not considered, but the focus is on fluid flow, and corresponding pressure gradients in the fracture (ECONOMIDES and NOLTE 2003). PERKINS and KERN (1961) assumed that a straight planar height-limited vertical fracture is propagating in a well-confined layer. Normal stresses in the layers above and below are large enough to prevent an out-of-zone growth of the fracture, and plane strain conditions in the vertical direction are assumed. However, fluid leakoff and storage of fluid within the fracture were neglected. To add leakoff and storage, NORDGREN (1972) introduced the

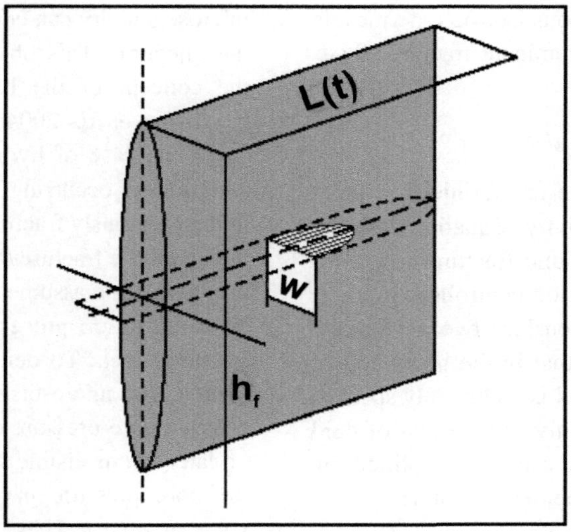

Figure 1
PKN 2-D hydraulic fracture geometry model, PERKINS and KERN (1961) and NORDGREN (1972)

equation of continuity (the conservation of mass) to the Perkins and Kern model.

To understand the dynamical behavior of induced microseismicity we apply an approximation of the process of fracture growth that results from a volume balance principle of an incompressible treatment fluid. The volume balance states that the total volume of the injected fluid, V_i, equals the sum of the fluid volume which is stored in the created fracture, V_f, and the fluid volume which is lost into the surrounding formation, V_L:

$$V_i = V_f + V_L. \tag{1}$$

We can rewrite Eq. 1 by replacing $V_i = Q_i t$, where Q_i is the average injection rate of the treatment fluid and t is the injection time, and $V_f = 2Lh_f w$, where L is the fracture half-length, h_f is the average fracture height and w is the average fracture width. The fluid-loss volume V_L can be approximated by $4LC_L h_f \sqrt{2t}$ (HARRINGTON and HANNAH 1975) which introduces the fluid-loss coefficient C_L and describes the fluid volume V_L which flows through the surface area of the fracture $S = 4Lh_f$ during the time period t. This formulation simplifies the integral of the Carter equation for leakoff (ECONOMIDES and NOLTE 2003). Then, the following expression for the half-length of a (symmetric) hydraulic fracture L as a function of injection time t is obtained (SHAPIRO *et al.*, 2006):

$$L(t) = \frac{Q_i t}{4h_f C_L \sqrt{2t} + 2h_f w}. \tag{2}$$

Here the first term in the denominator describes the total fluid-loss from the fracture walls into the surrounding formation. It is proportional to $\sqrt{t}$ and hence it has a diffusional character. The fluid-loss coefficient C_L is an important reservoir engineering parameter and is an active research topic. It is controlled by different mechanisms depending on several factors, including the wall-building characteristics of the treatment fluid, the relative permeability of the formation to the treatment fluid, the hydraulic diffusivity of the reservoir, the pressure difference between fracture and reservoir, total compressibility for the reservoir, and the viscosity of the reservoir fluid (see also Eq. 10). The second term in the denominator of Eq. 2 represents the contribution of the effective fracture volume and it depends on the geometry of the fracture vertical cross section. The shape of the hydraulic fracture, in particular the fracture width, is a major objective in reservoir engineering.

Equation 2 can be seen as a one-dimensional approximation for the triggering front of microseismicity in the case of a propagating hydraulic fracture. It is an alternative formulation to the triggering front equation of microseismicity induced by a borehole fluid injection in a homogeneous isotropic permeable medium without creating a dominant fracture (SHAPIRO *et al.*, 2002):

$$r(t) = \sqrt{4\pi D t}, \tag{3}$$

where r is the radial distance to the injection point and D is the medium diffusivity. Equation 2 shows that the growth of the hydraulic fracture is a combined geometry- and diffusion-controlled process. However, it is possible to consider two asymptotic situations. First, we assume that in the beginning of an injection the treatment fluid is completely spent to create new fracture volume only and the loss of fluid is insignificant. Then Eq. 2 can be simplified by neglecting the first term in the denominator:

$$L(t) = \frac{Q_i t}{2h_f w}. \tag{4}$$

Here the fracture half-length L is a linear function of the injection time t. The spatio-temporal characteristic of induced microearthquakes should obey this linear dependence. This means that microseismic events migrate roughly linear with time away from the injection well, indicating the rupturing process.

On the other hand, the loss of treatment fluid from the fracture walls into the surrounding formation is becoming large in the long-term limit of the injection. Therefore, the corresponding term in Eq. 2 is the dominant one and the geometry term can be neglected. The equation then becomes identical to the triggering front equation for a linear pore pressure diffusion (Eq. 3), with an apparent hydraulic diffusivity $D = D_{ap}$ given by:

$$r(t) \equiv L(t) = \frac{Q_i t}{4h_f C_L \sqrt{2t}} = \sqrt{4\pi D_{ap} t},$$
$$D_{ap} = \frac{Q_i^2}{128\pi h_f^2 C_L^2}. \tag{5}$$

The evolution of the microseismic event cloud in space and time should show the typical diffusional signature ($r \propto \sqrt{t}$). Equations 2–5 provide a model that describes the process of hydraulic fracture growth as long as the treatment fluid is injected into the reservoir rock. They relate the spatio-temporal dynamics of fracturing-induced microseismic clouds to the fracture propagation. However, fluid-induced microseismicity can continue to occur after shut-in of the injection. This observation can be explained by the concept of the back front of microseismicity (PAROTIDIS *et al.*, 2004; SHAPIRO *et al.*, 2006).

In the case of hydraulic fracturing, most microearthquakes occurring after the injection are located in the previously fractured area, including the body of the hydraulic fracture (the fracture can be filled with a proppant). We assume that these events are triggered by a relaxation process of the pore pressure in the fractured area. To define the back front, we further assume that microseismic events can be triggered by positive pore pressure changes only. This means that no later microseismic activity occurs in points where the pore pressure $p(x, t)$ has reached its maximal value at time t. These (x, t) coordinates of the pore pressure maximum define the minimal distance of an event to occur, even after the injection is stopped. This distance is the back front of induced seismicity $x_{bf}(t)$, and it is given by the following expression (SHAPIRO *et al.*, 2006):

$$x_{bf} = \sqrt{dD_{bf} t \left(\frac{t}{t_0} - 1\right) \ln\left(\frac{t}{t - t_0}\right)}, \tag{6}$$

where t_0 denotes injection end time. Equation 6 is identical with those equations valid for 2D and 3D poroelastic media (see PAROTIDIS *et al.*, 2004). The constant d is equal to a double dimensionality of the diffusion geometry, by hydraulic fracturing $d = 2$. The diffusion constant D_{bf} corresponds to the diffusivity of the fractured system. It can further be used to obtain permeability of the fracture κ_f by the following relation (SHAPIRO *et al.*, 2006):

$$\kappa_f = c_f \eta_f \phi_f D_{bf}. \tag{7}$$

Here η_f and c_f are viscosity and compressibility of the treatment fluid and ϕ_f is the porosity of the fracture.

3. *Extension of the Interpretational Model*

The main idea of our further development of the interpretational approach is to estimate fluid loss using engineering data and geometric parameters of the fracture obtained from microseismicity. Then, the fluid loss can be used to estimate the permeability of the reservoir. Equation 2 shows that in the beginning of a treatment new fracture volume will be opened and the fluid-loss can be neglected. With Eq. 4 the width of a fracture can be defined using the values of half-length $L(t)$ and time t from the spatio-temporal distribution of events induced during the fracture volume opening phase. Here we present an additional method to estimate geometric parameters of the fracture. We consider the tangent to a curve described by Eq. 2, which is assumed to represent the triggering front equation of fracturing-induced seismicity:

$$\frac{d\tilde{L}}{dt} = \frac{X + \frac{1}{2}Y\sqrt{t}}{(X + Y\sqrt{t})^2} \tag{8}$$

where $X = (2h_f w)/Q_i$, $Y = (\sqrt{32}h_f C_L)/Q_i$, and $\tilde{L} = L/Q_i$.

At time $t = 0$ the tangent coefficient is $1/X$. It provides the width of a fracture, or in combination with the fracture height, the cross-section area: $A = (X/2)Q_i$. The tangent coefficient can simply be obtained by data fitting. Knowing the geometric parameters, one can then consider the fluid loss from the fracture body using the volume balance principle (Eq. 1). A comparison of the injected fluid volume and the created fracture volume gives the lost fluid volume: $V_L = Q_i t - 2LA$. The fluid loss coefficient C_L can be calculated after rearranging Eq. 2:

$$C_L = \frac{Q_i t - 2LA}{4Lh_f\sqrt{2t}}. \tag{9}$$

Furthermore, one can now estimate the permeability κ of the virgin reservoir. If we consider the definition of reservoir leak-off coefficient given in Economides and Nolte (2003, chap. A9) and neglect fracture surface effects we can write (Shapiro *et al.*, 2006):

$$\kappa = \frac{{C_L}^2 \pi \eta}{\Delta p^2 \phi c_t}, \tag{10}$$

where η and c_t are viscosity and compressibility of the reservoir fluid, ϕ is the porosity of the unfractured reservoir, and Δp is the difference of the pressure in the fracture (downhole pressure) and the initial reservoir pressure. The pressure difference can be estimated from engineering data. Moreover, the fluid loss estimate permits calculation of the apparent hydraulic diffusivity according to Eq. 5. The so defined diffusivity can further be used in Eq. 3. The resulting triggering front can be compared with the microseismic data in the space-time domain for evaluation of our interpretational model.

With both the reservoir and the fracture permeability one can finally calculate the dimensionless fracture conductivity C_D, which is a key design parameter in hydraulic fracturing (Economides and Nolte 2003, chap. 12):

$$C_D = \frac{w\kappa_f}{L\kappa} = \frac{A\kappa_f}{h_f L\kappa} = \frac{4A\kappa_f}{S\kappa}, \tag{11}$$

with S being the surface of the fracture body. The geometric parameters, fracture width w and fracture half-length L, are included in this formulation. The dimensionless fracture conductivity is a measure of the ability of the fracture to transport fluid divided by the ability of the reservoir to feed the fracture.

4. Interpretation of Microseismic Signatures of Gel and Water Fracturing in the Cotton Valley Gas Field

In May and July 1997, a consortium of oil field operators and service companies carried out a series of hydraulic fracturing experiments in the Carthage Cotton Valley gas field in East Texas, USA (Walker 1997). The geological setting of the gas reservoir is defined by its location in the northern Gulf of Mexico sedimentary basin. The Cotton Valley formation consists of multiple, low-permeability sands within an interbedded sequence of sands and shales. The formation is approximately 325 m thick, its top is at a depth of about 2,600 m within the study area (Rutledge and Phillips 2003). A detailed description of the geology and of the stratigraphy of the Cotton Valley formation is given in Wescott (1984) and Walker (1997). During the operation of the experiments different target zones within the Cotton Valley formation were hydraulically fractured. The operational setup is shown in Fig. 2. In a first phase, fracturing was performed using well CGU21-10 as the treatment well. The treatment fluid was a cross-linked polymer gel with added proppant. The induced microearthquakes were detected and registered by borehole geophone arrays placed in two monitoring

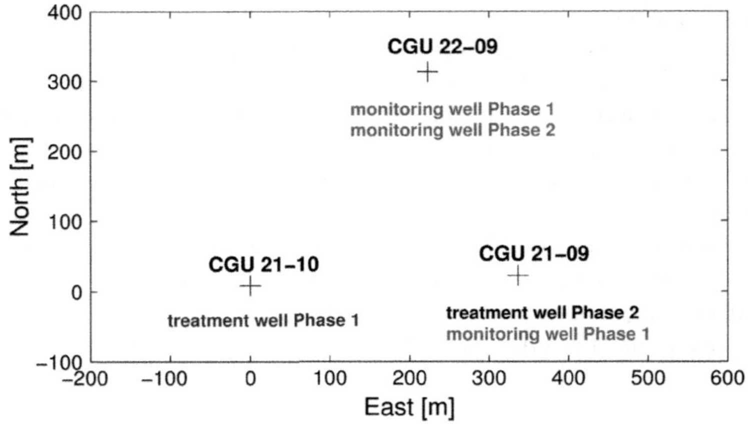

Figure 2
Well configuration and operational setup of the Carthage Cotton Valley gas field hydraulic fracturing experiments in 1997 (Phase 1: gel treatments, Phase 2: water treatments)

wells, CGU21-09 and CGU22-09. Initial source locations were determined using P- and S-wave traveltimes and P-wave particle motion data (Rutledge and Phillips 2003). Subsequently, the event locations were improved through high-precision repicking of P- and S-phase onsets (Rutledge and Phillips 2003). In a second phase, fracturing was performed via perforated domains of well CGU21-09 and monitored by the borehole sensors in well CGU22-09 only. Here a water-based fluid was used for the treatment. Microseismic events were located using the same approach as in phase 1.

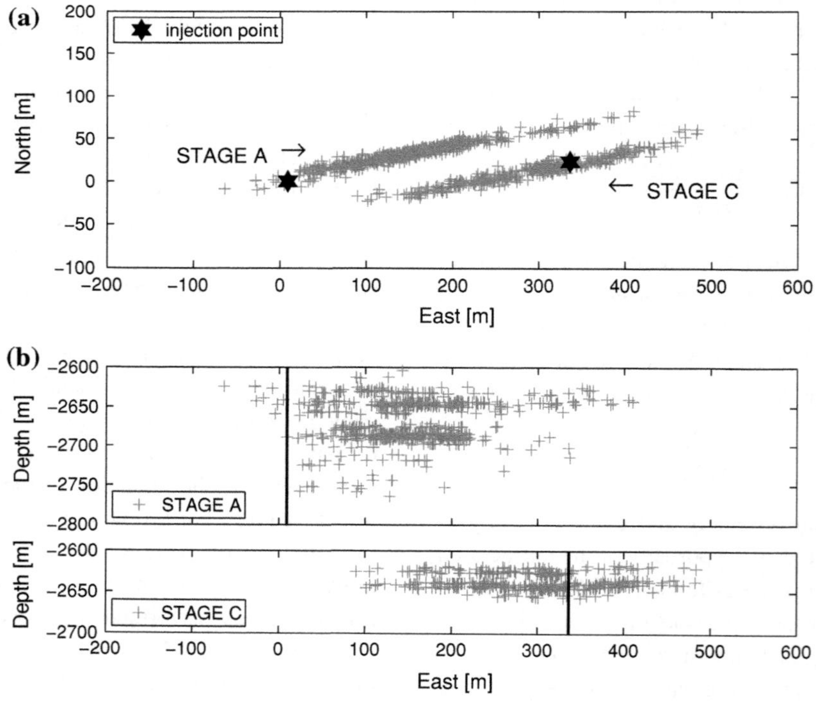

Figure 3
a Map view and **b** depth view of source locations of microearthquakes induced in Stage A gel treatment and in Stage C water treatment

In the following we consider two fracturing experiments in more detail. These are Stage A, a gel treatment, and Stage C, a water treatment. The precisely determined source locations of the two corresponding microseismic event clouds are shown in Fig. 3. The observed asymmetric distribution of Stage A microearthquakes with respect to the treatment well is likely to be caused by the position of receivers (Fig. 2). RUTLEDGE and PHILLIPS (2003) argued for a limited detection range of the monitoring system. They calculated a magnitude-distance threshold and could show that about 93% of microseismic events east of the treatment well would not have been detected if they had been triggered in same distance west of the well. Thus, the treatment could have resulted in a symmetric fracture.

4.1. A Gel Treatment

Here we present the interpretation of microseismicity recorded in the Stage A treatment. To study the dynamical behavior via the spatio-temporal

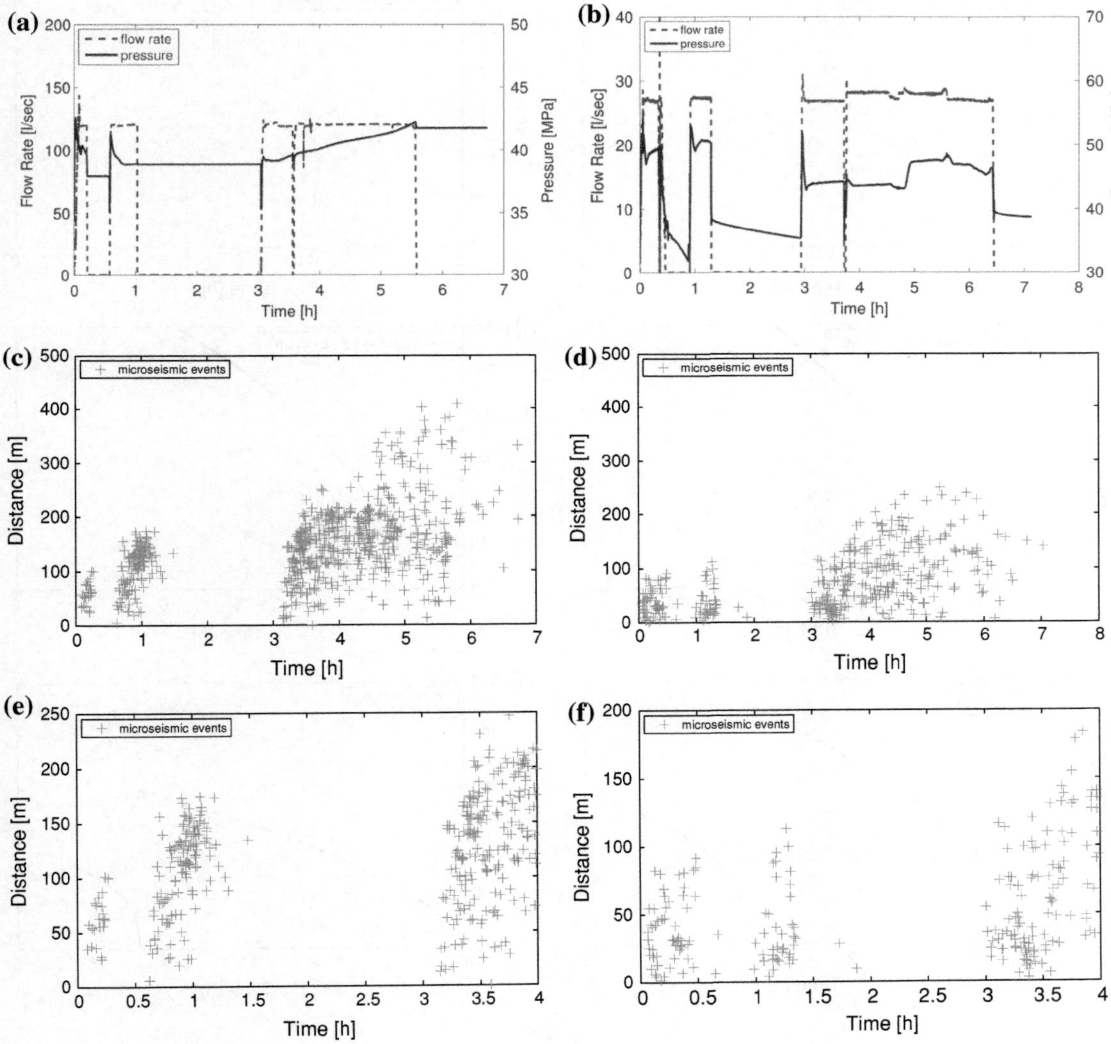

Figure 4

a Hydraulic fracturing treatment data, **c** $r - t$ diagram of induced microearthquakes of Stage A gel treatment. **e** Zoomed window of **c** to highlight the fracture volume opening. **b**, **d**, and **f** are the equivalent of Stage C water treatment evolution of the event cloud, we have computed the distance of an event to the nearest wellbore perforation point and plotted it as function of the occurrence time relative to the starting time of the treatment ($r - t$ diagram, Fig. 4). The fracturing was performed in multiple cycles of injection in a typical hydraulic fracturing sequence: mini-frac, step rate test and main treatment (Fig. 4). The cessations of injection allow the fracture to close if no proppant is added, and to reopen in the following injection cycle. Such behavior can be identified in the $r - t$ diagram of microseismicity (Fig. 4). If we consider only the first minutes of each of the injection cycles, one can see the quasi-linear event migration with time away from the treatment well. In our understanding, it represents opening, and in subsequent injection cycles, reopening, and further propagation of the fracture. The similarity in the velocities with which

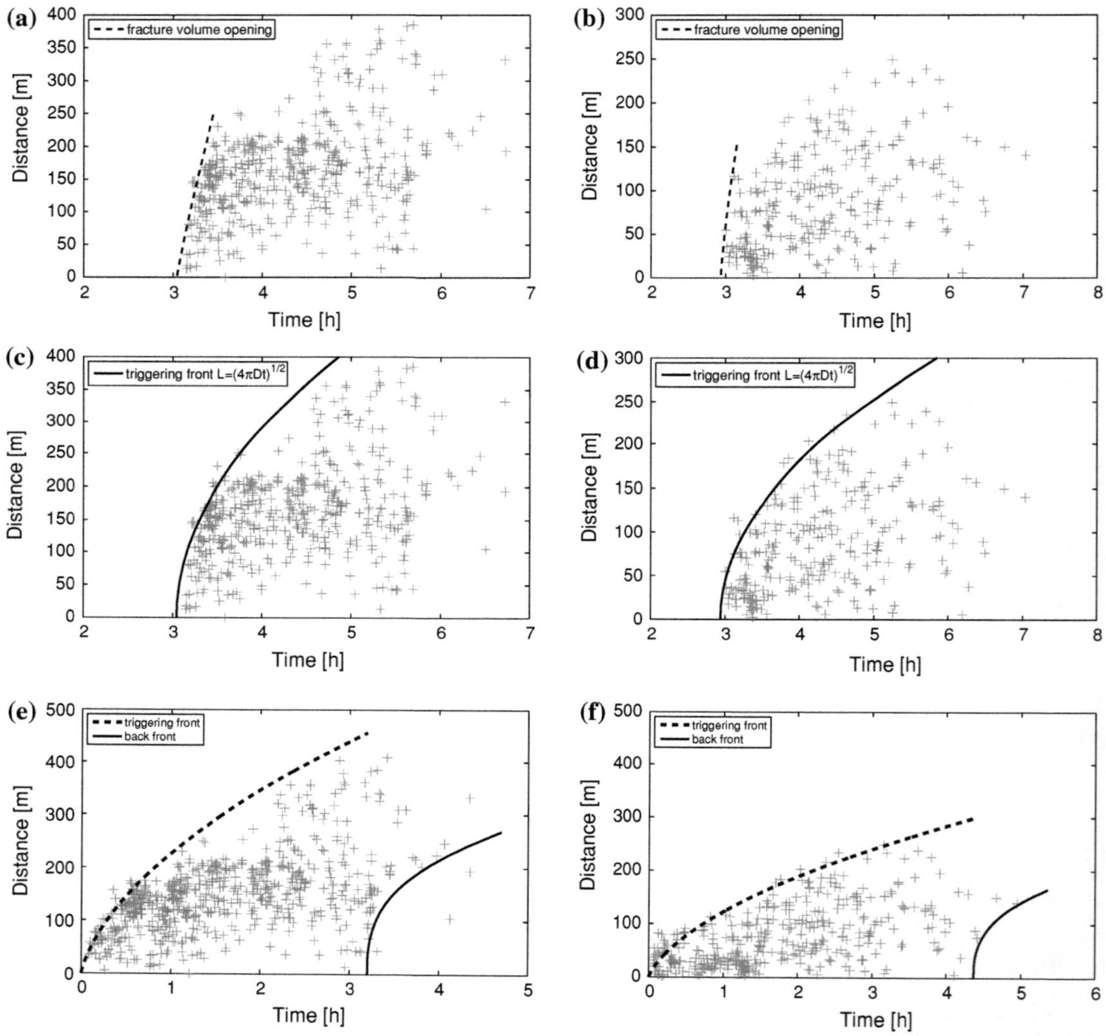

Figure 5

a $r - t$ diagram of microearthquakes induced in main treatment period of Stage A gel treatment, *dashed line* indicates fracture volume opening. **c** $r - t$ diagram as in (**a**) but with triggering front curve according to Eq. 3 to describe the diffusion-like propagation of seismicity. **e** Modified $r - t$ diagram with removed shut-in phases. Triggering and back front curves according to Eq. 2 and Eq. 6 well limit the microseismic event cloud. **b**, **d** and **f** are the equivalent of Stage C water treatment the fracture opens in the three individual injection cycles (velocity of fracture propagation, $v \sim 0.17$ m/s) indicates that reopening takes place with approximately the same geometry (width and height).

4.1.1 Estimate of Fluid Loss Volume

In the first step we have determined the volume of the newly created fracture. To estimate the cross-section area A according to Eq. 8 we have only used the microseismic events which were triggered in the main treatment. The tangent coefficient at time $t = 0$ s is obtained by data fitting (Fig. 5). It results in $A = 0.22$ m^2. Assuming the fracture height equal to the total perforated section of the treatment well, $h_f \sim 80$ m, then the average fracture width w is ~ 3 mm. The uncertainty in the estimate of fracture width is relatively high due to the simplified assumption on fracture height. However, with known cross-section area and length one can calculate the fracture volume and compare it with the injected volume: in Stage A the total injected fluid volume was $V_i = 1{,}340$ m^3, resulting in a fracture volume $V_f = 176$ m^3. With the volume balance, it means that $\sim 85\%$ of the injected volume was lost into the formation.

4.1.2 Estimate of Reservoir Permeability

The controlling parameter of fluid leakage is the fluid-loss coefficient C_L. According to Eq. 9 it is 5.2×10^{-5} m/s$^{1/2}$. In addition, a diffusion constant which is characterizing the diffusional fluid leakage process can be determined. Introducing C_L into Eq. 5, the apparent hydraulic diffusivity is $D_{ap} = 2$ m^2/s. However, this diffusion constant is apparent only and not directly related to the triggering of microearthquakes by a pore pressure diffusion. However for a quality control, we have plotted a triggering front curve according to Eq. 3 with $D = 2$ m^2/s to the microseismic data in Fig. 5. One can see that the growth of the event cloud following the fracture volume opening obeys a behavior similar to triggering by linear pore pressure diffusion. The fluid-loss coefficient is related to the reservoir permeability (Eq. 10). With typical values for the Cotton Valley tight gas reservoir (hydrocarbon gas at a temperature of 120°C and hydrostatic pressure of $p = 28$ MPa: a viscosity $\eta = 3 \times 10^{-5}$ Pa s and a compressibility $c_t = 3.5 \times 10^{-8}$ Pa^{-1}), porosity $\phi = 0.1$ (WALKER 1997), and pressure difference $\Delta p = 16$ MPa, the virgin reservoir permeability is $\kappa = 3 \times 10^{-7}$Darcy. To test our findings, we present a modified $r - t$ diagram where times of shut-in phases are removed, and plot the triggering front according to Eq. 2 using the parameters as estimated before (fracture cross-section area and fluid loss-coefficient) (Fig. 5). One can see that the calculated curve limit (as the upper envelope) the distribution of microseismic events in the spatio-temporal domain.

4.1.3 Estimate of Fracture Permeability and Conductivity

After completion of the hydraulic treatment, a seismicity back front can be identified in the $r - t$ diagram (Fig. 5). The signature provides $D_{bf} = 4$ m^2/s for the hydraulic diffusivity of the fractured area by fitting Eq. 6 to match the microseismic data (Fig. 5). Introducing the diffusion constant D_{bf} into Eq. 7 and assuming an increased porosity of $\phi_f = 0.3$ yields a fracture permeability κ_f in the order of some 10 Darcy. Thus, there is an increase in permeability about seven orders of magnitude due to the fracturing of the reservoir. Finally, we have introduced the results in Eq. 11 to obtain dimensionless fracture conductivity of $C_D = 250$.

4.2. A Water Treatment

Here the focus is on the interpretation of Stage C induced microseismicity. Additionally, we compare the results for water and gel treatment. Events in the distance–time domain are shown in Fig. 4. Contrary to the intuitive expectation that more fluid will be lost into the formation by leakage (because of the lower viscosity of the treatment fluid) the $r - t$ diagram does not show such an effect. An explanation for this observation could be the lower fluid volume that was injected at lower flow rates (Table 1), or a less dense population of pre-existing natural fractures at the location of Stage C.

Table 1

Summary and comparison of the results that were obtained by interpretation of fracturing-induced seismicity resulting from gel and water treatment

Stage A		Stage C
400 m	Fracture half-length L	250 m
0.22 m^2	Cross section A	0.13 m^2
80 m	Fracture height h_f	36 m
2.8 mm	Fracture width w	3.6 mm
176 m^3	Fracture volume V_f	65 m^3
1,340 m^3	Injected volume V_i	419 m^3
86%	Relative fluid loss V_L/V_i	85%
120 l/s	Flow rate	26 l/s
39–43 MPa	Bottom hole pressure	45–52 MPa

Despite different treatment parameters, the ratio of fractured volume and lost fluid volume is nearly identical in both type of treatments

4.2.1 Estimate of Fluid Loss Volume

The characteristic fracture opening and subsequent reopening phases are observable in the $r - t$ diagram (Fig. 4). The velocity of fracture propagation is $\sim$0.2 m/s. Despite the different treatment data (Table 1), the water treatment resulted in a similar fracture geometry. The estimation of the cross-section area yields $A = 0.13$ m^2. With the assumption that the fracture height h_f is equal to the total perforated domain of the treatment well ($\sim$36 m), the average fracture width is $w \sim 4$ mm. The water treatment resulted in a fracture volume $V_f = 65$ m^3. Thus, it is much less than the volume of the Stage A fracture. However, by taking into account that the volume of injected fluid was significantly lower ($V_i = 419$ m^3), then the ratio of fracture volume and loss volume is nearly identical in both treatments. Therefore, the fracturing process seems to be decoupled from the type of treatment fluid. Another point which supports this result is the seismic moment release. RUTLEDGE *et al.*, (2004) compared cumulative moment release versus cumulative injected volume for the various treatments. Because cumulative seismic moment describes total seismic deformation, it should be proportional to the created volume. The gel treatment and the water treatment have an identical moment release per unit volume injected (Fig. 6). Thus, it can be concluded that the fluid loss was similar in both treatments. If fluid leakage in water treatment was higher, it should be reflected by a smaller normalized moment release.

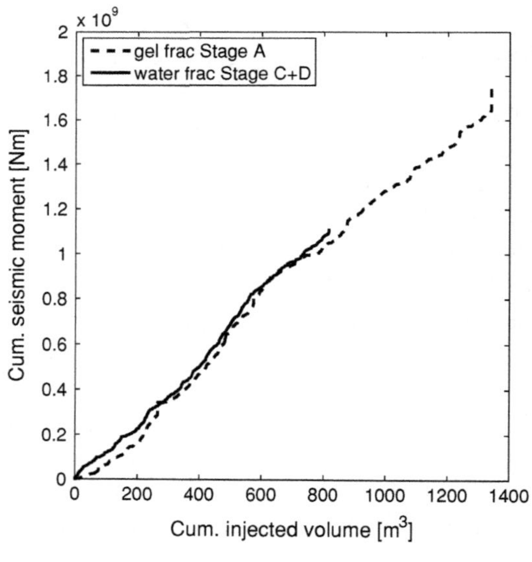

Figure 6
Cumulative seismic moment versus cumulative injected fluid volume. Stages C and D of water treatments were combined to correspond to the same depth interval as Stage A gel treatment (RUTLEDGE *et al.* 2004)

Table 2

Continuation of Table 1. Differences in the hydraulic characteristics of the two fractures are caused by the different volume of pumped proppant in gel and water treatment

Stage A		Stage C
5.2×10^{-5} m/s$^{1/2}$	Fluid loss coefficient C_L	4.6×10^{-5} m/s$^{1/2}$
3×10^{-7} Darcy	Reservoir permeability κ	8×10^{-8} Darcy
4 m^2/s	Fracture diffusivity D_{bf}	1.8 m^2/s
>10 Darcy	Fracture permeability κ_f	1 Darcy
250	Fracture conductivity C_D	160
230,000 kg	Proppant volume	15,000 kg

4.2.2 Estimate of Reservoir Permeability

The fluid-loss coefficient C_L is therefore similar to the one obtained in the gel treatment ($C_L = 4.6 \times 10^{-5}$ m/s$^{1/2}$). The coefficient is again transferred to an equivalent diffusion constant ($D_{ap} = 0.6$ m^2/s). A triggering front curve (Eq. 3) based on this hydraulic diffusivity limits the microseismicity apart from the volume opening phase (Fig. 5). The fluid loss coefficient is used to obtain a reservoir permeability $\kappa = 8 \times 10^{-8}$ Darcy, which is smaller

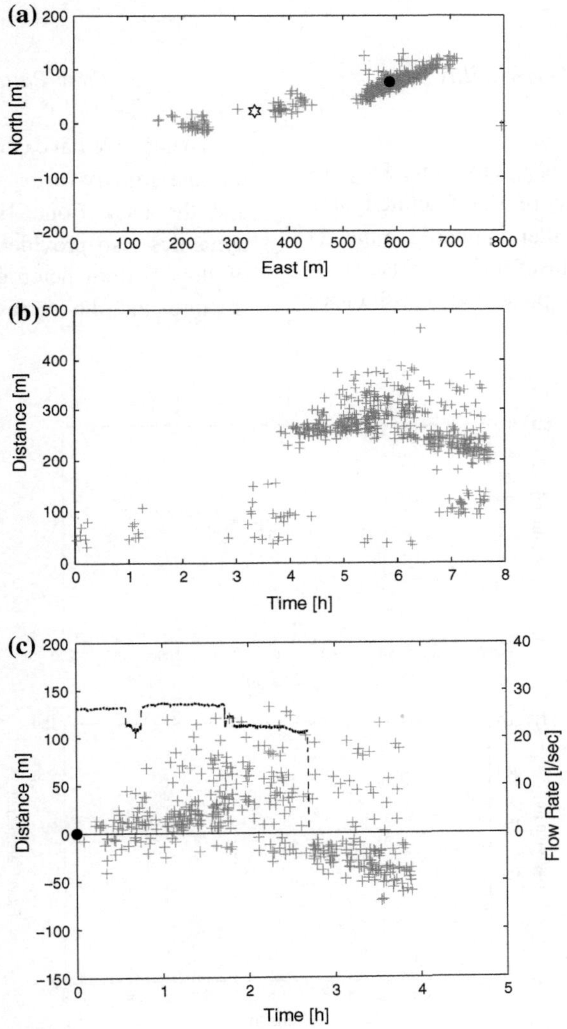

Figure 7

a Map view of induced microearthquakes in Stage E water treatment. The *black point marks* the first triggered event within an event cluster in a distance of ~250 m east of the injection well (*star*). **b** $r - t$ diagram of Stage E seismicity. **c** Directional $r - t$ diagram of the event cluster, *dashed line* denotes fluid flow rate. Note, that after shut-in seismicity migrates towards the treatment well by a factor four than the estimate from the gel treatment. One reason may be the higher pressure difference $\Delta p = 26$ MPa. Nonetheless in our opinion, the following aspect also has to be taken into account here. If proppant is pumped into a fracture it will be transported to the fluid-invaded zone around the fracture body as well. As a consequence, it affects fluid loss, pressure field, and hence the estimate of reservoir permeability using induced seismicity. The water treatment was performed without a significant volume of proppant (Table 2), which can be an explanation for the different estimates of reservoir permeability.

4.2.3 Estimate of Fracture Permeability and Conductivity

The back front of induced seismicity (Fig. 5) gives for the hydraulic diffusivity of the fractured area $D_{bf} = 1.8\ m^2/s$, which is smaller compared to the gel treatment. Despite rather insufficient statistics of post-treatment seismicity, the presented best-fit curve is still reasonable. The fracture permeability κ_f is approximately 1 *Darcy* and thus, it is one order of magnitude lower than in the gel treatment. We assume that this difference is caused by the strongly reduced volume of pumped proppant. It results in a decline of the fracture width after shut-in due to the missing resistance against horizontal stresses. Consequently, the fracture permeability will also be affected. A decrease in fracture width also affects the resultant dimensionless fracture conductivity (see Eq. 11). Taking this into account, the obtained value of $C_D = 160$ here should be seen as an upper limit.

5. Flow Pattern Heterogeneities

To date, we have considered envelopes of induced microseismicity in $r-t$ diagrams, the triggering and the back front. However, the spatio-temporal dynamics also provide information on the influence of flow pattern heterogeneities on the growth of a fracture and the corresponding microseismic event

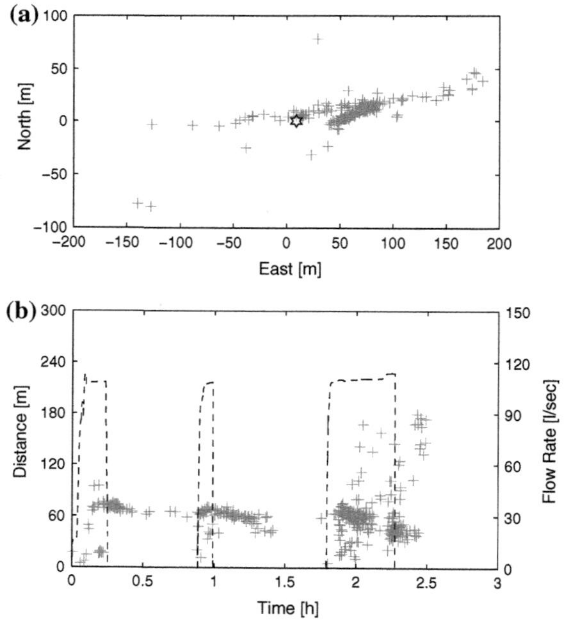

Figure 8

a Map view of induced microearthquakes in Stage B gel treatment. Shown are the first 3 h of the treatment only. **b** Corresponding $r-t$ diagram, *dashed line* denotes fluid flow rate. Note, that most of the seismicity occurs in a natural fracture system which intersects the hydraulic fracture trajectory. In the $r-t$ diagram, these events migrate towards the treatment well cloud. Two different phenomena, the opening of a pre-existing fracture system and the back-flow of the injected fluid, can be identified in the $r - t$ diagram of the water treatment Stage E (Fig. 7). The observations are explained by the following considerations. The treatment fluid flows over a pre-existing, thin aseismic channel to a natural weakness or fractured zone at a distance of about 250 m east of the treatment well (Fig. 7). The forced fluid opens a fracture in this zone mainly in a direction away from the treatment well, but also in a direction towards the well. This will be more intuitive by assuming the microearthquake that occurred first in this region to be the location of a secondary injection source (Fig. 7). Now, representing the induced events of this cluster in a separate directional $r - t$ diagram shows that the microseismicity (and hence the fluid) migrates in both directions from the assumed secondary fluid source (Fig. 7). Interestingly, a clear intensification of the flow in direction back to the treatment well can be observed after the injection is stopped. We propose that the back-flow is caused by some manner of "air balloon" effect. It is caused by an increase of the pressure gradient directed towards the treatment well: zero injection rate and, consequently, a pressure drop at the perforated injection well but remaining high pressure in the fracture. It forces the fluid flow back in the direction of the treatment well.

A slightly different phenomenon is observed in gel treatment Stage B. Figure 8 clearly demonstrates that part of the microseismic events migrates also in a direction back to the borehole. Note, that the propagation in backward direction occurs here during the injection period as well as after termination of the injection. These events occur within a fracture off-trend cluster that accounts for about 40% of detected events in Stage B (Rutledge *et al.*, 2004). This means that the injected fluid opens a pre-existing fracture which intersects the hydraulic fracture trajectory. Because both the Stage E and the Stage B treatments were performed in the same sand horizon, the distribution of microseismicity reflects the influence of a more heterogeneous natural fracture system compared to the targeted sand layers in the Stage A and Stage C treatments.

6. Conclusions

The space-time diagrams of induced microseismicity show signatures of fracture volume growth, loss of treatment fluid, diffusion of injection-induced pore pressure perturbations within the fracture and from the fracture walls into the surrounding formation, and heterogeneities in flow pattern. The interpretation of envelopes of the spatio-temporal evolution of fracturing-induced seismicity allows determination of geometrical parameters of the fracture. By assuming a conservation of the injected fluid volume, hydraulic properties of the fractured area and of the virgin reservoir can be estimated. We propose here a further development of this approach. We estimate the volume of the newly created hydraulic fracture using microseismic parameters. Thereafter, the volume of treatment fluid which was lost into formation by filtration through the fracture walls can be computed. Knowing the fluid loss, we define the virgin reservoir permeability. Additionally, we calculate an apparent diffusivity of the leakage, and compare it to observations from the microseismic data for quality control. A comparison of microseismicity resulting from a gel and a water treatment shows that the fracturing resulted in similar characteristics of the fractures: the ratio of new fracture volume to the total injected volume is nearly identical in both treatments. We therefore conclude that the fracturing process is decoupled from the type of treatment. This is confirmed by comparing the released seismic energy as a function of the cumulative injected volume. The permeability of the hydraulic fractures is influenced by the volume of pumped proppant which keeps the fracture open after shut-in of the injection. Signatures of heterogeneous flow pattern are also present in the dynamics of microseismicity. They provide information pertaining to interactions of the forced fluid with the pre-existing fracture system. The heterogeneities can be explained by the orientation of pre-existing natural fractures in respect to the hydraulic fracture trajectory, and pressure gradients within the fracture after termination of the injection.

Acknowledgments

The research presented in this paper is supported by the sponsors of the PHASE University research project. We express our thanks to Scott Phillips and to one anonymous reviewer for their valuable comments and suggestions.

References

ALBRIGHT, J.N. and PEARSON C.F. (1982), *Acoustic emissions as a tool for hydraulic fracture location: Experience at the Fenton Hill Hot Dry Rock site*, Soc. Petro. Eng. August, 523–530.

ARCO EXPLORATION and PRODUCTION TECHNOLOGY (1997), *Cotton Valley production stimulation hydrofracture Phase II*, Technical Report, UPR Consortium.

BLOCK, L.V., CHENG, C.H., FEHLER, M.C., and PHILLIPS, W.S. (1994), *Seismic imaging using microearthquakes induced by hydraulic fracturing*, Geophysics *59*, 102–112.

ECONOMIDES, M.J. and NOLTE, K.G., (eds.) *Reservoir Stimulation* (Wiley, Chichester, 2003).

FEHLER, M., HOUSE, L., and KAIEDA, H. (1987), *Determining planes along which earthquakes occur: Method and application to earthquakes accompanying hydraulic fractures*. J. Geophys. Res. *92*, 9407–9414.

FISCHER, T., HAINZL, S., EISNER, L., SHAPIRO, S.A., and LE CALVEZ, J. (2008), *Microseismic signatures of hydraulic fracture growth in sediment formations: Observations and modeling*. J. Geophys. Res. *113*, doi:10.1029/2007JB005070.

HARRINGTON, L. and HANNAH, R.R. *Fracturing design using perfect support fluids for selected fracture proppant concentrations in vertical fractures*. In *SPE Annual Meeting* (Dallas, September 1975).

HUBBERT, M.K. and WILLIS, D.G. (1957), *Mechanics of hydraulic fracturing*. J. Pet. Tech. *9*(6), 153–168.

KECK, R.G., and WITHERS, R.J. *A field demonstration of hydraulic fracturing for solid waste injection with real-time passive seismic monitoring*. In *Proc 1994 Soc Petro. Eng. Annu. Tech. Conf., Paper 28495* (New Orleans, September 1994).

NORDGREN, R.P. (1972), *Propagation of a vertical hydraulic fracture*, SPE J. *12*(8), 306–314.

PAROTIDIS, M., SHAPIRO, S.A., and ROTHERT, E. (2004), *Back front of seismicity induced after termination of borehole fluid injection*. Geophys. Res. Lett. *31*, doi:10.1029/2003GL018987.

PEARSON, C. (1981), *The relationship between microseismicity and high pore pressures during hydraulic stimulation experiments in low permeability granitic rocks*, J. Geophys. Res. *86*, 7855–7864.

PERKINS, T.K. and KERN, L.R. (1961), *Width of hydraulic fractures*, J Petro. Tech. *13*(9), 937–949.

PHILLIPS, W.S., HOUSE, L.S., and FEHLER, M.C. (1997), *Detailed joint structure in a geothermal reservoir from studies of induced microearthquake clusters*, J. Geophys. Res. *102*, 11745–11763.

PINNACLE TECHNOLOGIES, *Cotton Valley fracture imaging project Phase III—Final Report. Technical Report (JIP Consortium, 1999)*.

ROWE, C.A., ASTER, R.C., PHILLIPS, W.S., JONES, R.H., BORCHERS, B., and FEHLER, M.C. (2002), *Using automated, high-precision repicking to improve delineation of microseismic structures at the Soultz geothermal reservoir*. Pure Appl. Geophys. *159*, 563–596.

RUTLEDGE, J.T. and PHILLIPS, W.S. (2003), *Hydraulic stimulation of natural fractures as revealed by induced microearthquakes, Carthage Cotton Valley gas field, east Texas*, Geophysics *68*, 441–452.

RUTLEDGE, J.T., PHILLIPS, W.S., and MAYERHOFER, M.J. (2004), *Faulting induced by forced fluid injection and fluid flow forced by faulting: An interpretation of hydraulic-fracture microseismicity, Carthage Cotton Valley gas field, Texas*, Bull. Seismol. Soc. Am. *94*, 1817–1830.

SHAPIRO, S.A. and DINSKE, C. (2007), *Violation of the Kaiser effect by hydraulic-fracturing-related microseismicity*. J. Geophys. Eng. *4*, doi:10.1088/1742-2132/4/4/003.

SHAPIRO, S.A., DINSKE, C., and ROTHERT, E. (2006), *Hydraulic-fracturing controlled dynamics of microseismic clouds*, Geophys. Res. Lett. *33*, doi:10.1029/2006GL026365.

SHAPIRO, S.A., HUENGES, E., and BORM, G. (1997), *Estimating the permeability from fluid-injection induced seismic emission at the KTB site*, Geophys. J. Int. *131*, F15–F18.

SHAPIRO, S.A., ROTHERT, E., RATH, V., and RINDSCHWENTNER, J. (2002), *Characterization of fluid transport properties of reservoirs using induced microseismicity*, Geophysics. *67*, 212–220.

SILENY, J., EISNER, L., HILL, D.P., and CORNET, F.H. (2008), *Non-doubel-couple mechanisms of microearthquakes induced by hydraulic fracturing*. In *70th EAGE Conference and Exhibition*, page PO21 (Rome, June 2008).

URBANCIC, T.I. and RUTLEDGE, J.T. (2000), *Using microseismicity to map cotton valley hydraulic fractures*. In *SEG Expanded Abstracts 19*, doi:10.1190/1.1815676.

WALKER, R.N. *Cotton Valley hydraulic fracture imaging project*. In *Proc 1997 Soc. Petro. Eng. Annu. Tech. Conf.* 38577 (San Antonio, October 1997).

WARPINSKI, N.R., ENGLER, B.P., YOUNG, C.J., PETERSON, R., BRANAGAN, P.T., and FIX, J.E. *Microseismic mapping of hydraulic fractures using multi-level wireline receivers*. In *Proc 1995 Soc. Petro. Eng. Annu. Tech. Conf.* 30507 (Dallas, October 1995).

WESCOTT, W.A. (1984), *Diagenesis of a tight gas formation: Jurassic Cotton Valley sandstone, East Texas Basin, AAPG Bull. 68*, 118–125, doi:10.1306/AD4618C2-16F7-11D7-8645000102C1865D.

(Received August 30, 2008, revised April 1, 2009, accepted April 9, 2009, Published online November 18, 2009)

Pure Appl. Geophys. 167 (2010), 183–213
© 2009 Birkhäuser Verlag, Basel/Switzerland
DOI 10.1007/s00024-009-0012-5

Pure and Applied Geophysics

Hydromechanics of the Koyna–Warna Region, India

INMACULADA DURÁ-GÓMEZ[1] and PRADEEP TALWANI[1]

Abstract—Continuous reservoir-induced seismicity has been observed in the Koyna–Warna region in western India following the beginning of impoundment of Koyna and Warna Reservoirs in 1961 and 1985, respectively. This seismicity includes 19 events with $M \geq 5.0$ which occurred in 7 episodes (I–VII) between 1967 and 2005 at non-repeating hypocentral locations. In this study, we examined the first six episodes. The seismicity occurs by diffusion of pore pressures from the reservoirs to hypocentral locations along a saturated, critically stressed network of NE trending faults and NW trending fractures. We used the daily lake levels in the two reservoirs, from impoundment to 2000, to calculate the time history of the diffused pore pressures and their daily rate of change at the hypocentral locations. The results of our analysis indicate that Episodes I and IV are primarily associated with the initial filling of the two reservoirs. The diffused pore pressures are generated by the large (20–45 m) annual fluctuations of lake levels. We interpret that critical excess pore pressures $>\sim$300 kPa and $>\sim$600 kPa were needed to induce Episodes I–III and Episodes IV–VI, respectively, suggesting the presence of stronger faults in the region. The exceedance of the previous water level maxima (stress memory) was found to be the most important, although not determining factor in inducing the episodes. The annual rise of 40 m or more, rapid filling rates and elevated d*p*/d*t* values over a filling cycle, contributed to the rapid increase in pore pressure.

Key words: Reservoir-induced seismicity, Koyna reservoir, pore pressure diffusion, stress memory, strain hardening.

1. *Introduction*

Continuous reservoir-induced seismicity (RIS) has been observed since 1963 in the Koyna–Warna region (KWR) after the impoundment of the Koyna Reservoir (in western India) in 1961 and Warna Reservoir in 1985. The seismicity covers a volume $\sim$30 km long, $\sim$20 km wide and $\sim$10 km deep. Through 2006, more than 100,000 earthquakes have been recorded, including more than 170 events with $M \geq 4$ and 19 events with $M \geq 5$ (henceforth referred as M 5 events) with non-repetitive epicenters (GUPTA, 2002; GUPTA *et al.*, 2007). Over the years, the KWR has been the subject of numerous studies (see e.g., in the last two decades, GUPTA, 1992, 2002; TALWANI *et al.*, 1996; TALWANI, 1997b, 2000; RASTOGI *et al.*, 1997; MANDAL *et al.*, 1998; GUPTA *et al.*, 2002, 2005, 2007; SATYANARAYANA *et al.*, 2005; KUMAR *et al.*, 2006; SARMA and SRINAGESH, 2007; CHADHA *et al.*, 2008). A few studies have been aimed at understanding the mechanism of RIS at the KWR (e.g., SIMPSON *et al.*, 1988; KALPNA and CHANDER, 2000; TALWANI 1995, 1997a; RAJENDRAN and HARISH, 2000; PANDEY and CHADHA, 2003).

A unique observation in the region is the continuing occurrence of M 5 earthquakes, 44 and 20 years after the impoundment of the Koyna and Warna Reservoirs, respectively. In this paper, we combine our improved understanding of both the seismotectonic framework of the KWR and the mechanism of RIS in order to explain the RIS and hydromechanics of the KWR, by a study of the pore pressure history associated with the spatial and temporal evolution of M 5 earthquakes. We chose M 5 events because of their non-repeating hypocentral locations, each of which was associated with a $\sim$10 km^2 rupture area. Understanding the origin of these earthquakes is challenging and fascinating, and is the subject of this study.

We first present the revised seismotectonic framework of the KWR in light of results of studies of recent seismicity and geophysical investigations. We suggest the association of various fractures and faults with the M 5 earthquakes. Our analysis suggests that the saturated fractures in the KWR are critically stressed (Sect. 2). Next we describe our understanding of the predominant role of pore pressure diffusion (CHEN and

[1] Department of Geological Sciences, University of South Carolina, Columbia, SC 29208, USA. E-mail: talwani@geol.sc.edu

TALWANI, 2001) in isolated, discrete fractures (CORNET and YIN, 1995; TALWANI *et al.*, 1999; EVANS *et al.*, 2005; TALWANI *et al.*, 2007) in the onset of RIS due to lake level fluctuations. We address (a) the role of long-period lake level fluctuations (ROELOFFS, 1988) on the diffusion of pore pressure to long distances; (b) the increase in pore pressures to threshold levels to induce M 5 earthquakes; (c) the observation that large dp/dt changes help trigger RIS (NUR and BOOKER, 1972; TALWANI *et al.*, 2007) (Sect. 3). Then we present observations that show that some of the episodes of M5 earthquakes display evidence of stress memory in this large, critically stressed volume (Sect. 4). The specific goals of this study to understand the hydromechanics of the KWR are described in Sect. 5. To achieve them, we use the daily lake-level data at Koyna and Warna Reservoirs up to 2000 and the assumption of one-dimensional pore pressure diffusion through fractures comprising the seismotectonic framework to calculate the pore pressure (p) and the rate of pore pressure change (dp/dt) histories at hypocentral locations of M 5 earthquakes (Sect. 6).

We conclude that the continued occurrence of seismicity in the KWR for more than 45 years is a result of a combination of factors, which include (a) the renewed generation of pore pressure ($\sim$100 s kPa) due to large (20–45 m) annual fluctuations of lake levels of the Koyna and Warna Reservoirs; (b) the availability of a network of fractures with permeabilities greater or equal than the seismogenic permeability ($k \geq k_s$) to diffuse pore pressure to long distances in a (c) critically stressed volume (30 × 20 × 10 km^3); (d) the cumulative effect of annual cycles of lake-level changes contributes to an increase of the pore pressure to threshold levels; (e) the exceedance of lake levels over the previous maxima; (f) large dp/dt amplitudes and filling rates, which appear to contribute to higher pore pressures; (g) elastic–plastic deformation with strain hardening, which possibly strengthens the fractures, accounts for the observation of stress memory in this critically stressed volume (Sect. 7).

2. *Seismotectonic Framework in the KWR*

In the KWR the seismicity is associated with lake-level changes in the two reservoirs. Figure 1a compares the daily histogram of the number of $M \geq 4.0$ recorded earthquakes (catalog from Gupta 2002) with the lake levels in the two reservoirs since their impoundment in 1961 (KR) and 1985 (WR) to 2000. Figure 1b compares the times of occurrence of M 5 events with the two lake levels in and their locations are given in Fig. 2.

TALWANI (1997b) integrated seismological data from 1962 to 1995 with geomorphic, geological and geophysical data to describe the seismotectonic framework of the Koyna–Warna area. In the revised seismotectonic framework (Fig. 2), we use additional information based on the analysis of over 600 digitally recorded earthquakes from April 1996 to December 1997 (SRINAGESH and SARMA 2005; SARMA and SRINAGESH 2007) and magnetotelluric surveys in the Koyna area (SARMA et al. 2004; HARINARAYANA et al. 2007).

The revised seismotectonic framework (Fig. 2) consists of a pattern of crisscrossing steeply dipping, NE trending faults, associated with primarily left-lateral strike-slip motion; and a system of sub-parallel, NW trending near vertical fractures, associated with primarily normal faulting; in response to a $\sim$N-S oriented stress field (TALWANI, 1997b; GUPTA *et al.*, 2002; SARMA and SRINAGESH, 2007). The former include the Koyna River fault zone (KRFZ), Donachiwada, Patan, and P1 faults, and the latter include linear features L1–L7 (Fig. 2). (We have renamed the lineaments given in TALWANI (1997b) to follow the nomenclature provided by GUPTA *et al.*, 2002). The steep NW-SE trending faults divide the area between the Koyna and Warna rivers into distinct blocks, providing conduits for fluid pressure flow to hypocentral depths.

L3 is a major NW-SE lineament that crosses the Koyna–Warna region and extends up to the Warna dam. It was found by LANGSTON (1981) from the inspection of Landsat images. L1 and L2 are based on the analysis of seismicity data by TALWANI *et al.* (1996) and TALWANI (1997b). L4 and L5 are based on aeromagnetic data (TALWANI et al., 1996; PESHWA, 1991). The location of the Patan fault is based on geomorphological features and 1994 seismicity data (TALWANI *et al.*, 1996, TALWANI, 1997b). P1 is exclusively based on the analysis of seismicity data (TALWANI et al., 1996; TALWANI, 1997b). L6 and L7 are based on the revised hypocentral locations obtained by SRINAGESH and SARMA (2005) and SARMA

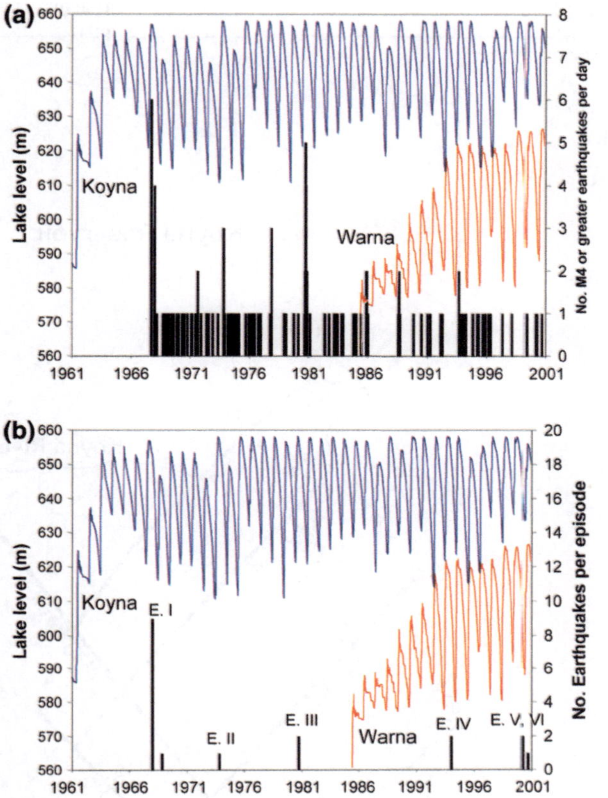

Figure 1
a Koyna and Warna water level fluctuations from impoundment to 2001, and daily histogram of the number of $M \geq 4.0$ recorded earthquakes (catalog from GUPTA, 2002). **b** Koyna and Warna water level fluctuations from impoundment to 2001, and times of occurrence of $M \geq 5.0$ earthquakes (Episodes I–VI)

and SRINAGESH (2007). They divided the revised hypocentral locations into three zones, one of which lies along the Donachiwada fault, the other two NW-SE trending zones were labeled SEZ and WSZ. Cross sections across WSZ and SEZ show that earthquakes lie along four vertical ~parallel NW-SE trending planes, two of which coincide with L4 and L5 described earlier, and the other two we have named L6 and L7 (Fig. 2). The locations of the 19 M 5 earthquakes in the KWR are shown in Fig. 2, and their causative faults are listed in Table 1a and b.

3. *Mechanisms of RIS Applied to the KWR*

TALWANI (1997a) determined that the protracted induced seismicity in the Koyna–Warna region is a consequence of the large annual water level fluctuations associated with the monsoon season. The study of the spatio-temporal pattern of seismicity in the KWR for over four decades, and comparison with the anomalously large (25–45 m) annual lake-level fluctuations reveals some unique features. The seismicity is primarily associated with, and delayed with respect to, the large annual filling between June and August, with about two-thirds of the annual seismic energy release occurring from September to December, including the continuing occurrence of M4 and 5 events long after the initial impoundment (TALWANI, 1997b, 2000; TALWANI *et al.*, 1996). The seismicity is induced by excess pore pressures associated with the filling, as they diffuse through saturated, critically stressed fractures (TALWANI, 1997b; TALWANI *et al.*, 2007). It is confined to a ~30 km long, 20 km wide and 10 km deep seismogenic volume (Fig. 2).

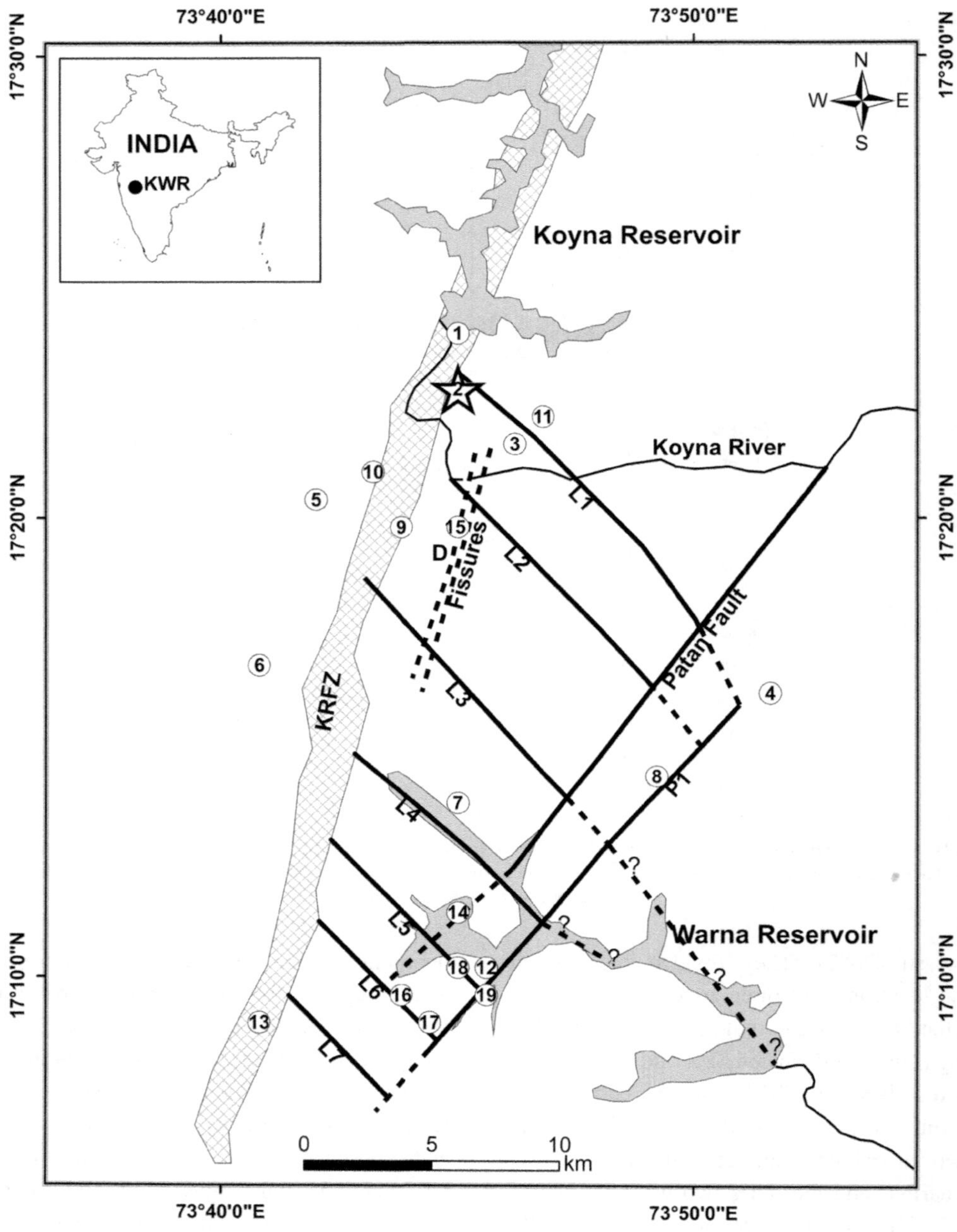

Figure 2
Location of Koyna–Warna region in western India, (*inset*), showing the chronological locations of 19 $M \geq 5$ events, taken from Talwani et al. (1996), Gupta et al. (2002 and 2005) superposed on the revised seismotectonic framework (this study) modified from Talwani (1997a) (see text for details). D is the Donachiwada fault, with fissures associated with the 1967 shock

Several other factors are known to be associated with reservoir-induced seismicity (RIS). These include the total height of the water column, the filling rate, the duration of the highest lake level and the amplitude and frequency of the lake level fluctuations (see e.g., Gupta, 2002). Another observation, the occurrence of large magnitude events when the lake level exceeds the previous maximum lake-level (Gupta and Combs, 1976; Simpson and Negmatullaev, 1981) was recognized to occur in the

Table 1

Lake levels and episodes of $M \geq 5.0$ events near (a) Koyna Reservoir[1], *(b) Warna Reservoir*[2]

Previous H_{max} exceeded on Y M D	New H_{max} reached on Y M D	New H_{max} (m)	Δh (m)	Episode	Eq. #	Event Date Y M D	Magnitude	Days after exceeded H_{max}	Fault mechanism	Causative fault
(a) Koyna										
	61 07 11	624.68								
62 07 11	62 08 15	636.65								
63 07 09	63 08 14	654.10								
64 08 01	64 08 13	654.94								
65 07 22	65 07 22	655.16								
67 08 30	67 10 04	656.99	1.83	I	1	67 09 13	5.2	14	Strike-slip (Langston 1981)	KRFZ
					2	67 12 10	6.3		Strike-slip (Talwani 1997b)	Near the intersection of KRFZ and L1
					3	67 12 10	5.0			L1
					4	67 12 11	5.2			P1
					5	67 12 12	5.4		Normal (Langston and Franco-Spera 1985)	L3
					6	67 12 12	5.0		Strike-slip (Rao et al. 1975)	KRFZ
					7	67 12 12	5.0			L4
					8	67 12 24	5.5			P1
					9	67 12 25	5.1			KRFZ
					10	68 10 28	5.4			KRFZ
73 09 02	73 09 27	657.98	0.99	II	11	73 10 17	5.1	45	Strike-slip? (Gupta *et al.* ,1980) CFPS	L1
80 09 01	80 09 03	658.13	0.15	III	12	80 09 02	5.5	1		L5
					13	80 09 20	5.3	19	Normal (Dziewonski *et al.*, 1988)	KRFZ/L7
94 09 02	94 09 02	658.2	~0.1			No EQ.				
02 09 07	02 09 07	658.3	~0.1			No EQ.				
(b) Warna										
	85 08 01	581.30								
86 06 25	86 06 30	586.15								
87 07 08	87 07 09	588.50								
88 07 14	88 07 19	592.65								
89 07 05	89 07 24	601.50								
90 07 09	90 07 16	604.15								
91 07 15	91 07 28	609.06								
92 07 29	92 08 19	617.90								
93 07 03	93 08 04	621.80	3.9	IV	14	93 12 08	5.2	158	Normal (Talwani 1997b)	Patan or L5
					15	94 02 01	5.4	213	Normal (Talwani 1994)	Patan
94 09 02	94 09 02	622.05	0.25			NO EQ				
98 08 28	98 09 28	625.85	3.8	V	16	00 03 12	5.2	562	Normal (Gupta *et al.* 2002)	L6 or Patan
					17	00 04 06	5.1		Normal (Gupta *et al.* 2002)	L6 or P1
00 09 11	00 10 16	626.30	0.45	VI	18	00 09 05	5.3			L5
03* 09 10 (est)	03 09 13	627.88	1.58	VII	19	05 03 14	5.2	551	Normal (Kumar *et al.* 2006)	L5

[1] Data compiled from Talwani et al. (1996), Talwani (2000), Gupta et al. (2002)

[2] Data compiled from Talwani et al. (1996), Gupta et al. (2002, 2005) and Satyanarayana et al. (2005)

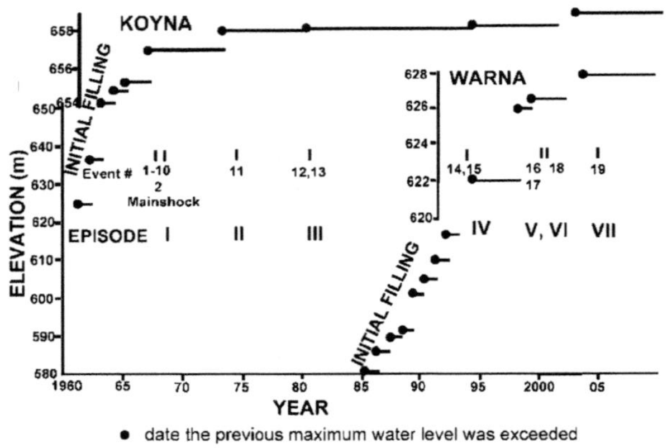

Figure 3

Temporal distribution of $M \geq 5$ events and the seven episodes of stress memory, compared with the lake levels. The *solid dots* show the date when the previous maximum water level was exceeded, and the horizontal bars when it was not. The initial fillings of the Koyna and Warna Reservoirs (1961–1965 and 1985–1992) were followed by small rises in those two reservoirs (shown with an exaggerated scale). $M \geq 5$ events, constituting the episodes were found to occur only when the previous lake level was exceeded, displaying stress memory in the rocks of the KWR

KWR for three M 5 events, and the phenomenon, stress memory, to be a crustal manifestation of the Kaiser effect observed in the laboratory (Simpson *et al.*, 1988; Talwani *et al.*, 1996; Talwani, 1996, 2000). More recent studies show that most of the M 5 events in the KWR manifest evidence of stress memory (next section).

Assuming a simple 2-D reservoir, Roeloffs (1988) calculated that the stress and pore pressure fields produced by time oscillatory reservoir loading are governed by a dimensionless frequency Ω given by $\Omega = \omega L^2/2c$, where ω is the loading frequency, L is the width of the reservoir and c is the hydraulic diffusivity. The depth below which diffusion associated with cyclic pore pressure changes at the surface is negligible, is given by $z = \pi(2c/\omega)^{1/2}$. Talwani (1997a) calculated this distance for a range of c and ω. For the Koyna and Warna Reservoirs (loading frequency is 1/year) he found it to be 25 km for $c = 1$ m^2/s and 78 km for $c = 10$ m^2/s. Kessels and Kück (1995) showed the applicability of the 1-D diffusion analysis along horizontal fractures. The observation of RIS more than 30 km south of the Koyna Reservoir suggests that the pore pressures associated with the annual filling cycles diffused along fractures connecting the reservoirs to the hypocenters.

Nur and Booker (1972) showed that when aftershocks could be attributed to pore pressure changes, their frequency was proportional to the rate of pore pressure increase. Talwani et al. (2007, Fig. 3a and b in their paper) demonstrated that the time of occurrence of induced earthquakes at the Monticello Reservoir coincided with a peak in the rate of pore pressure change, dp/dt, with time.

In summary, earlier work suggests that RIS is associated with highly diffused pore pressures (in accordance with the Coulomb failure criterion) and that it is triggered by high dp/dt values. We will test these ideas with over 40 years of lake level and seismicity data from the KWR.

4. *Stress Memory*

Stress memory of rocks is the ability to accumulate, retain and, under certain conditions, to reproduce information on the stresses experienced in the past (Lavrov, 2003). The Kaiser effect and strain hardening are manifestations of stress memory in various types of materials (e.g., metals and rocks), and are called stress memory effects (see Lavrov, 2003, for a review). The Kaiser-effect was discovered in the laboratory by Kaiser (1953). During cyclic loading with increasing stress, he observed a marked increase in acoustic emission activity (or microseismicity) when the largest stress level of the previous cycle was exceeded.

Another possible stress memory effect is strain hardening. Strain hardening is a strengthening mechanism that increases the yield point of a material, i.e., upon reloading a material will not deform plastically further unless the load is increased above its previous level (HILL, 1950; MENDELSON, 1983). Strain hardening is associated with plastic contraction of rocks, their subparticles become irreversibly more densely packed, which strengthens their elastic interactions that better withstand the subparticles from being moved apart (COUSSY, 1995). Seminal work on elastic–plastic deformation of rocks was carried out by Rice (1975, 1977) and RUDNICKI and RICE (1975). Strain hardening can only take place during elastic–plastic deformation of rocks. Given the stress-strain curve of a material, its behavior can be characterized as elastic, elastic–plastic or perfectly plastic, depending on how large the plastic (permanent) strain is with respect to the elastic strain (MENDELSON 1983). In the elastic case, the elastic strain is dominant and the stress-strain curve follows the generalized Hooke's law. In the rigid-plastic or perfectly plastic case, the plastic strain is dominant and the theory of plasticity is applicable. Finally, in the elastic–plastic case, the elastic and plastic strains are comparable, and both the generalized Hooke's law and the theory of plasticity have to be applied. Recent work by TEMPLETON and RICE (2008) and VIESCA *et al.*, (2008) examine off-fault elastic–plastic deformation for dry and fluid-saturated materials, respectively, focusing mainly on the undrained (no fluid diffusion) effects. We suggest from our study that elastic–plastic deformation with strain hardening in drained conditions might explain the phenomena of stress memory observed in the KWR.

4.1. *Stress Memory in the KWR*

Figure 2 shows the location of the 19 events ($M \geq 5$) plotted on an outline of the revised seismotectonic framework. These 19 events, which include the largest reservoir induced event, M 6.3 in December, 1967, occurred in seven episodes (Table 1a and b). An episode is defined as a sequence of one or more $M \geq 5.0$ events that follows the exceedance of a lake-level elevation above the previous highest value (H_{max}). (Different magnitudes have been listed for the larger events by various workers. In this paper, the magnitudes of these events have been taken from GUPTA *et al.*, 2002).

Episode I which includes the M 6.3 main shock in December, 1967, consists of ten $M \geq 5$ events, while the other episodes consist of one or two such events (Table 1a and b). The temporal distribution of these events, vis-a-vis the water levels in the Koyna and Warna Reservoirs is shown in Fig. 1b.

Evidence of stress memory was first realized with the comparison of the daily water level changes with the times of onset of the first three episode of $M \geq 5.0$ (TALWANI, 1996, 2000; TALWANI *et al.*, 1996). The highest water level and filling rate in any year depends on the annual rainfall between June and August, and on the draw-down governed by engineering considerations. In Fig. 3 we show those times when the previous maximum water levels (H_{max}) are exceeded (solid dots), and the durations when they are not (horizontal lines). We compare the times of maximum water levels (H_{max}) for the Koyna and Warna Reservoirs with the times of Episodes I–VII.

Filling of the Koyna Reservoir started in 1961 and H_{max} was exceeded on July 11, 1962, and the new H_{max} (636.65 m) was reached on August 15, 1962. The initial filling stage lasted until 1965, with H_{max} value of 655.16 m. This value of H_{max} was first exceeded on August 30, 1967. It was followed by a M 5.2 event (#1) on September 13, 1967. The highest water level observed that year was on October 4, 1967 (656.99 m) and the largest reservoir induced event, M 6.3, occurred on December 10,1967 (UTC), (4.21 a.m. on December 11,1967 local time). This event was followed by five aftershocks with $M \geq 5.0$ in the following 2 days, and three later, the last one on October 29, 1968 (GUPTA *et al.*, 2002).

The farthest event (#8 on December 24, 1967) was located $\sim$20 km to the southeast. Other $M \sim 4$ aftershocks were located near the present-day Warna Reservoir, which had not been impounded at that time (TALWANI, 1997b). The locations of the aftershocks are probably good to better than $\sim$5 km. These ten events constituting the foreshock-main shock-aftershock sequence constitute Episode I, and their pattern is similar to RIS at other locations following the initial impoundment (TALWANI, 1997a).

The 1967 H_{max} (656.99 m) was exceeded on September 2, 1973 and the new H_{max} (657.98 m) was reached on September 27, 1973. It was followed by a magnitude 5.1 event on October 17, 1973, (earthquake #11) which constitutes Episode II.

The 1973 H_{max} was exceeded on September 1, 1980 and was followed by an M 5.5 event the next day. H_{max} (658.13 m) was reached on September 3, 1980. It was followed by an M 5.3 event (#13) on September 20, 1980. These two events, which were located more than 25 km from the Koyna Reservoir, constitute Episode III. Earthquake #13 was associated with the Koyna River fault zone, and earthquake #12 was located on L5. TALWANI (1997b) speculated that these locations were connected to the Koyna Reservoir by a network of cross-cutting faults and fractures (Fig. 2).

The next episode did not occur until after the impoundment of the Warna Reservoir, which began in 1985. Following its impoundment there were four more episodes of $M \geq 5$ events (Table 1b).

After the initial filling of the Warna Reservoir (1985–1992), the 1992 H_{max} (617.90 m) was exceeded on July 3, 1993 and the new H_{max} (621.80 m) was reached on August 4, 1993. Two $M \geq 5$ events (#14 and #15, constituting Episode IV) occurred on December 8, 1993 and on February 1, 1994 at distances of ~13 and ~25 km from the deepest part of the Warna Reservoir (near the dam) respectively, the latter event was associated with normal faulting on the Patan fault. Its aftershocks lie along L2 (Fig. 2). These events, which occurred 158 and 213 days after the 1992 H_{max} was exceeded, showed evidence of pore pressure diffusion to the northwest of the Warna Reservoir, possibly along the same sets of fractures that were associated with southward pore pressure diffusion from the Koyna Reservoir during earlier episodes. We attribute these events to the initial filling of the Warna Reservoir.

The H_{max} in the Warna Reservoir was next exceeded in September 2, 1994 by 0.25 m and then again in 1998 by 3.8 m (Table 1b). The next Episode, V, consists of two events that occurred during a period of lake-level decline in March and April 2000, about 10 km from the deeper part of the Warna Reservoir. The March 2000 event occurred after a lag of 562 days after the 1994 H_{max} was exceeded on August 28, 1998. These two events were the farthest south of the $M \geq 5$ events in the KWR, and occurred in a previously aseismic area.

Earthquake #18 on September 5, 2000 occurred 6 days before the 1998 H_{max} (625.85 m) was exceeded. Although not strictly consistent with our definition of an episode, we consider earthquake #18 as a new Episode, VI, because it was associated with a water level rise and the elevation was a few cm of the ultimate H_{max} (626.30 m). This event was located closer to the reservoir than the previous two.

The 2000 H_{max} was exceeded on ~September 10, 2003, and earthquake #19 constituting Episode VII, occurred on March 03, 2005, 551 days later, during a period of lake-level decline.

Summarizing, there have been seven episodes of $M \geq 5$ events in the KWR in the last four decades. The 19 events that form these episodes did not repeat their hypocentral locations, and are contained in a critically stressed $30 \times 20 \times 10$ km^3 volume. Comparing the temporal patterns of these episodes with laboratory data, we can think of Episodes I and IV, which followed the impoundment of the Koyna and Warna Reservoirs, as being associated with the initial filling. We suggest that Episodes II, III, IV, VI and plausibly also V and VII, are manifestations of stress memory on a crustal scale. That is, a ~6,000 km^3 seismogenic volume remembers the water level maximum H_{max} that preceded a $M \geq 5$ event, and when that water level is exceeded, even by 0.15 m, a new episode occurs.

If the excess lake load above the previous H_{max} was the only contributing factor, it would imply very small triggering stresses associated with these M 5 events. Next we examine other possible causes of additional stress perturbations, which are summarized in the next section.

5. *Goals of the Study*

Several factors have been suggested as possible contributors to the development of the critical pore pressure (p_c) and the triggering of M 5 events. To obtain an estimate of p_c values associated with various episodes and the relative contributions of different factors, we analyzed the excess pore pressures associated with the filling histories of the two reservoirs.

Specifically, we evaluated the contributions of the following factors: (a) the annual cycle of lake-level fluctuations; (b) the amplitude of the annual increase in lake level, ΔH, especially for those years when it exceeds 40 m; (c) the maximum rate of pore pressure (dp/dt) increase over a filling cycle, A; (d) exceedance of the lake level over a previous maximum level (identified as a manifestation of stress memory in rocks); (e) the cumulative effect of annual cycles of filling and emptying of the reservoir between episodes; (f) complete leakage or no leakage of built-up pore pressures between episode and (g) the combined effect of the two reservoirs.

To study the *relative* contributions of these factors, we calculated the excess pore pressure by assuming simple one-dimensional diffusion of pore pressure through a network of fractures based on our current understanding of the seismotectonic framework. Although the fractures are two-dimensional features, hypocentral locations suggest that only finite strips of the fractures are seismogenic and conduits for pore pressure diffusion. As the details of these are not known, we assumed simple one-dimensional pore pressure diffusion between the reservoirs and the hypocentral locations. The calculated pore pressures may overestimate the true excess pore pressures nonetheless they give meaningful estimates of the relative contributions of the various factors.

6. *Calculation of Diffused Pore Pressure Histories at Hypocentral Locations*

6.1. *Theory*

Roeloffs (1988) calculated the pore pressure at time $t > 0$ and a depth z due to a change in the level of a reservoir lake behind a dam at time $t = 0$. The total pore pressure at a depth z and time t, $p(z,t)$ can be expressed as the coupled response of diffused pore pressure and undrained pore pressure.

$$p(z,t) = (1-\alpha)\cdot p_0 \cdot \mathrm{erfc}\left[\frac{z}{(4ct)^{1/2}}\right] + \alpha \cdot H(t)\cdot p_0,$$
$$\alpha = B\cdot(1+\nu_u)/[3\cdot(1-\nu_u)] \qquad (1)$$

where erfc is the complementary error function, $H(t)$ is the Heaviside unit step function, B is the Skempton's coefficient, ν_u is the undrained Poisson's ratio, z is the depth beneath the reservoir, p_o is the vertical stress at a depth $z = 0$, c is the hydraulic diffusivity and t is time.

The undrained pore pressure term in Eq. (1) builds-up instantaneously in the vicinity of a reservoir via the Skempton effect due to the load of water. For a point load, the undrained pore pressure effect is contained within a cone whose horizontal and vertical dimensions are about 3–4, and 6–7 times the width of the reservoir (Talwani et al., 2007). For the Koyna Reservoir, assuming the deepest part to be adjacent to the dam and a maximum width of ~2.5 km, the undrained pore pressure is contained within a horizontal distance of ~10 km and a vertical distance of ~17.5 km. Our study focuses on M 5 earthquakes, some of which occurred in the vicinity of the Koyna Reservoir within the cone shaped volume with a non-zero undrained effect. We estimate this undrained effect at hypocentral locations within this volume.

The pore pressure increase due to compression (undrained response) at a distance r from the reservoir is

$$\Delta p_u(r) = -B\Delta\sigma_{kk}/3, \qquad (2)$$

where $\Delta\sigma_{kk}/3$ is the change in the mean stress at the given distance r. Considering a uniform circular load, Ahlvin and Ulery (1962) tabulated the vertical, radial and transverse stresses at a point at any radial distance from the center of the circular load and at a given depth. Assuming $B \sim 0.7$ for crystalline rocks (Talwani *et al.*, 1999) and water height of ~70 m, for an earthquake at a depth of 4 km and a radial distance of 2 km, the undrained response is of the order of 20 kPa (Appendix 1), one order of magnitude less than the diffused pore pressure values calculated in this study (see next section). Therefore, we disregard the undrained effect in our calculations and consider only the changes in pore pressure due to diffusion. Substituting $\alpha = 0$ (no undrained effect) in Eq. (1), the diffused pore pressure component can be written as

$$p(z,t) = p_0 \cdot \mathrm{erfc}\left[\frac{z}{(4ct)^{1/2}}\right] \qquad (3)$$

Rajendran and Talwani (1992) and Chen and Talwani (2001) calculated the diffused pore pressures away from a reservoir. We have revised the formulation following the methodology given in those papers and, applying the principle of superposition to the equation obtained by ROELOFFS (1988). The following mathematical expression can be written for one-dimensional pore pressure diffusion in any direction:

$$P_n = \delta p_1 \cdot \mathrm{erfc}\left[\frac{r}{(4cn\delta t)^{1/2}}\right] + \delta p_2 \cdot \mathrm{erfc}\left[\frac{r}{(4c(n-1)\delta t)^{1/2}}\right] + \delta p_3 \cdot \mathrm{erfc}\left[\frac{r}{(4c(n-2)\delta t)^{1/2}}\right] + \ldots + \quad + \delta p_n \cdot \mathrm{erfc}\left[\frac{r}{(4c\delta t)^{1/2}}\right] \qquad (4)$$

where δt is a fixed increment of time (i.e., 1 day, see next section), n is the number of time increments considered in our calculation (i.e., number of days), δp_n is the water load change for the n-th day, r is the distance from the reservoir, c is the hydraulic diffusivity and erfc is the complementary error function. The equation states that the diffused pore pressure n days after the filling of a reservoir at a distance r from the reservoir is the sum of the diffused pore pressures generated by discrete water load changes in a given time step δt. The first discrete water load change δp_1 contributes for a total of n days to the total diffused pore pressure, while the last discrete water load change δp_n contributes only for one day.

We use Eq. (4) to calculate the diffused pore pressure associated with the impoundment and annual filling cycles of the Koyna and Warna Reservoirs at the times and locations of episodes of M 5 earthquakes. The water load changes δp correspond to time steps δt of 1 day. The rate of pore pressure change dp/dt is calculated by subtracting the pore pressure values in two consecutive days and dividing them by the time step of 1 day. Next we describe how we estimate the distances from the reservoirs to the earthquakes and the hydraulic diffusivity values along the diffusion paths.

6.2. Data and Assumptions

Water-level data cover the period from 1961 (impoundment) to 2000 for the Koyna Reservoir (14,606 data points) and 1985 (impoundment) to 2000 for the Warna Reservoir (5,685 data points). Figure 1b shows the daily water-level curves for the Koyna and Warna Reservoirs, and the times of the episodes (vertical lines). The hypocentral distance from the reservoir, r, assuming that pore pressure diffusion is restricted to faults and fractures (CORNET and YIN, 1995; TALWANI *et al.*, 1999; EVANS et al., 2005), was estimated from the inferred paths of pore pressure diffusion along them. The values of hydraulic diffusivity, c, along these paths were obtained from TALWANI *et al.* (2007, Tables A.1 and A.3 in that paper), and vary between ~1 and 10 m^2/s. Based on this range of values, we calculate diffused pore pressures in the Koyna–Warna region for three values of c, i.e., 1, 5 and 10 m^2/s.

The distance to the hypocenter for pore pressure diffusion, r, was calculated from the dams because they represent the deepest part of the reservoir and water levels are measured next to the dams. In calculating this distance, r, we first estimated the total horizontal distance (x) along the chosen surface path. Then, taking the hypocentral depth (z), we estimated $r = (x^2 + z^2)^{1/2}$. Table 2a–c list the estimated values of x, z and r for the different earthquakes. Because of the uncertainties in the estimation of distances, and to obtain a meaningful order of magnitude estimates of the diffused pore pressures, we used a range of representative distances for r, rounded to the nearest 5 km (shaded in Table 2a–c).

Given the uncertainties in the horizontal distances along fractures, depths and hydraulic diffusivity values, we calculated the diffused pore pressures for a range of distances and hydraulic diffusivity values. For example, we assigned earthquake #11 to occur on L1. We assumed the path of pore pressure diffusion from the Koyna Reservoir to the hypocenter of earthquake #11 to be along KRFZ and L1 (Fig. 2). Taking the horizontal distances along KRFZ and L1 to be ~2.0 and 4.3 km we obtain $x = 6.3$ km. With the hypocentral depth $z = 4.0$ km, we obtain the hypocentral distance $r = 7.5$ km. We used 5 km and 10 km as representative distances in our calculations of diffused pore pressures. In common with other earthquakes, the pore pressure for earthquake #11

Table 2

Distances from Koyna dam to Episodes (a) I, II and III, (b) IV and V, (c) Warna dam to Episode IV, V and VI

Earthquake # (Date)	Route followed	Depth, z (km)	Horizontal distance along fractures, x (km)	Hypocentral Distance, r (km)	Representative r (km)
(a)					
Episode I					
2 (67 12 10)	KRFZ	10.0 (TALWANI *et al.*, 1996)	2.7	10.4	5, 10
Episode II					
11 (73 10 17)	KRFZ, L1	4.0 (TALWANI *et al.*, 1996)	6.3	7.5	5, 10
Episode III					
12 (80 09 02)	KRFZ, L5	8.0 (GUPTA *et al.*, 2002)	29.4	30.5	30
13 (80 09 20)	KRFZ	12.0 (GUPTA *et al.*, 2002)	28.8	31.2	
(b)					
Episode IV					
14 (93 12 08)	KRFZ, L4 and Patan	8.2 (TALWANI *et al.*, 1996)	27.3	28.5	30
15 (94 02 01)	KRFZ, L3 and Patan	8.8 (TALWANI *et al.*, 1996)	7.4	11.5	15
Episode V					
16 (00 03 12)	KRFZ and L6	5.0 (GUPTA *et al.*, 2002)	28.2	28.6	30
17 (00 04 06)	KRFZ and L6	7.8 (GUPTA *et al.*, 2002)	30.3	31.3	
Episode VI					
18 (00 09 05)	KRFZ and L5	6.9 (GUPTA *et al.*, 2002)	28.2	29.0	30
(c)					
Episode IV					
14 (93 12 08)	Lake, L4 and Patan	8.2 (TALWANI *et al.*, 1996)	15.0	17.1	15
15 (94 02 01)	L3 and Patan	8.8 (TALWANI *et al.*, 1996)	25.0	26.5	30
Episode V					
16 (00 03 12)	Lake, L4, P1 and L6	5.0 (GUPTA *et al.*, 2002)	20.0	20.6	20
17 (00 04 06)	Lake, L4, P1 and L6	7.8 (GUPTA *et al.*, 2002)	17.7	19.3	
Episode VI					
18 (00 09 05)	Lake, L4, P1 and L5	6.9 (GUPTA *et al.*, 2002)	16.0	17.4	15

was calculated for three values of c, i.e. 1, 5 and 10 m^2/s.

6.3. *Pore Pressure Changes Associated with M 5 Earthquakes*

To obtain realistic values of diffused pore pressure we incorporated in our calculations how M 5 earthquakes affect the pore pressure field. MUIR-WOOD and KING (1993) found that the character of the hydrological changes which follow major earthquakes depend on the style of faulting. Earthquakes associated with normal faulting expel sizeable amounts of water however earthquakes associated with reverse faulting do not. Their coseismic strain model explains the differences between the hydrological signatures of normal and reverse fault earthquakes. For normal faulting, cracks open and the effective porosity increases during the interseismic period; at the time of an earthquake, cracks close and water is expelled. We interpret that, after the normal faulting earthquake, the pore pressure values are lower than those before the earthquake, there is a leakage (decrease) of pore pressure during normal faulting. For reverse faulting—according to their model—cracks close and water is expelled during the interseismic period; at the time of an earthquake, cracks open and water is drawn in. We interpret that, after the reverse faulting earthquake, the pore pressure values are higher that those before the earthquake. For strike-slip faulting we have a mixture of the two. Pore pressure values after an earthquake can drop (normal faulting) or increase (reverse faulting) depending on the style of faulting. As normal faulting is the dominant focal mechanism in the Koyna–Warna region (except for the main shock in December, 1967), we expect that there is at least some release of build-up pore pressures after each event. In calculating the pore pressure increase following each episode, we considered that a) there is no loss of pore pressure and that it continuously builds-up after impoundment and b) all the built-up pore pressure is released (complete leakage) during each episode, and starts to increase anew in response to the total water column in the reservoir. These two extreme cases comprise any other possibility, where there is a partial release of pore pressures. In the absence of reverse faulting mechanism for any M 5 earthquake, we do not consider any pore pressure increase after an episode.

6.4. *Observed Duration Between Episodes*

From the Coulomb failure criterion assuming no cohesion, a new episode can occur when the pore pressures reach a threshold or critical value (p_c) that depends on the pre-existing stress state and strength of faults,

$$p_c = \sigma_n - \tau/\mu, \tag{5}$$

where σ_n and τ are the normal and shear stresses, and μ is the coefficient of friction. The durations of 5–7 years between consecutive episodes (except between V and VI) suggests that there is a partial or complete leakage of pore pressure after each episode, and that this duration reflects the time needed to raise the pore pressure again to seismogenic levels. The following calculation illustrates how the cumulative effect of several years of annual lake-level fluctuations can result in a diffused pore pressure increase to critical levels. From Fig. 1a and b we note that the annual lake-level changes in the two reservoirs vary from 20 to 45 m. These fluctuations occur above a 'base' water level of about 20–30 m in the two reservoirs. Accordingly the contributions of the water levels in the reservoirs can be divided into two parts. The first is the contribution of a constant head of water (20–30 m) above the pre-impoundment level, and the second is due to the annual lake-level fluctuations above that level. The contribution of a constant head of water above pre-impoundment levels can be calculated using Eq. (4). For $c = 1\ \text{m}^2/\text{s}$ and at a 5-km distance, the diffused pore pressures reach ~80% of the load 6 years after impoundment (Fig. 4a). For ~20–30 m of water, corresponding to a load of 0.2–0.3 MPa at the surface, the pore pressure increase is 0.16–0.24 MPa after 6 years. Using Eq. (4) we evaluate pore pressure changes due to diffusion associated with six annual filling-emptying cycles, at a distance of 5 km from a reservoir, assuming $c = 1\ \text{m}^2/\text{s}$. We note that the maxima in pore pressures lag behind the annual maxima in lake levels, and that the pore pressures increase from about 35% of the load after the first year, to ~55% after 6 years (Fig. 4b). For an annual fluctuation of the Koyna Reservoir between ~30–40 m, excluding the small stress release associated with M 4 and smaller events, the excess pore pressure at 5 km from the reservoir by diffusion alone would be ~0.17–0.22 MPa after 6 years. In summary, the two contributions to the pore pressures at hypocentral depths are of the same order (~0.15–0.2 MPa). Both effects need to be taken into consideration when attempting to calculate diffused pore pressures generated after each episode.

6.5. *Combined Effects of the Two Reservoirs*

Annual seismicity maps of the KWR (Talwani *et al.*, 1996; Talwani, 1997b), depict a migration of seismic activity from the Koyna Reservoir towards the Warna River region from 1967 to 1983 and then a decrease of activity near the Warna River in 1984. This observation suggests that pore pressures generated by Koyna from 1967 to 1983 caused earthquakes along the weakest faults of the Warna region. Pore pressures from Koyna were not sufficient to destabilize the stronger faults (as per the Coulomb failure criterion) and seismicity decreased dramatically. After the impoundment of the Warna Reservoir, diffused pore pressures began to increase again in the region, amounting to the pore pressures generated by Koyna. The sudden increase of seismicity in 1993 suggests that the combined diffused pore pressure (from Koyna and Warna) was able to destabilize stronger faults in the KWR. Episode IV consists of earthquakes #14 and #15 which occurred in December 1993 and February 1994 (Table 1a and b). While earthquake #14 occurred near the Warna Reservoir, earthquake #15 occurred near Koyna Reservoir (Fig. 2). Interestingly, earthquake #15 was the first M 5 earthquake to occur near Koyna after 1973 (Episode II). This observation suggests that it required the combined effect of the Koyna and

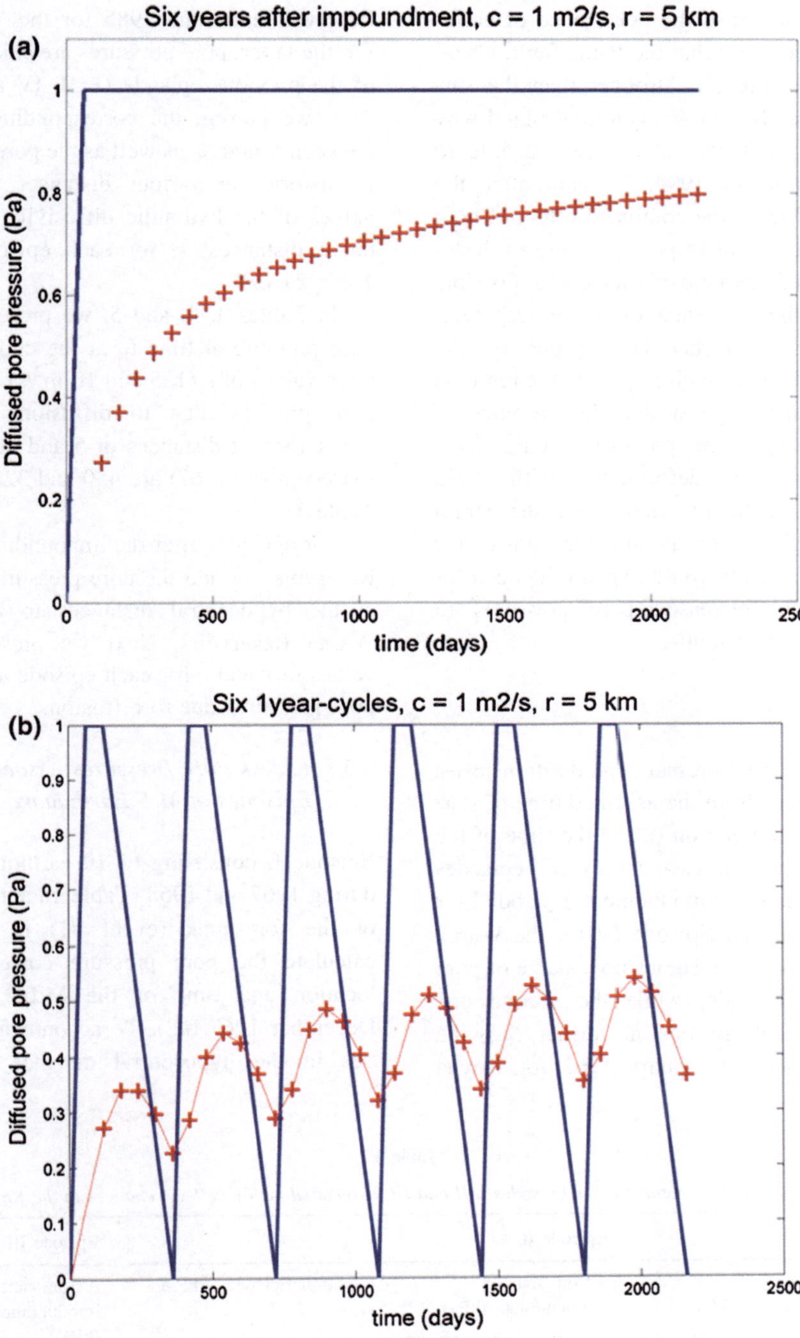

Figure 4

a Effect of maintaining a constant level of water above pre-impoundment levels, calculated for a 6-year period. The pore pressures (shown in *red* crosses) have been normalized with respect to the lake levels. **b** Effect of six loading and unloading 1 year-cycles (*blue*) on the diffused pore pressure (*red*). The pore pressures have been normalized with respect to the lake levels

Warna Reservoirs to cause the occurrence of earthquake #15. If we consider that the Patan fault, where earthquake #15 occurred, is stronger than the surrounding faults near Koyna, we can understand why pore pressures from Koyna alone were unable to destabilize the fault. In 1994, 9 years after the impoundment of Warna, the combined effect of both reservoirs provided adequate pore pressure to destabilize the fault and generate earthquake #15. Possible explanations for the inference of a stronger fault include differences in the rheological properties and the orientation of the fault with respect to the regional stress field. We also suggest that the presence of fluids and changing pore pressures could have facilitated elastic–plastic deformation with strain hardening in the fault. In such case, the strain hardening provided a strengthening mechanism for rocks in the KWR, which would explain the need for increased time and increased pore pressures to destabilize the causative faults.

6.6. *Results*

We calculate the pore pressure and dp/dt histories for different distances from the assumed time of start of pore pressure accumulation (t_0) to the time of the episode (t_E). We present two curves for episodes occurring after the initial impoundment, Episode I for the Koyna Reservoir and Episode IV for the Warna Reservoir. The first curve assumes no leakage of pore pressure after an episode, while the second one assumes complete leakage. For the former, t_0 is the start of impoundment in January 1961 for Koyna Reservoir and June 1985 for the Warna Reservoir. For the latter, pore pressures are calculated at the end of the previous episode (I, II, IV and V). For each case, we present the corresponding lake-level data between t_0 and t_E, as well as the pore pressure and dp/dt histories at various distances for the different values of the hydraulic diffusivity c. The representative distances, r, for each episode are given in Table 2a–c.

In Tables 3, 4, and 5, we present the calculated pore pressure at time t_E at the chosen distances for three values of c (1, 5 and 10 m^2/s). For example, the pore pressure due to diffusion for an assumed $c = 1\ m^2/s$ at distances of 5 and 10 km for Episode I (December, 1967) are 450 and 320 kPa (Fig. 5 and Table 3).

For episodes after the impoundment of the Warna Reservoir, we add the pore pressure due to diffusion at the hypocentral distances to both Koyna and Warna Reservoirs. Next we present our selected values of r and c for each episode and the calculated pore pressures due to diffusion.

6.6.1 *Excess Pore Pressures Associated with Episodes of M 5 Earthquakes in the KWR*

Episode I, consisting of 10 earthquakes, took place during 1967 and 1968 (Table 1a and b). As the depth of the foreshock (event #1) is not known, we calculate the pore pressure corresponding to the location and time of the M 6.3 main shock in December 1967 (#2). To account for any uncertainties in the hypocentral distance r (estimated at

Table 3

Calculated values of pore pressure, p, for Episodes I, II and III, associated with lake level changes in the Koyna Reservoir

	Episode I		Episode II				Episode III	
	p from start of impoundment (kPa)		p from start of impoundment (kPa)		p after Episode I (kPa)		p from start of impoundment (kPa)	p after Episode II (kPa)
	p ($r = 5$ km)	p ($r = 10$ km)	p ($r = 5$ km)	p ($r = 10$ km)	p ($r = 5$ km)	p ($r = 10$ km)	p ($r = 30$ km)	p ($r = 30$ km)
$c = 1\ m^2/s$	**450**	**320**	410	340	380	270	200	90
$c = 5\ m^2/s$	**560**	**470**	**540**	**430**	**530**	**400**	**380**	**290**
$c = 10\ m^2/s$	590	520	580	490	580	470	**420**	**360**

Table 4

Calculated values of pore pressure, p, for Episodes IV, V and VI, associated with lake level changes in the Koyna Reservoir

	Episode IV				Episode V		Episode VI	
	p from start of impoundment (kPa)		*p* after Ep. III (kPa)		*p* from start of impoundment (kPa)	*p* after Ep. IV (kPa)	*p* from start of impoundment (kPa)	*p* after Ep. V (kPa)
	p ($r = 15$ km)	*p* ($r = 30$ km)	*p* ($r = 15$ km)	*p* ($r = 30$ km)	*p* ($r = 30$ km)	*p* ($r = 30$ km)	*p* ($r = 30$ km)	*p* ($r = 30$ km)
KOYNA								
$c = 1\ m^2/s$	410	280	340	170	300	70	300	0.026
$c = 5\ m^2/s$	**510**	**420**	**480**	**360**	**450**	**290**	**450**	**4**
$c = 10\ m^2/s$	**540**	**460**	**530**	**420**	**490**	**380**	**490**	**30**

Table 5

Calculated values of pore pressure, p, for Episodes IV, V and VI, associated with lake level changes in the Warna Reservoir

	Episode IV		Episode V		Episode VI	
	p from start of impoundment (kPa)		*p* from start of impoundment (kPa)	*p* after Ep. IV (kPa)	*p* from start of impoundment (kPa)	*p* after Ep. V (kPa)
	p ($r = 15$ km)	*p* ($r = 30$ km)	*p* ($r = 20$ km)	*p* ($r = 20$ km)	*p* ($r = 15$ km)	*p* ($r = 15$ km)
WARNA						
$c = 1\ m^2/s$	160	50	210	140	280	1
$c = 5\ m^2/s$	**290**	**180**	**370**	**330**	**380**	**70**
$c = 10\ m^2/s$	**350**	**240**	**420**	**390**	**400**	**140**

~10 km, Table 2a) and hydraulic diffusivity c (estimated at 1.7 m^2/s; TALWANI, 1981; TALWANI *et al.*, 2007), we calculate the diffused pore pressure at distances of 5 and 10 km, and for hydraulic diffusivities of 1 and 5 m^2/s (Table 3, Figs. 5 and 6). Note that Fig. 6 covers the period from initial impoundment to the time of Episode I. The largest value of pore pressure, 560 kPa, corresponds to $r = 5$ km and $c = 5\ m^2/s$, whereas the smallest value, 320 kPa, corresponds to $r = 10$ km and $c = 1\ m^2/s$ (Table 3). We treat this range, 320–560 kPa, as a plausible range of excess pore pressure required to induce Episode I.

For subsequent episodes, we consider two scenarios. In the first, there is no leakage of build-up pore pressure after an episode; in the second, there is complete leakage after an episode and the build up pore pressure starts new.

For Episode II, the selected hypocentral distances are 5 and 10 km, while the hydraulic diffusivity is 5 m^2/s in view of the increased hydraulic diffusivities (TALWANI, 1981; TALWANI *et al.*, 2007) immediately after the 1967 main shock, and ~1 m^2/s (TALWANI, 1981; TALWANI *et al.*, 2007) immediately preceding the 1973 earthquake (Episode II). For hypocentral distances of 5 and 10 km, assuming $c = 5\ m^2/s$ and no leakage of pore pressure, we obtain diffused pore pressures of 540 and 430 kPa (Table 3, Fig. 7, vertical yellow line). In the case of complete leakage of pore pressure after Episode I (Fig. 8), the time for the start of building up a pore pressure is immediately after Episode I, and the duration of pore pressure increase is from 1968 to 1973. For the same range of values of r and c, the corresponding pore pressures are 530 and 400 kPa. Thus we find the plausible range of excess pore pressure between 400 kPa (with complete leakage of pore pressure after Episode I, and a hypocentral distance of 10 km) and 540 kPa (for no leakage and a hypocentral distance of 5 km).

Episode III occurred 7 years after Episode II, 30 km away from the Koyna Reservoir (Table 2a). Estimates of the hydraulic diffusivity in the region

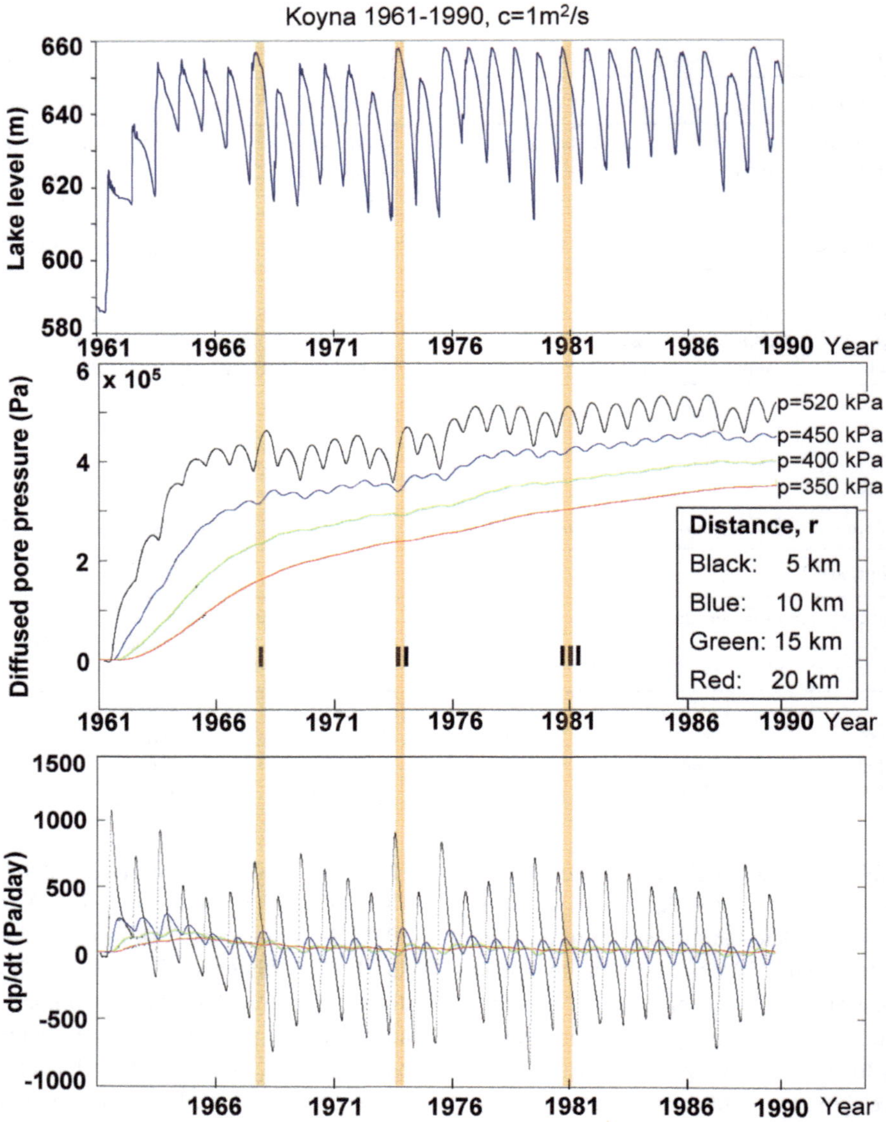

Figure 5

The top panel shows the water levels in Koyna Reservoir from the beginning of impoundment in 1961–1990. The middle panel shows the calculated build-up of diffused pore pressure for $c = 1$ m^2/s at different distances (5, 10, 15, 20 km), and the times of Episodes I, II and III. The accumulated pore pressures on 1/1/1990 for various distances are given for each curve in kPa. The bottom panel shows the daily rate of pore pressure change in Pa/day

suggest values of c between 5 and 10 m^2/s (Rastogi *et al.*, 1997; Talwani *et al.*, 2007). For this range of c and a 30 km hypocentral distance, we obtain pore pressure values ranging from 380 to 420 kPa for the case of no leakage, and 290 to 360 kPa for the case of complete leakage (Table 3, Figs. 7 and 9). Thus for Episode III, the plausible range of excess pore pressure by diffusion ranges between 290 kPa and 420 kPa.

For Episodes IV, V and VI, we add the combined build-up of pore pressure at hypocentral locations due to the water level fluctuations in both reservoirs. Estimates of the hydraulic diffusivity from July to August 1993 in the area close to the Warna Reservoir

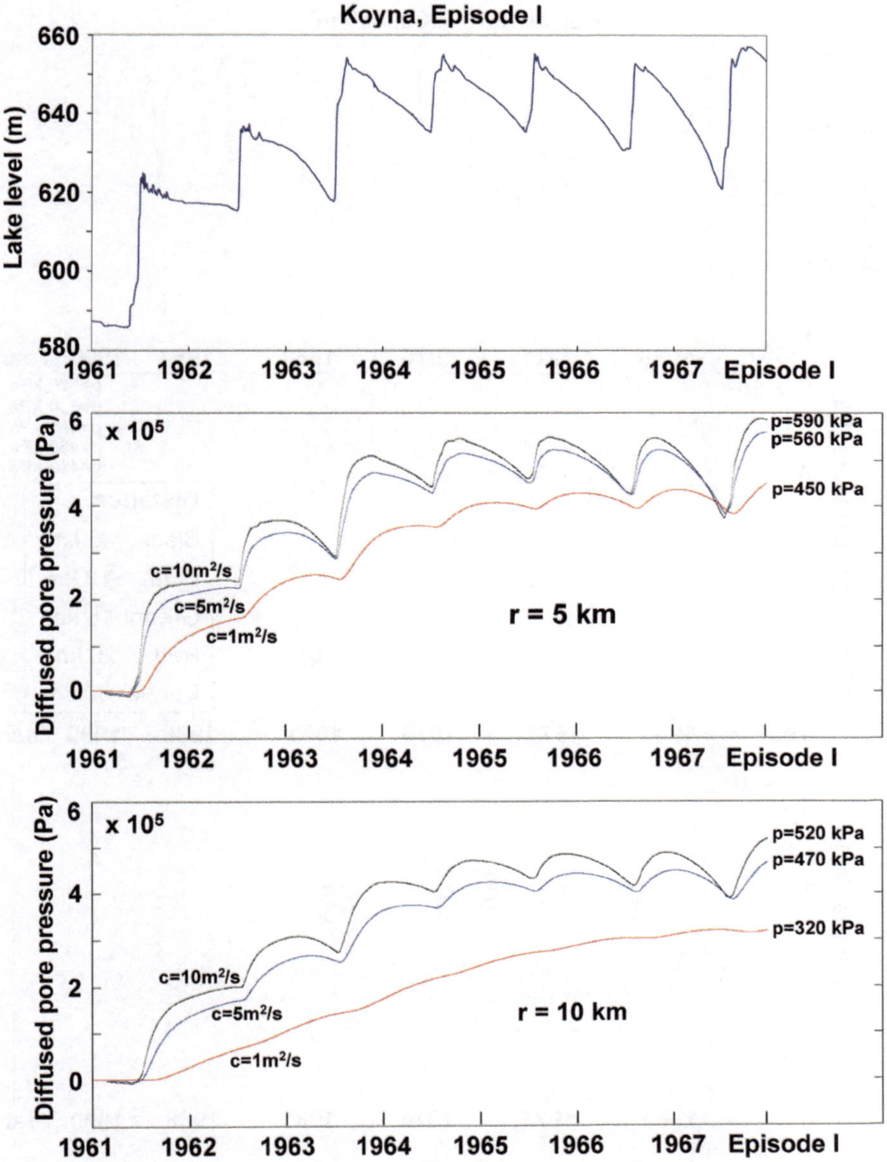

Figure 6

Build-up of diffused pore pressures from beginning of impoundment of the Koyna Reservoir to the time of the main shock in 1967 (earthquake #2) at distances of 5 and 10 km for different values of *c* (*middle* and *bottom panels*). The pore pressures are compared with the daily water level (*top panel*)

(Rastogi *et al.*, 1997) suggest $c = 6.9\ \mathrm{m^2/s}$. Correspondingly, we selected two values of c to estimate pore pressures, 5 and 10 $\mathrm{m^2/s}$. Figures 10 and 11 show the diffused pore pressure histories at 10, 15 and 20 km for $c = 5$ and 10 $\mathrm{m^2/s}$ assuming no leakage. The first part of Figs. 10 and 11 has been replotted for $r = 15$ km-corresponding to earthquake #14 in Episode IV—and different values of c (Fig. 12).

Earthquake #14 (Episode IV) occurred ~15 km away from Warna (Table 2c) and ~30 km away from Koyna (Table 2b). Pore pressure values due to Koyna at the hypocentral location range from 360 kPa (with complete leakage of pore pressure

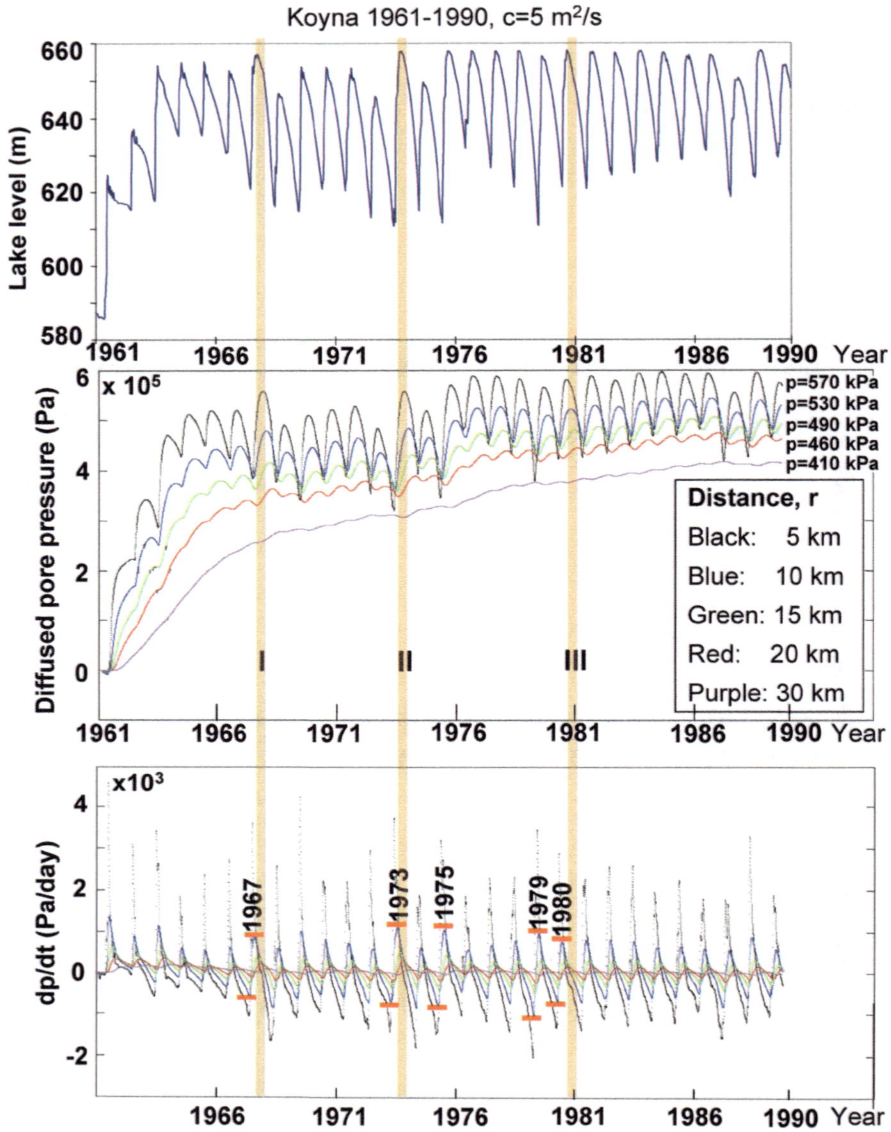

Figure 7
The *top panel* shows the water levels in the Koyna Reservoir from the beginning of impoundment in 1961–1990. The *middle panel* shows the calculated build-up of diffused pore pressure for $c = 5$ m^2/s at different distances (5, 10, 15, 20, 30 km), and the times of Episodes I, II and III. The accumulated pore pressures on 1/1/1990 for various distances are given for each curve in kPa. The *bottom panel* shows the daily rate of pore pressure change in Pa/day. In the curve showing dp/dt, the horizontal ticks define the amplitude of annual dp/dt increase (A) for the 1973 cycle (Pa/day) for $c = 5$ m^2/s and $r = 10$ km, i.e. 1900 Pa/day. A is listed in Table 6 for various filling cycles

during Episode III and $c = 5$ m^2/s; Table 4) to 460 kPa (for no leakage of pore pressure and $c = 10$ m^2/s; Table 4). Pore pressure values due to the Warna Reservoir at the hypocentral location range from 290 kPa (for $c = 5$ m^2/s; Table 5; Fig. 12) to 350 kPa (for $c = 10$ m^2/s; Table 5; Fig. 11). Therefore, adding the effects of the two reservoirs, the plausible excess pore pressure values could vary between 650 kPa and 810 kPa.

Earthquake #15 (Episode IV) took place ~15 km away from the Koyna Reservoir (Table 2b) and ~30 km away from Warna Reservoir (Table 2c).

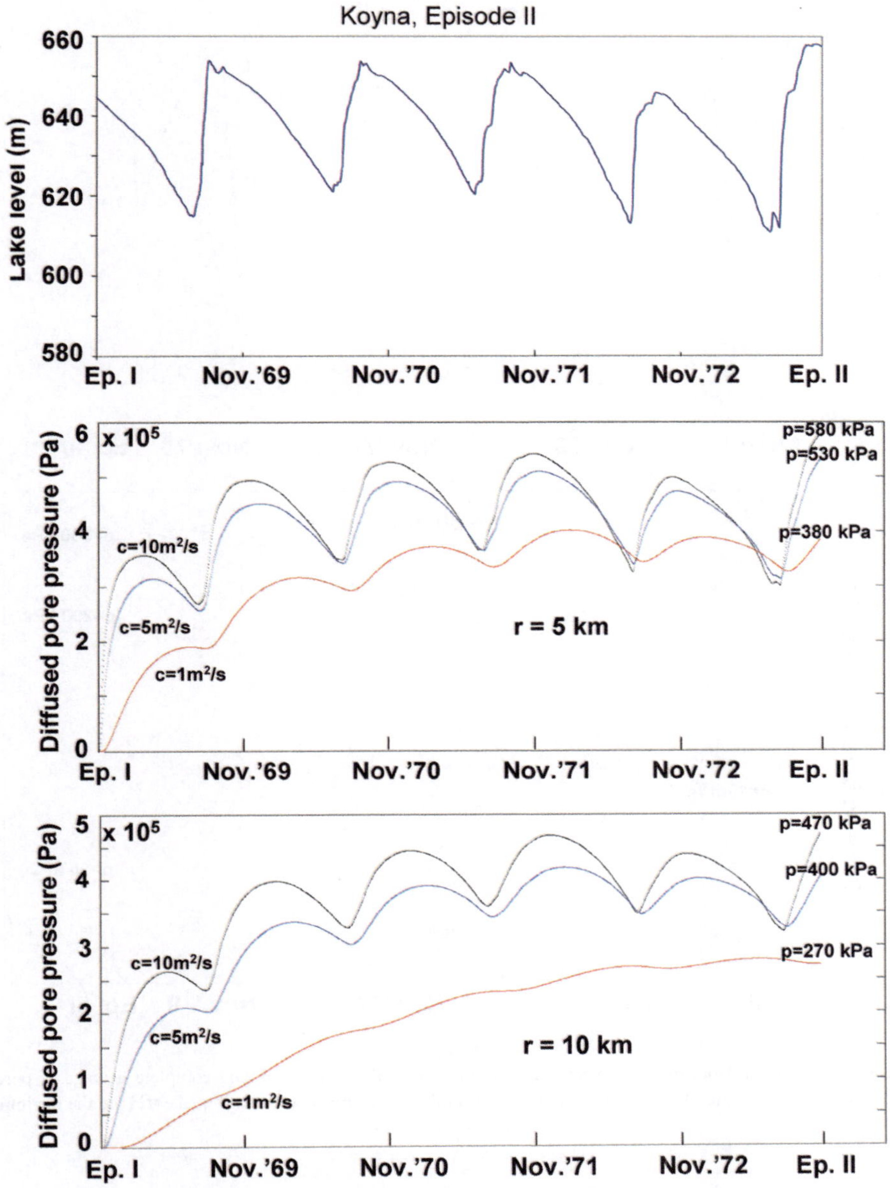

Figure 8

Build-up of diffused pore pressure from the Koyna Reservoir at a distances of 5 and 10 km, assuming complete leakage of pore pressure during Episode I. Pore pressures are calculated for different values of c after the end of Episode I (earthquake #10) up to the beginning of Episode II (earthquake #11)

Pore pressure values generated by the Koyna Reservoir at these distances and time of the earthquake vary from 480 kPa (with complete leakage of pore pressure during Episode III and $c = 5$ m^2/s; Table 4) and 540 kPa (for no leakage of pore pressure and $c = 10$ m^2/s; Table 4). For this earthquake, pore pressure values due to the Warna Reservoir range from 180 kPa (for $c = 5$ m^2/s; Table 5) to 240 kPa (for $c = 10$ m^2/s; Table 5; Fig. 11). The combined effect of both reservoirs generated the earthquake, with plausible excess pore pressures ranging from 660 kPa to 780 kPa.

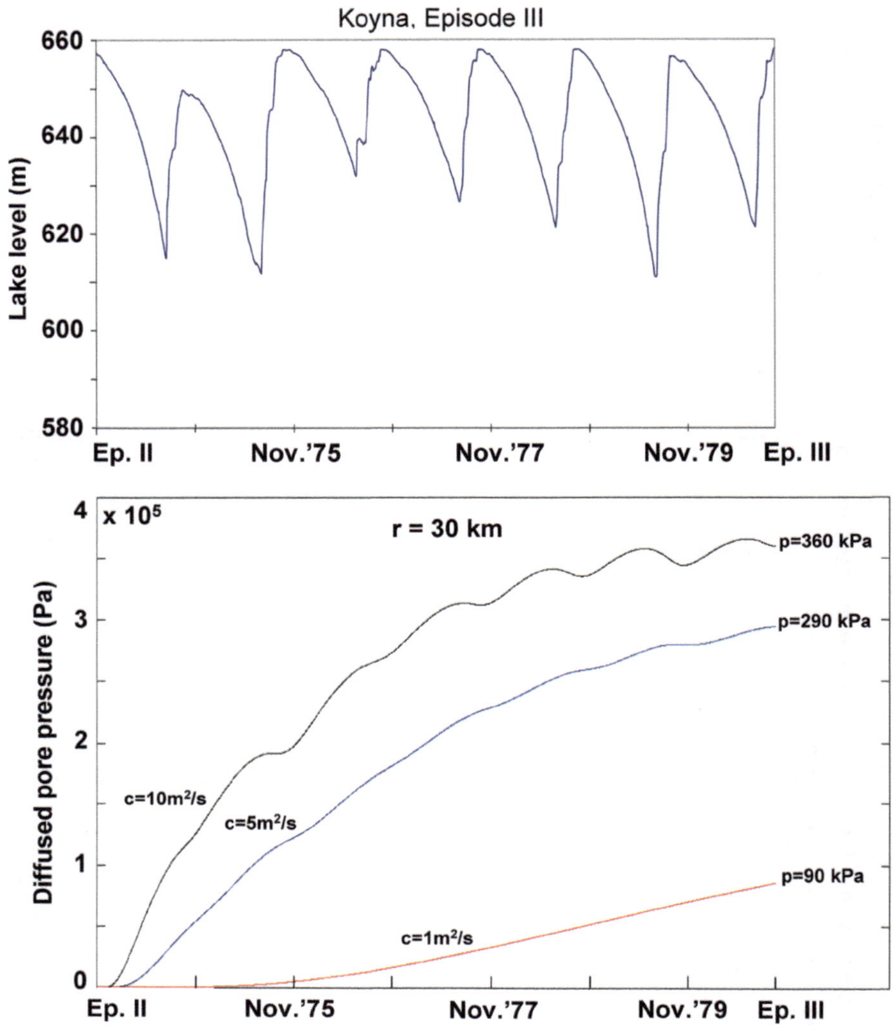

Figure 9

Build-up of diffused pore pressure from the Koyna Reservoir at a distance of 30 km, assuming complete leakage of pore pressure during Episode II. Pore pressures are calculated for different values of *c* from the end of Episode II (earthquake #11) to the beginning of Episode III (earthquake #12)

Episode V (earthquakes #16 and #17) occurred ~30 km away from the Koyna Reservoir (Table 2b) and ~20 km away from the Warna Reservoir (Table 2c). The range of pore pressures generated by the Koyna Reservoir at the hypocentral location and time of the earthquake range from 290 kPa (with complete leakage of pore pressure during Episode IV and $c = 5$ m^2/s; Table 4) and 490 kPa (for no leakage of pore pressure and $c = 10$ m^2/s; Table 4). Similarly, the range of pore pressures due to the Warna Reservoir varies from 330 kPa (with complete leakage of pore pressure during Episode IV and $c = 5$ m^2/s; Table 5; Fig. 13) to 420 kPa (for no leakage of pore pressure and $c = 10$ m^2/s; Table 5; Fig. 11). Therefore, after combining the effect of the two reservoirs, plausible excess pore pressure values range from 620 kPa to 910 kPa.

Episode VI (earthquake #18) occurred after the monsoon rains, 5 months after Episode V, at 15 km distance from Warna Reservoir (Table 2c) and 30 km distance from the Koyna Reservoir (Table 2b). The two previous earthquakes in March and April of 2000

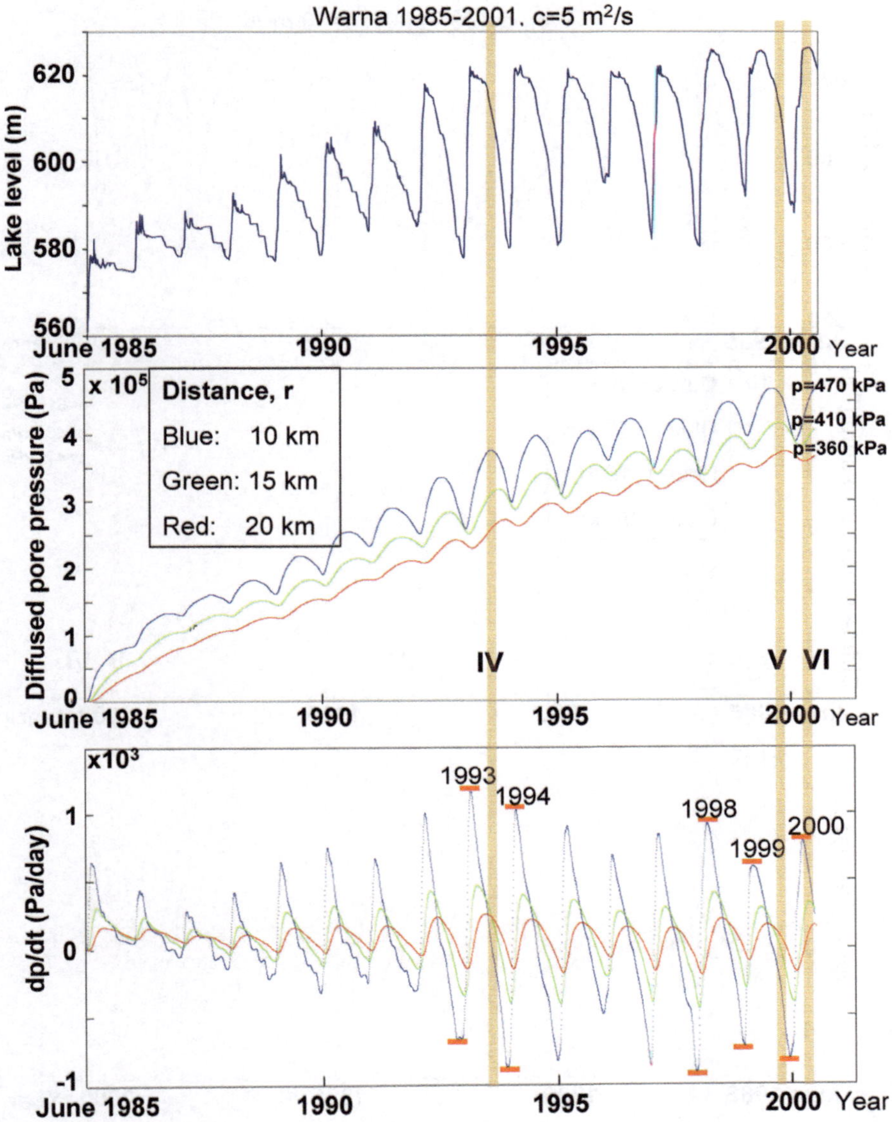

Figure 10
The *top panel* shows the water levels in the Warna Reservoir from the beginning of impoundment in 1985 to 2000. The *middle panel* shows the calculated build-up of diffused pore pressure for $c = 5$ m^2/s at different distances (10, 15, 20 km), and the times of Episodes IV, V and VI. The accumulated pore pressures on 1/1/1990 for various distances are given for each curve in kPa. The *bottom panel* shows the daily rate of pore pressure change in Pa/day

were associated with normal faulting and therefore we should expect to have leakage of pore pressure in the hypocentral area associated with Episode VI. The range of pore pressures generated by the Koyna Reservoir at the hypocentral location and time of the earthquake range from 4 kPa (with complete leakage of pore pressure during Episode V and $c = 5$ m^2/s; Table 4) to 490 kPa (for no leakage of pore pressure and $c = 10$ m^2/s; Table 4). Similarly, the range of pore pressures due to the Warna Reservoir varies from 70 kPa to 400 kPa (Table 5). The combined excess pore pressures range from 74 kPa to 890 kPa.

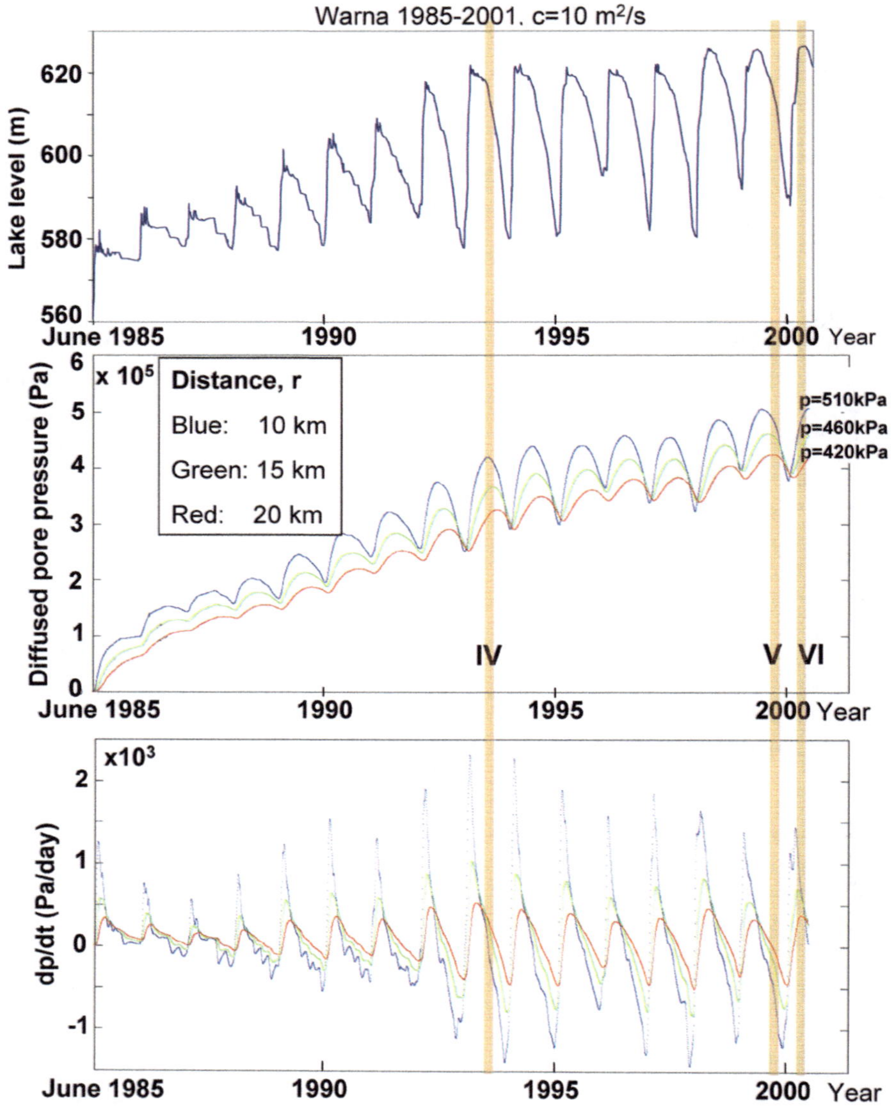

Figure 11
The *top panel* shows the water levels in the Warna Reservoir from the beginning of impoundment in 1985–2000. The *middle panel* shows the calculated build-up of diffused pore pressure for $c = 10\ m^2/s$ at different distances (10, 15, 20 km) and the times of Episodes IV, V and VI. The accumulated pore pressures on 1/1/1990 for various distances are given for each curve in kPa. The *bottom panel* shows the rate of pore pressure change in Pa/day

This large range is not consistent with those obtained for the previous episodes, the reason being the short period to build-up pore pressures after Episode V. Bearing in mind that the interval of time required to build-up pore pressures from episode to episode has been normally 5–7 years, we do not fully understand how Episode VI followed Episode V in such a brief amount of time. Pore pressures might not have dropped as expected during the two previous normal faulting earthquakes (Episode V) in the absence of fractures extending to the surface (Muir-Wood and King 1993). This is the subject of future research.

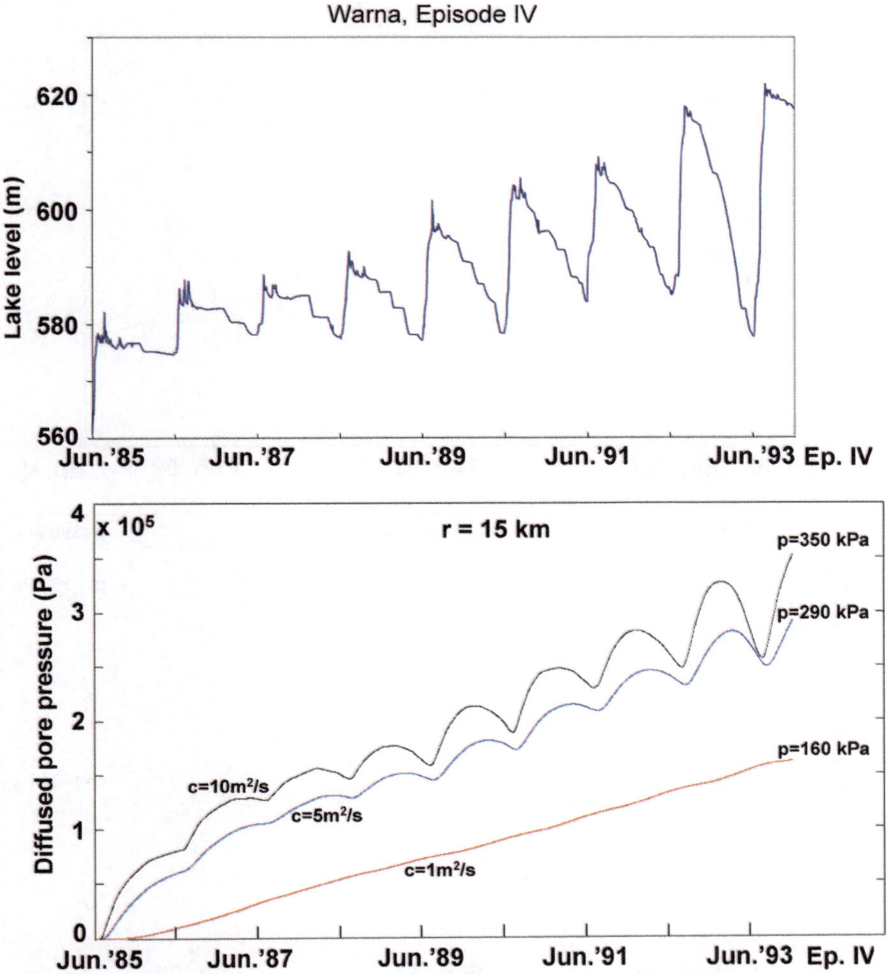

Figure 12
Build-up of diffused pore pressure from the Warna Reservoir at a distance of 15 km from the beginning of impoundment to the beginning of Episode IV (earthquake #14)

6.6.2 Excess Pore Pressure Associated with M ~ 4.0 Earthquakes in the KWR

For comparison, we calculate the required diffused pore pressure to generate an M 4 earthquake in the Warna and Koyna areas.

Given the scarcity of seismic stations near Koyna in the early years of seismicity (before 1967), earthquakes were poorly located and any attempt to calculate the diffused pore pressures is subject to the error associated with the earthquake location. The first reported M 3.8 earthquake near the Koyna Reservoir occurred in 1965/11/06 (lat. 17.4202°, long. 73.7623°) (Talwani *et al.*, 1996) at a depth of 5 km. Considering the horizontal distance along the main fractures to be ~2.6 km, we obtain a hypocentral distance r of ~6 km from the Koyna dam. Assuming the hydraulic diffusivity to be $c = 0.1$ m^2/s (Talwani *et al.*, 2007) and using Eq. (4), pore pressures of the order of ~160 kPa are needed to generate a M 3.8 earthquake near Koyna.

The first M 4.0 earthquake generated after the impoundment of the Warna Reservoir occurred on January 6, 1991 near Warna Reservoir at a depth of

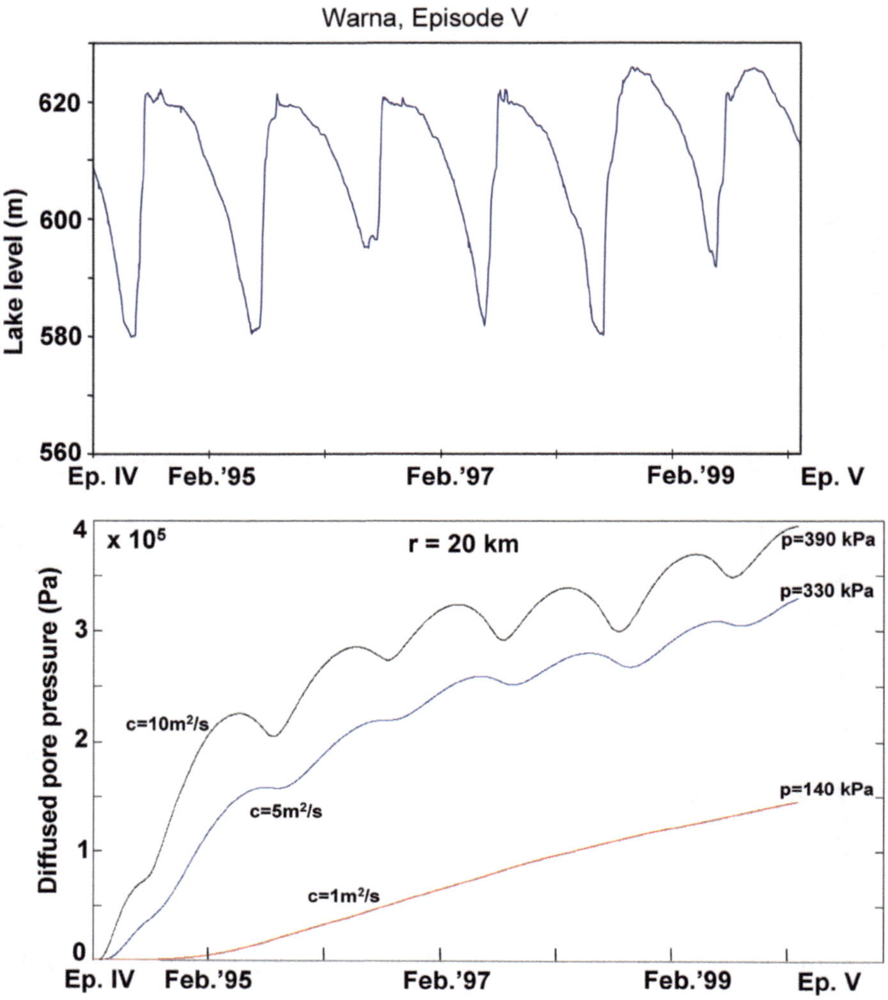

Figure 13
Build-up of diffused pore pressure from the Warna Reservoir at a distance of 20 km assuming complete leakage of pore pressure during Episode IV (earthquakes #14 and #15). Pore pressures are calculated from the end of Episode IV (earthquake #15) to the beginning of Episode V (earthquake #16)

6.3 km (lat. 17.1973°, long. 73.7355°) (TALWANI *et al.*, 1996). Following the argument given earlier, both Koyna and Warna Reservoirs contributed to the diffused pore pressure values and their combined effect caused the earthquake. From the Warna dam, the hypocentral distance is ~18 km (the horizontal distance along the main fractures is ~17 km, considering that pore pressure flows through the lake, L4, P1 and L5). For a hydraulic diffusivity $c = 5$ m^2/s (RASTOGI *et al.*, 1997 obtained $c = 6.9$ m^2/s near Warna from July to August 1993, the time at which the seismicity near Warna increased dramatically), applying Eq. (4) we derive that Warna contributed with diffused pore pressures of the order of ~180 kPa. From Koyna dam, the hypocentral distance is about 26 km (assuming that pore pressures flow through KRFZ and L5). For $c = 5$ m^2/s, the pore pressure values from Koyna vary from 340 kPa (assuming complete leakage of pore pressure after Episode III) to 420 kPa (without leakage). Therefore, for $c = 5$ m^2/s, the first M 4 earthquake associated with the combined effect of Koyna and Warna Reservoirs occurred at diffused pore pressure values of 520–600 kPa.

6.6.3 *Amplitude of the Annual dp/dt Increase*

We compare the excess pore pressure (in 10^5 Pa or 100 s of kPa—middle panel) with its daily rate of change (dp/dt, in 10^3 Pa/day or kPa/day—bottom panel) at Koyna (1961–1990) and Warna Reservoirs (1985–2001) for two diffusivity values each, in Figs. 5, 7, 10 and 11, respectively. Note that the calculated excess pore pressure values reflect the cumulative effect of water level fluctuations from t_0 to a given time t. The calculated dp/dt values reflect the changes in those cumulative pore pressures, and are not directly related to the lake-level changes in that cycle. We estimated the amplitude of the annual dp/dt increase, A (Fig. 7). It is those changes in dp/dt for a given cycle that we compare with values of $\Delta H(m)$ and excess diffused pore pressures.

6.6.4 *Comparison of the Roles of Various Factors Influencing RIS*

Next, for various filling cycles, we compare those factors that have been related to RIS in Tables 6 and 7. In deciding the filling cycles for comparison, we chose those that included the occurrence of an episode, or in which the lake level exceeded the previous maximum, i.e., displayed stress memory; or the amplitude of the annual increase in water level (ΔH) exceeded 40 m. As we are interested in the relative values of A, for various filling cycles, we calculated it assuming $c = 5$ m^2/s at a 10 km distance.

For those filling cycles associated with an episode, the excess pore pressures were calculated at the hypocentral locations and times of the given episodes as described in Sect. 6.6.1.

For those filling cycles associated with large ΔH and A, but not accompanied by an episode, we calculated the excess pore pressure at the hypocentral location of the next known episode on December 31 of the given filling cycle (middle of the cycle). For all those filling cycles after Episode II, we calculated these pore pressures assuming both complete leakage of pore pressure following an episode and $c = 5$ m^2/s and no leakage with $c = 10$ m^2/s. The calculated values of the excess pore pressure define the range of possible values for that particular case. We account for the effect of both reservoirs for filling cycles after 1985.

We summarize our findings in Table 6. For a given filling cycle (column 1), if the maximum water level exceeds the previous maximum water level from Tables 1a and 1b (column 2) and an episode follows, we consider it as evidence of stress memory (we highlight and label the corresponding cell as 'yes'). If the annual lake-level increase (ΔH, column 3) exceeds 40 m, we consider it anomalous and highlight the corresponding cell. If the amplitude of the annual dp/dt increase, A (column 4) is more than 1.80 kPa/day, we consider it anomalous and highlight the corresponding cell. We note the occurrence of an episode (column 5), the estimated minimum excess pore pressure, with complete leakage (column 6) and maximum pore pressure with no leakage (column 7). Next we compare these sets of parameters for the various filling cycles.

From Sect. 6.6.1 and Table 6 we notice that after the first episode which was associated with initial filling, Episodes II and III occurred when p_c was greater than $\sim$300 kPa, whereas the subsequent episodes occurred when p_c exceeded $\sim$600 kPa.

For Episode *I* the values of $\Delta H(m)$ and A(kPa/day) were not anomalous, but the previous water level was exceeded. The pore pressure increased to $\sim$560 kPa and induced a M 6.3 earthquake and its aftershocks. This behavior of the lake level and seismicity is characteristic of RIS associated with initial filling (Talwani, 1997a). For Episode II, we note that the previous H_{max} was exceeded, ΔH and A were anomalously high, and the minimum induced pore pressure (assuming leakage) was $\sim$400 kPa. In 1975, although ΔH and A were anomalously high, the minimum excess pore pressure $\sim$130 kPa was less than the values obtained for Episodes I, II and III. Only after the 1979–1980 filling cycle did the excess pore pressures at 30 km distance increase enough to cause Episode III. However, Episode III did not occur until the 1980–1981 filling cycle when the lake level exceeded the previous maximum. For the 1992–1993 filling cycle, ΔH and A were anomalously large at Koyna and large excess pore pressures were reached without inducing an episode. Episode IV occurred the following year when H_{max} was exceeded at Warna and ΔH and A were anomalously high. For the 1995–

Table 6

Comparison of factors contributing to the generation of $M \geq 5$ earthquakes

Year of filling cycle	Water level exceeds H_{max}. (evidence of stress memory)	ΔH (m)	A for $c = 5$ m^2/s at 10 km distance (kPa/day)	$M \geq 5$	p_c (kPa)	
					Minimum value with leakage	Maximum value without leakage
Koyna						
6/1967–5/1968	**Yes**	36.11	1.48	**Yes, Episode I.**	–	560
6/1973–5/1974	**Yes**	**47.09**	**1.90**	**Yes, Episode II.**	400	540
6/1975–5/1976	No	**46.18**	**1.9**	No	130	380
6/1979–5/1980	No	**45.41**	**2.24**	No	280	420
6/1980–5/1981	**Yes**	36.96	1.48	**Yes, Episode III**	290	420

Year of filling cycle	Water level exceeds H_{max}. (evidence of stress memory)		ΔH (m)		A for $c = 5$ m^2/s at 10 km distance (kPa/day)		$M \geq 5$	p_c (kPa)	
	Koyna	Warna	Koyna	Warna	Koyna	Warna		Minimum value with leakage	Maximum value without leakage
Koyna + Warna									
6/1992–5/1993	No	No	**43.90**	32.60	**1.98**	1.30	No	620	780
6/1993–5/1994	No	**Yes**	36.15	**44.11**	1.56	**1.85**	**Yes, Episode IV**	(Eq. 14) 650	810
								(Eq. 15) 660	780
6/1994–5/1995	**Yes**	**Yes**	36.00	**41.88**	1.62	**1.90**	No	160	860
6/1995–5/1996	No	No	36.70	**40.46**	1.64	1.75	No	310	840
6/1997–5/1998	No	No	23.25	**40.15**	1.62	1.70	No	480	880
6/1998–5/1999	No	**Yes**	30.80	**45.04**	0.98	**1.85**	No	540	900
6/1999–5/2000	No	No	22.40	33.77	0.96	1.40	**Yes, Episode V**	620	910
6/2000–5/2001	No	**Yes**	22.50	38.09	1.02	1.60	**Yes, Episode VI**	74?	890

Table 7

Comparison of filling rates, duration of high water levels and episodes

Filling cycle	Episode	Koyna		Warna	
		Filling rate (m/day)	Duration water level > 651 m (days)	Filling rate (m/day)	Duration water level > 615 m (days)
1967–1968	I	**0.95**	**173**		
1973–1974	II	0.52	142		
1975–1976		0.58	155		
1979–1980		0.89	153		
1980–1981	III	**0.76**	128		
1992–1993		0.73	120	0.56	52
1993–1994	IV	0.53	**168**	**0.83**	148
1994–1995		0.83	143	0.77	165
1995–1996		0.47	127	0.53	158
1997–1998		0.48	163	0.88	146
1998–1999		0.30	168	0.46	205
1999–2000	V	**0.82**	**190**	0.34	**215**
2000–2001	VI	0.23	112	0.36	

1996 and the 1996–1997 filling cycles, ΔH was anomalously high and the minimum excess pore pressures were less than 500 kPa and there were no episodes. For the 1998–1999 filling cycle, there was evidence of exceedance of H_{max}, large A and ΔH at Warna, and the minimum excess pore pressure was $\sim$540 kPa and no earthquake occurred, suggesting that p_c had not been reached. Episode V occurred during the 1999–2000 filling cycle which showed no evidence of stress memory or anomalously high values for ΔH and A, but for which the excess minimum pore pressure exceeded $\sim$600 kPa. For the 2000–2001 filling cycle, the lake level at Warna exceeded the previous maximum and there were no anomalous changes in ΔH or A, and the minimum excess pore pressure exceeded $\sim$74 kPa and episode VI followed. This low value of the minimum excess pore pressure suggests that the excess pore pressure may not have completely leaked following Episode V. The exact cause of this earthquake is not understood.

In order to evaluate the role of filling rate, dH/dt, and the duration of the highest lake level in a filling cycle, D in days, in inducing RIS (see e.g., Gupta, 2002) we examined these parameters for the filling cycles examined in Table 6. In some years the lake level begins to rise following the onset of the monsoon in June, and rises to a maximum level and then decreases; we used the dates and heights of the lowest and the highest water levels to calculate the filling rate, dH/dt. In other years, the lake level rises following the initial rains, and then decreases and increases again to the highest level in that filling cycle; this usually happens following delayed rains in September. We used the water level attained after the initial filling to calculate dH/dt, and compare it with other years (Table 7). We counted the number of days, D, that the water level stayed above elevations of 651 m and 615 m (arbitrarily decided) for the Koyna and Warna Reservoirs. These levels are within 10 m of the maximum levels attained by the reservoirs. As we were interested in the relative roles of these factors for the different filling cycles, we arbitrarily chose >0.75 m/d for the filling rate, dH/dt, and >160 days for D, to define as "high" (Table 7).

In Table 7, using data for the two reservoirs, we note that four of the six episodes occurred when the filling rates were high, and that there were four instances of these high filling rates without an episode. Of the seven cycles when D was high, only three were accompanied by an episode. These observations suggest that high filling rates and longer duration of high water levels are not the primary factors in inducing an episode, but play a supporting role, if the diffused pore pressures have nearly reached threshold levels.

Next we summarize these observations of the comparative role stress memory and anomalous values of ΔH, A, dH/dt and D play in the triggering of episodes (Table 8), and discus them in the next section.

7. *Discussion and Conclusions*

In this paper we have attempted to examine the hydromechanics related to the occurrence of six of the seven episodes of reservoir-induced M 5 events from 1967 to 2005 in the Koyna–Warna region. The seismicity is related to the increase in pore pressures to threshold levels, with the subsequent slippage along the fractures and fault planes identified in the seismotectonic framework. The increase in pore pressure at hypocentral locations large enough to induce the episodes results from changes in surface loads due to lake-level fluctuations. We assumed 1-D pore pressure diffusion from the reservoirs to hypocentral locations through a network of fractures. Using the daily lake level data, from the time of impoundment up to 2000, we calculated a history of diffused pore pressure, p, and dp/dt at hypocentral locations. For our calculations, we used a range of values of the hypocentral distance, r, and the hydraulic diffusivity of the involved fractures, c, so as to bracket the true conditions. To account for the release of accumulated pore pressures at the time of M 5 events, we calculated the pore pressure increase for the subsequent events by assuming both a complete leakage of pore pressure and a complete absence of leakage at the time to the preceding event, thus further bracketing the real situation.

The episodes occur when the excess pore pressures at the hypocentral locations attained a critical value, p_c, in accordance with the Coulomb failure criterion. In the Koyna–Warna region, the annual cycle of reservoir's filling and emptying contributed to the diffusion of fluid pressures to tens of kms. The amplitude of lake-level fluctuations, ΔH, and the pre-fluctuation lake levels control these excess pore pressures. For Episodes II and III, the critical excess pore pressures were found to be $\sim$300 kPa or more, whereas for the subsequent episodes they were $\sim$600 kPa or more. The need for higher pore pressures to induce Episodes IV, V and possibly VI could be because of rheological differences or unfavorable orientation of the fault planes with respect to the regional stress field.

Most of these episodes were associated with normal faulting, which suggested leakage of the excess pore pressures following these events. This assumption was found to be consistent with the observation that 5–7 years elapsed between episodes. Our calculations further support this assumption, that 5–7 years are needed to build-up the excess pore pressures to critical levels.

The role of five factors was examined in the triggering of these episodes after the excess pore pressures had reached critical levels (Table 8). We note that water level exceeded the previous maximum for five of the six episodes. For two cases where the previous H_{max} was exceeded, no episode followed in that cycle. Additionally, for the 2003–2004 filling cycle in Warna, the lake level also exceeded the previous H_{max} and the M 5 event did not occur until the following filling cycle (Table 1b). This observation suggests that in these cases the excess pore pressure had not reached critical values. We conclude that the exceedance of ΔH (stress memory) plays an important but not decisive role in the triggering of these episodes.

For the cases in which ΔH exceeded 40 m and A exceeded 1.80 kPa/day, only two were associated

Table 8

Comparison of triggering factors

	Evidence of stress memory		$\Delta H > 40$ m		$A > 1.8$ kPa/day		Filling rate > 0.75 m/day		Duration > 160 days	
	Observed		Observed		Observed		Observed		Observed	
	Yes	No	Yes	No	Yes	No	Yes	No	Yes	No
Episode	5	1	2	4	2	4	4	2	3	3
No episode	2		7		5		4		4	

with episodes. There were seven cases when ΔH exceeded 40 m and five cases when A exceeded 1.80 kPa/day and no episode occurred. This suggests that large values of ΔH and A contribute towards the general increase of excess pore pressures but do not induce the episodes. Also for four episodes the filling rate was >0.75 m/d, and there were four occasions during which these high filling rates were not accompanied by an episode. Anomalous durations, D, were found to occur before three episodes, while on four occasions there were no episodes. Faster filling rates (see e.g., Fig. 5 in Talwani *et al.*, 2007) and longer durations of D contribute to higher diffused pore pressures and at earlier times. We estimate that these two factors contribute a few additional kPa of pore pressure in the diffusion process. They are inadequate of themselves to induce the episodes.

The words 'induced' and 'triggered' have been used interchangeably to describe the seismicity that follows reservoir impoundment or lake-level fluctuations. Our results at KWR allow us to distinguish between them. We suggest that the seismicity results from a two-step process. The impoundment of a reservoir (in the case of initial seismicity), or large annual lake-level fluctuations (in the case of protracted seismicity) over a prolonged period of time, raises the pore pressure by $\sim$100 s kPa, which induces the seismicity. Short-term factors (such as the exceedance of the lake level over the previous maxima, large values of H, ΔH, dH/dt, D, dp/dt and A), with pore pressure changes of 10 s kPa or less, are the end products of the induction process—advancing the timing of the earthquakes and triggering them.

We did not have daily lake levels after 2000 to analyze Episode VII in 2005. However, based on the limited data, we note that the H_{max} was exceeded in the Warna reservoir in the third filling cycle (September 2003) after Episode VI. These three cycles were inadequate to generate excess pore pressures to induce any event. It was in the next filling cycle that Episode VII occurred in 2005, 4.5 years after episode VI. Episode VII was preceded by high filling rates in both reservoirs (Gupta *et al.*, 2005; Satyanarayana *et al.*, 2005).

We expect that the seismicity will expand to the southwest of the Warna Reservoir, where another M 5 earthquake might occur in the future, and whose timing may be advanced if the current H_{max} is exceeded.

Acknowledgments

This paper benefited from critical comments by David Simpson, Hari Rajaram and an anonymous reviewer. We appreciate the invitation of the Editor Cezar Trifu to submit this paper, and his patience and input in its publication.

Appendix 1

The following expression (6) relates the increase of pore pressure due to compression (undrained response), $\Delta p_u(r)$, at a distance r from the reservoir and the change in the mean stress $\Delta\sigma_{kk}/3$ at the given distance r

$$\Delta p_u(r) = -B\Delta\sigma_{kk}/3 \qquad (6)$$

where B is the Skempton's coefficient.

An estimate of the change in the mean stress can be obtained following the procedure in Ahlvin and Ulery (1962). For a uniform circular load, the authors tabulated the vertical, radial and transverse stresses at a given depth and radial distance d from its center using elliptic functions. Considering the radius R of the circular loaded area to be 1.25 km, for an earthquake located at a depth $z = 4$ km and a radial distance $\rho = 2$ km (i.e., $r = (4^2 + 2^2)^{1/2} \sim 4.5$ km), we obtain $\sigma_z = -0.09193p$, $\sigma_\rho = -0.009830p$ and $\sigma_\theta = 0.006325p$, where p is the surface contact pressure. Therefore, the mean stress at a distance of 5 km is

$$\Delta\sigma_{kk}/3 \sim -0.032p. \qquad (7)$$

The maximum water height in Koyna is $\sim$70 m, providing a surface contact pressure p of about 700 kPa

$$p = \rho gh = 1000 \times 9.8 \times 70 \sim 700\,\text{kPa}$$

and, substituting p in (7), a mean stress value of about -22.4 kPa.

Assuming $B \sim 0.7$ for crystalline rocks (Talwani *et al.*, 1999), substituting in (6) we obtain

$$\Delta p_u(5\,\mathrm{km}) \sim 16\,\mathrm{kPa}.$$

The undrained response is one order of magnitude less than the diffused pore pressure values calculated in this study.

References

Ahlvin, R.G. and Ulery, H.H. (1962), *Tabulated values for determining the complete pattern of stresses, strains, and deflections beneath a uniform circular load on a homogeneous half space*, Highw. Res. Board Bull. *342*, 1–13.

Chadha, R.K., Kuempel, H.-J., and Shekar, M. (2008), *Reservoir Triggered Seismicity (RTS) and well water level response in the Koyna-Warna region, India*, Tectonophysics *456*, 94–102.

Chen, L. and Talwani, P. (2001), *Mechanism of initial seismicity following impoundment at the Monticello Reservoir, South Carolina*, Bull. Seismol. Soc. Am. *91*, 1582–1594.

Cornet, F.H. and Yin, J. (1995), *Analysis of induced seismicity for stress field determination and pore pressure mapping*, Pure Appl. Geophys. *145*, 677–700.

Coussy, O., *Mechanics of Porous Continua* (Wiley, Chichester, New York, 1995) 472 pp.

Dziewonski, A.M., Ekstrom, M.G., Franzen, J.E., and Woodhouse, J.H. (1988), *Global seismicity of 1980: Centroid-moment tensor solutions for 515 earthquakes*, Phys. Earth Planet. Inter. *50*, 127–154.

Evans, K.F., Genter, A., and Sausse, J. (2005), *Permeability creation and damage due to massive fluid injections into granite at 3.5 km at Soultz: 1. Borehole observations*, J. Geophys. Res., B04203, doi:10.1029/2004JB003168.

Gupta, H.K., *Reservoir Induced Earthquakes* (Elsevier Publishers, Amsterdam, 1992).

Gupta, H.K. (2002), *A review of recent studies of triggered earthquakes by artificial water reservoirs with special emphasis on earthquakes in Koyna, India*, Earth Sci. Rev. *58*, 279–310.

Gupta, H.K. and Combs, J. (1976), *Continued seismic activity at the Koyna Reservoir site, India*, Eng. Geol. *10*, 307–313.

Gupta, H.K., Ram Krishna Rao, C.V., Rastogi, B.K., and Batia, S.C. (1980), *An investigation of earthquakes in Koyna Region, Maharashtra, for the period October 1973 through December 1976*, Bull. Seismol. Soc. Am. *70*, 1833–1847.

Gupta, H.K., Mandal, P., and Rastogi, B.K. (2002), *How long will triggered earthquakes at Koyna, India continue*?, Curr. Sci. *82*, 202–210.

Gupta, H.K., Mandal, P., Satyanarayana, H.V.S., Shashidhar, D., Sairam, B., Shekar, M., Singh, A., Devi, E.U., Kousalya, M., Rao, N.P., and Dimri, V.P. (2005), *An earthquake of M ~ 5 may occur at Koyna*, Curr. Sci. *89*, 747–748.

Gupta, H., Shashidhar, D., Pereira, M., Mandal, P., Rao, N.P., Kousalya, M. Satyanarayana, H.V.S., and Dimri, V.P. (2007), *Earthquake forecast appears feasible at Koyna, India*, Cur. Sci. *93*, 843–848.

Harinarayana, T., Patro, B.P.K., Veeraswamy, K., Manoj, C., Naganjaneyulu, K., Murthy, D.N., and Virupakshi, G. (2007), *Regional geoelectric structure beneath Deccan Volcanic Province of the Indian subcontinent using magnetotellurics*, Tectonophysics *445*, 66–80.

Hill, R., *The Mathematical Theory of Plasticity* (Oxford, Clarendon Press, 1950).

Kaiser, J. (1953), *Erkenntnisse und Folgerungen ayus der Messung von Geräuschen bei Zugbeanpruchung von metallischen Werstoffen*, Archiv fur das Eisenhüttenwesen *24*, 43–45. (in German).

Kalpna and Chander, R. (2000), *Green's function based stress diffusion solutions in the porous elastic half space for time varying finite reservoir loads*, Phys. Earth Planet. Inter. *120*, 93–101.

Kessels, W. and Kück, J. (1995), *Hydraulic communication in crystalline rock between the two boreholes of the continental deep drilling project in Germany*, Int. J. Rock Mech. Min. Sci. Geomech. Abstr. *32*, 37–47.

Kumar, B.A., Ramana, D.V., Kumar, Ch.P., Rani, V.S., Shekar, M., Srinagesh, D., and Chadha, R.K. (2006), *Estimation of source parameters for 14 March 2005 earthquake of Koyna-Warna region*, Curr. Sci. *91*, 526–529.

Langston, C.A. (1981), *Source inversion of seismic wave form : The Koyna India earthquake of 13, September 1967*, Bull. Seismol. Soc. Am. *71*, 1–24.

Langston, C.A. and Franco-Spera, M. (1985), *Modeling of the Koyna, India, aftershock of 12 December 1967*, Bull. Seismol. Soc. Am. *75*, 651–660.

Lavrov, A. (2003), *The Kaiser effect in rocks; principles and stress estimation techniques*, Intl. J. Rock Mech. Min. Sci. *40*, 151–171.

Mandal, P., Rastogi, B.K., and Sarma, C.S.P. (1998), *Source parameters of Koyna earthquakes*, India, Bull. Seismol. Soc. Am. *88*, 833-842.

Mendelson, A., *Plasticity: Theory and Application* (Krieger, Malabar, Florida, 1983).

Muir-Wood, R., and King, G.C.P. (1993), *Hydrological signatures of earthquake strain*, J. Geophys. Res. *98*, 22035–22068.

Nur, A. and Booker, J. R. (1972), *Aftershocks caused by pore fluid flow*?, Science *175*, 885–887.

Pandey, A.P. and Chadha, R.K. (2003), *Surface loading and triggered earthquakes in the Koyna-Warna region, western India*, Phys. Earth Planet. Inter. *139*, 207–223.

Peshwa, V.V. (1991), *Geological Studies of Chandoli Dam Site Area Warna Valley, Sangli Dist., Maharashtra State. Studies Based on Remote Sensing Techniques*. Unpublished Report to Maharashtra Engineering Research Institute, Nashik, Department of Geology, University of Pune, 45 pp.

Rajendran, K. and Talwani, P. (1992), *The role of elastic, undrained, and drained responses in triggering earthquakes at Monticello Reservoir, South Carolina*, Bull. Seismol. Soc. Am. *82*, 1867–1888.

Rajendran, K. and Harish, C.M., (2000), *Mechanism of triggered seismicity at Koyna: an assessment based on relocated earthquake during 1983–1993*, Curr. Sci. *79*, 358–363.

Rao, B.S.R., Prakasa Rao, T.K.S., and Rao, V.S. (1975), *Focal mechanism study of an aftershock in the Koyna Region of Maharashtra State, India*, Pure Appl. Geophys. *113*, 483–488.

Rastogi, B.K., Chadha, R.K., Sarma, C.S.P., Mandal, P., Satyanarayana, H.V.S., Raju, I.P., Kumar, N., Satyamurthy, C., and Rao, A.N. (1997), *Seismicity at Warna Reservoir (near Koyna) through 1995*, Bull. Seismol. Soc. Am. *87*, 1484–1494.

Rice, J.R. (1975), *On the stability of dilatant hardening for saturated rock masses*, J. Geophys. Res. *80*, 1531–1536.

RICE, J.R. (1977), *Pore pressure effects on inelastic constitutive formulations for fissured rock masses, Advances in Civil Engineering Through Engineering Mechanics*, Proc. 2nd ASCE Eng. Mech. Div. Specialty Conf., N.C. Raleigh, 1977, pp. 295–297, Am. Soc. Civ. Eng., New York.

ROELOFFS, E.A. (1988), *Fault stability changes induced beneath a reservoir with cyclic variations in water level*, J. Geophys. Res. *83*, 2107–2124.

RUDNICKI, J.W. and RICE, J.R. (1975), *Conditions for localization of deformation in pressure-sensitive dilatant materials*, J. Mech. Phys. Solids *23*, 371–394.

SARMA, S.V.S., PRASANTA, B., PATRO, K., HARINARAYANA, T., VEERASWAMY, K., SASTRY, R.S., and SARMA, M.V.C. (2004), *A magnetotelluric (MT) study across the Koyna seismic zone, western India: Evidence for block structure*, Phys. Earth Planet. Int., *142*, 23–36.

SARMA, P.R. and SRINAGESH, D. (2007), *Improved earthquake locations in the Koyna-Warna seismic zone*, Nat. Hazards *40*, 563–571.

SATYANARAYANA, H.V.S., SHASHIDHAR, D., SAGARA RAO, T., and KOUSALYA, M. (2005), *An M 5.2 earthquake occurs in Koyna region after 4 1/2 years*, Curr. Sci. *89*, 612–613.

SIMPSON, D.W. and NEGMATULLAEV, S.K. (1981), *Induced seismicity at Nurek Reservoir, Tadjikistan, USSR*. Bull. Seismol. Soc. Am. *71*, 1561–1586.

SIMPSON, D.W., LEITH, W.S., and SCHOLZ, C.H. (1988), *Two types of reservoir-induced seismicity*, Bull. Seismol. Soc. Am. *78*, 2025–2040.

SRINAGESH, D. and SARMA, P.R. (2005), *High precision earthquake locations in Koyna-Warna seismic zone reveal depth variations in brittle-ductile transition zone*, Geophys. Res. Lett. *32*, 8310–8313.

TALWANI, P. (1981), *Hydraulic diffusivity and reservoir induced seismicity*, Final Tech. Rep., 48 pp., U.S. Geol. Surv., Reston, VA.

TALWANI, P. (1994), *Ongoing Seismicity in the Vicinity of the Koyna and Warna Reservoirs*, A Report to the United Nations Development Programme and Department of Science and Technology, Government of India, August, 85 pp.

TALWANI, P. (1995), *Speculation on the causes of continuing seismicity near Koyna Reservoir, India*, Pure Appl. Geophys. *145*, 167–174.

TALWANI, P. (1996), *Kaiser effect and reservoir induced seismicity*, EOS Trans. AGU, 77, F500.

TALWANI, P. (1997a), *On the nature of reservoir-induced seismicity*, Pure Appl. Geophys. *150*, 473–492.

TALWANI, P. (1997b), *Seismotectonics of the Koyna-Warna Area, India*, Pure Appl. Geophys. *150*, 511–550.

TALWANI, P. (2000), *Seismogenic properties of the crust inferred from recent studies of reservoir-induced seismicity—Application to Koyna*, Curr. Sci. *79*, 1327–1333.

TALWANI, P., KUMARA SWAMY, S.V., and SAWALWADE, C.B. (1996), *Koyna Revisited: The Reevaluation of Seismicity Data in the Koyna-Warna Area, 1963–1995*, Univ. South Carolina Tech. Report (Columbia, South Carolina) 343 pp.

TALWANI, P., COBB, J.S., and SCHAEFFER, M.F. (1999), *In situ measurements of hydraulic properties of a shear zone in northwestern South Carolina*, J. Geophys. Res. *104*, 14993–15003.

TALWANI, P., CHEN, L., and GAHALAUT, K. (2007), *Seismogenic permeability, k_s*, J. Geophys. Res. *112*, B07309, doi: 10.1029/2006JB004665

TEMPLETON, E.L. and RICE, J.R. (2008), *Off-fault plasticity and earthquake rupture dynamics: 1. Dry materials or neglect of fluid pressure changes*, J. Geophys. Res. *113*, B09306, doi:10.1029/2007JB005529.

VIESCA, R. C., TEMPLETON, E.L., and RICE, J.R. (2008), *Off-fault plasticity and earthquake rupture dynamics: 2. Effects of fluid saturation*, J. Geophys. Res. *113*, B09307, doi10.1029/2007JB005530.

(Received November 18, 2008, revised March 14, 2009, accepted March 26, 2009, Published online November 18, 2009)

GPSR Compliance

The European Union's (EU) General Product Safety Regulation (GPSR) is a set of rules that requires consumer products to be safe and our obligations to ensure this.

If you have any concerns about our products, you can contact us on ProductSafety@springernature.com

In case Publisher is established outside the EU, the EU authorized representative is:

Springer Nature Customer Service Center GmbH
Europaplatz 3
69115 Heidelberg, Germany

Batch number: 10578397

Printed by Printforce, the Netherlands